Psychobiology of Aggression and Violence

Psychobiology of Aggression and Violence

Luigi Valzelli, M.D.

Chief, Section of Neuropsychopharmacology
Mario Negri Institute of Pharmacological Research
Milan, Italy

Raven Press ■ New York

Raven Press, 1140 Avenue of the Americas, New York, New York 10036

Made in the United States of America

Library of Congress Cataloging in Publication Data

Valzelli, Luigi, 1927–
  Psychobiology of aggression and violence.

  Includes bibliographical references and index.
  1. Aggressiveness (Psychology) 2. Violence.
3. Psychobiology. I. Title. [DNLM: 1. Aggression.
2. Violence. BF 575.A3 V215p]
BF575.A3V3   155.2′32   78–55807
ISBN 0–89004–403–1

Great care has been taken to maintain the accuracy of the information contained in the volume. However, Raven Press cannot be held responsible for errors or for any consequences arising from the use of the information contained herein.

This book is dedicated to my wife, Liliana

Foreword

Aggressive behavior is one of the most important problems facing the world today. Hostility within the family creates unhappiness, crime in the streets promotes social disruption. The potential for the use of the sophisticated weapons of the atomic age may result in the demise of all of mankind. If these problems are not solved, it makes very little difference what the other problems are.

The *Psychobiology of Aggression and Violence* covers a vast literature on this topic and will certainly contribute to our understanding of the phenomenon of aggression. It is an important book that merits the attention of all individuals concerned with the problem of aggressive behavior.

Although there are hundreds of books currently in print that are devoted to some aspect of aggressive or violent behavior, only a limited number have been concerned primarily with the organic basis of these behaviors, and only two, the present one and my own,[1] have attempted to provide an integration of the vast number of studies which are frequently islands of knowledge in an unintegrated sea of information.

Integration is a theme of this book. There is enough information on the thousands of animal studies to cast light on the origins of human aggression. Valzelli's integration of findings on animals with the results from human studies is a strong point of this book.

The data from a wide variety of fields from anatomy to zoology are integrated. A perusal of the bibliography will show that works from innumerable fields of scholarly endeavor have been used.

The topic of aggression, from its least complicated early beginnings to its most complex manifestations, is covered. The book begins with the origin of life and closes with the aggression of man and how it might be controlled. In this last chapter on the control of aggression, the author draws on the evidence that has been presented throughout the book and attempts to formulate potential controls that might be applied to individuals or to groups. He gives, of course, no final answers. The state of the art does not permit it. However, the world needs individuals who will take a critical look at what is known and attempt to apply that to what is needed.

---

[1] Moyer, K. E. *The Psychobiology of Aggression.* New York, Harper and Row, 1976.

Every phrase and every concept in this volume is documented in detail. Valzelli has been remarkably thorough in his attempts to discover what is known about every phase of aggression and violence. The more than two thousand references will be of great value to any one who is concerned with the physiology of this behavior.

K. E. Moyer
Carnegie-Mellon University
Pittsburgh, Pa.

Preface

The need to integrate animal and human studies was a major reason for writing this volume. I began these studies sixteen years ago while evaluating my own animal experiments on aggression, which are based on specific training in psychology and psychiatry. Over the years, an ever-increasing bulk of evidence has developed, thus permitting a path to be traced, in common with both animal and man, through the jungle of aggression.

There has been some opposition to the integration of animal and human data concerning aggression. This opposition, which disregards neuroanatomical, neurochemical, genetic, and behavioral data, has attempted to depict violent man as the exclusive and pitiable product of an alien and wickedly obscure entity—the society.

The purpose of this volume is to offer an acceptably wide range of data concerning aggression and violence, and to present an easily comprehensible basis for an understanding of the continuity of these phenomena from animals to man. The implications of increasing human violence are explored. Preliminary data concerning the evolution of life and the brain, as well as neuronal function, are provided.

This volume represents an effort to construct a framework within which significant and converging information from different research sources, such as zoology, ethology, neurochemistry, neuroanatomy, neurophysiology, psychology, psychiatry, genetics, endocrinology, psychopharmacology, and medicine are included.

This volume will be of interest to students and professionals in the biomedical sciences fields.

Luigi Valzelli, M.D.

Acknowledgments

There are many people who have helped me with the production of this volume. Foremost, I want to express my love and appreciation to my dear wife Liliana, to whom this book is dedicated. There are other people who deserve my gratitude, and a complete list would require considerably more space than these few lines. It is not possible to express my acknowledgments to each one of them, but I would like to express my appreciation to my secretary, Mrs. Loredana Morgese, for her constant help and outstanding professional performance. Last, but not least, I want to give my grateful thanks to Dr. Walter B. Essman, who applied his considerable talents to make the original manuscript acceptable.

Contents

FIGURES INDEX

TABLES INDEX

1

The Brain

EXCITABILITY AND THE ORIGIN OF LIFE

The fundamental feature of the brain and of any kind of nervous tissue is excitability. Nerve impulses are propagated along nerves and muscle fibers by electric current, and bioelectricity can consequently be considered one of the vital functions of the body. Apart from the ever-debated question of the marine or volcanic origin of life (Sylvester-Bradley, 1976), it is interesting to remember that protoplasmic material is formed by organic and inorganic components, both consisting of the same elementary particles. Organic matter is formed by a series of different inorganic substances having a particular and highly organized structure that give protoplasm particular features and states with manifestations that obey biological laws (Macovschi, 1966).

An essential and indispensable characteristic of any living organism, even the most primitive, is its capacity to respond adaptively to the stimuli produced by its environment. As a consequence, it must also possess the capacity to become excited; in this way, the concept of life can be conceived of as coinciding with the concept of excitability. As suggested by Gabel in 1965, excitability must be an inherent property of an organized group of particles to transfer information from the environment through its own components with the preservation of its structural integrity. This last condition has to be regarded as essential, since if information or stimuli could not be properly transferred and utilized, the structural organization of the entire system would disintegrate, undergoing a loss of structure and then being disrupted by the stimulus.

In this context, the substances that comprise excitable tissues also have the same properties that occur in an inorganic system. What makes them biologically viable is their structural organization and their relationship to the organization of other systems. The structuring of even the most elementary living system is dependent on the formation of an excitable membrane. This must have evolved from chemicals present on the earth before the existence of any kind of life (Gabel, 1965). This dates back 1.5 miliard years. Such a proto-living membrane must have been formed by inorganic cations, anions, and organic molecules, and it is believed that the *primaeval broth* (Oparin, 1964) was abundant in inorganic and organic material, including amino acids, purines, pyrimidines, imidazoles, pentoses, polypeptides, and porphyrins (Oro, 1960; Oro and Guidry, 1960; Oro and Kamat, 1961; Oro and Kimball, 1961; Scott, 1956; Wilson, 1960). Thus, when the primitive earth was lifeless, it still constituted a place for the abiogenic synthesis of organic substances, the evolution of which concluded in the primitive living structures (Oparin, 1976). It is noteworthy that the crust of such a primitive universe consisted of large phosphate deposits (Van Wazer, 1961).

This is certainly a crucial point of our story, since the universal occurrence and fundamental importance of organic phosphates and polyphosphates in the bioelectrical phenomena of excitable tissues preclude the role of a template for an excitable proto-living structure for all materials except phosphates. Thus, polyphosphates must be regarded as essential in the origin of life. It has been shown that under exceedingly mild conditions and in the presence of a small amount of water, it is possible to prepare polypeptides from amino acids, polyglucoside from glucose, polyribosides from ribose, adenosine and adenylic acid from adenine, and polyadenylic acid and polydeoxythmidylic acids from their respective monomers through the use of a high molecular weight polyphosphate (Schramm, 1960; Schramm et al., 1962). In the primaeval scene, significant amounts of water-soluble polyphosphate salts would have been continuously leached from the igneous phosphate rocks and fed into the streams and runoff water flowing toward the shores (Gabel, 1965). Since the monomers of several biologically significant polymers were present in the primaeval broth, and since the polymerization of these monomers occurs in the presence of small amounts of water, the polymerization process could have been initiated and then continued within the matrix of the proto-living filament or membrane. According to Gabel, as the concentration and the chain length of polymers increased, they would have also assumed increasing importance in the structural maintenance of the evolving proto-life membrane. Moreover, due to the arrangement of the membrane components (Gabel, 1965; Macovschi, 1966; Oparin, 1976), hydrophobic material would tend to accumulate in the interior and hydrophilic material on the exterior surface, with a tendency for the formation of a hydrophobic-hydrophilic double layer when the proto-life membranes were evolving at a liquid-gas or liquid-solid interface.

These observations relate to the notion of the *unit membrane* model, proposed by Danielli and Davson in 1935 for cell membranes. In this model, proteins were viewed as attached to the inside and outside of a bimolecular leaflet of phospholipids by ionic forces. More recently, cell membranes were depicted as a mosaic of functional units, formed by lipoprotein complexes, in which proteins formed the core while phospholipids were attached on the outside, probably by Van der Waals and coulombic forces and by hydrophobic bonds (Benson, 1966).

If a proto-life membrane were folded upon itself to form a vescicle or micelle (Kavanau, 1965), such a structure would grow and reproduce, providing that the internal materials could be replenished from a metastable outer surface. Material exchange between the micelle and the environmental solvent could occur either passively by virtue of pores or actively through energy-providing metabolic events. The polyphosphates, being amphoteric, would be capable of complexing with external cations, and this would result in an altered acid-base equilibrium or an excitation phenomenon; those proto-life membranes and micelles which had incorporated proteins and porphyrins and in which synthetic photochemical processes had evolved would survive. The initiation of protein synthesis (Crick et al., 1976) then represented another important step in this sequence, due to the role of these compounds in the excitable cell membrane (Danielli and Davson, 1935; Kennedy, 1967; Robertson, 1960) and to their function in bioelectrical phenomena (Nachmansohn, 1970). However, other viable proto-living structures could have developed anaerobic metabolism, utilizing organic materials through the photosynthetic process of other proto-life forms. Most probably, the first organization of life did not acquire excitability but was excitable by the very nature of its organization; during the evolutionary process, the proto-living structures which survived would be those chemically or biochemically most capable of utilizing resources of their environment (Gabel, 1965).

This consideration implies that biological evolution took over from chemical evolution as soon as life had been formed, contemporaneously evolving in the principle of natural selection as result of competitive growth (Sylvester-Bradley, 1976) and adaptation. According to Oparin (1976), life precursors, as represented by high molecular weight compounds and primitive living entities, repeatedly developed, disintegrated, and emerged again in different areas of the earth and at various times. Therefore, primitive organisms must have coexisted for a long time, and probably for several hundred million years, with simpler earlier life stages. In this general framework, the survival of any living structure must have depended on plasticity and adaptability to environmental stimuli.

As a consequence, life can be considered as the expression of a metastable and dynamically evolving equilibrium that continuously undergoes environmental interaction. Any alteration of such a dynamic equilibrium implies the concept of a decreased adaptational state, or a *disease;* if the alteration is severe and adaptation is irreversibly lost, the concept of disintegration or *death* emerges (Valzelli, 1977*a*). Thus, in all systems or organisms, the presence of life must coincide with the presence of nervous activity or of a nervous system that permits the organism to respond, within certain limits, to the needs of survival. The distinction between *nervous activity* and *nervous system* appears useful in that some living systems, such as vegetables, do not show any apparent organized nervous structure, but are capable of clear nervous activities and reactions.

PRIMITIVE NERVOUS SYSTEM

The relationship between the chemical composition of living organisms and that of natural environments strongly supports the idea that life began in a water-rich environment interfacing with the primitive atmosphere of the earth (Banin and Navrot, 1975). This does not justify or impose a requirement on existing nonterrestrial or cosmic explanations for the origin of life (Crick and Orgel, 1973). Conversely, the aqueous initiation and development of life can add further value to those studies with elementary marine organisms concerned with the primitive organization of the nervous system.

Considering the evolutionary alternatives of proto-living vescicles, only two possibilities for the development of new living entities appear probable; first, an increase of the primitive and microscopic size to improve original performance with better probabilities of adaptative survival; second, the multiplicative assemblage of several proto-elements to evolve into new forms of life, more suitable to cope with environmental demands and future adaptative changes.

Living and growing processes of vesicular or unicellular organisms initially consisted of biochemical transformations and an assimilation of nonliving proto-elements; it is quite likely that, in parallel, a "cannibalic" predatory activity of the most viable elements toward others also developed. An increased size of *predators* would be of advantage for survival and natural selection. However, the increased size of unicellular element is finite. These limits are defined by a positive relationship between increased volume and decreased exchange; such exchange across square areas of surface varies negatively with cubic size of volume. This probably explains why living "gigantic" amoeboid organisms cannot exist, and why this theoretical evolutionary route did not consistently emerge. Thus, the exclusive unidirectional evolution for life on earth has been that founded on the assembly of several cells, eventually requiring a regulatory system for the coordination of functional components of the new organism.

In 1872, after the discovery of the neuromuscular cell in Hydra, Kleinenberg suggested that nervous activity and reflexes originated from the evolutionary division of what had once been a single cell. Later Hertwig and Hertwig (1879) identified the neuromuscular cell as an epitheliomuscular element, supporting the view of a simultaneous evolution of nerve and muscle cells from separate epithelial cells, as they are formed in coelenterate development. Afterward, and for nearly 50 years, the dominant theory of the evolution of the nervous system had been that of Parker (1919), who based it on three successive phylogenetic stages. According to this theory, there were initially only *independent effectors,* as is currently represented in sponge myocytes or in coelenterate nematocysts; second, receptor cells evolved from undifferentiated epithelium near the muscle cells, so that the most primitive nerve cells would have been sensory, receptive cells, as occur in coelenterate sensory epithelium; finally, *protoneurons* evolved between receptor and effector, eventually originating the reflex triad of receptor, adjuster, and effector.

Although widely accepted for many years, Parker's theory of independent effectors has undergone recent neurophysiological (Bishop, 1956; Grundfest, 1959) and biological (Pantin, 1956; Passano, 1963; Passano and McCullough, 1962) criticisms. A fundamental idea is that behavioral mechanisms of the newly assembled organisms, the *metazoans,* must have involved the structure of the entire animal in order to be sufficiently organized to meet various behavioral requirements. Then, just as in coelenterates of today, the effector can be represented not by a mosaic of single cells acting individually, but by an entire muscle sheet, since innervation of a single muscle cell by a receptor cell could not effect any meaningful response unless all of the units of an entire responsive field were involved (Passano, 1963). The single conductor or protoneuron of Parker's triad cannot function to assure *nervous system integration;* since integration is as fundamental as conduction to any nervous system (Sherrington, 1906), inputs from several receptors must reach common "coordinators" before integration can be achieved. As a consequence, since metazoan behavior is not cellularly organized, the nerve net has probably originated as a supplement to a conducting and integrating muscle sheet, with the primary purpose of allowing the evolution of specialized reflex responses (Pantin, 1956).

Accordingly, it is almost inconceivable that any organized nervous system could have evolved prior to the evolution of muscles or other effectors such as cilia or flagella and pseudopodia (Parker, 1919; Pantin, 1956); consequently, primitive metazoans must have had receptors before they had conductors or protoneurons.

Insofar as the phylogenetic sequence of nervous system structuring is concerned, Passano (1963) suggests that individual protomyocytes (comparable to proto-living excitable vesicles) first evolved into assemblages of independently contractile cells. This allowed the colony more movement than resulted from the contraction of individual myocytes, especially since some cells become activity regulators, or *pacemakers;* this probably occurs through the use of unstable and specialized areas of their excitable membrane, capable of active depolarization. In Passano's theory, such local pacemakers should have synchronized the simultaneous contractions of adjacent cells by passive depolarization spreading throughout the entire contractile field, perhaps utilizing intercellular bridges.

This pacemaker organization would permit the evolution of recurrent activities and feeding movements. The differentiation of what was becoming muscle, specialized for contractions, and what was becoming nerve, specialized for activity initiation, proceeded together, whereas the specialization of nerve cells or neurons for conduction rather than repetitive initiation of activity may be viewed as a secondary development in neuronal evolution (Passano, 1963). In a further sequence, a hierarchy of pacemakers emerged, in that some of them are specialized for the overall control of the organism, whereas others are subordinate centers for controlling specific activities; such development would mark the acquisition of synapses.

In parallel with the development of primitive nerve nets, with their sensory receptors, nerve cells, nerve bundles, and effectors which provide for the *conduction* of impulses, specialized subcellular devices also formed; these are capable of ensuring the *transmission* of nervous activity throughout the entire system. It seems reasonable to suppose that those remote synaptic junctions do not substantially differ from the structural and functional schema that applies to the synapses known today. This implies that primitive synaptic neurons contained neurochemical transmitters, including molecules similar to acetylcholine and possibly octopamine and others.

On this primitive organization, the emerging nervous system would have developed increased and differential sensitivities to various environmental stimuli, and thus would integrate recurring internal activity with rhythmic external events. The further evolution of the nervous system would involve the development of conducting pathways for specialized reflexes, the concentrations of nerve cells as ganglia or nerve rings, and the concentration of receptors into sense organs, as may be observed in the coelenterates (Passano, 1963). Finally, the tendency in primitive metazoans toward a reduction in the number of ganglia and the concentration and cephalization of neurons suggest a centralization process of the primitive nervous system, which occurs later in vertebrates.

VERTEBRATE EVOLUTION

Based on the hypothesis that competition among proto-living organisms created a new evolutionary process, that of natural selection (Sylvester-Bradley, 1976), we can easily imagine that in a primitive environment two opposite poles of feeding behavior could occur; one manifests itself through predatory aggression and the other by defensive strategies. This premise appears warranted in that life in an exclusively liquid world, such as a marine environment, does not offer consistent nutritional alternatives to predation. Hence to some primitive aquatic metazoans the best means for survival was the production of dermal armoured defenses, such as plaques, shells, and chelae; to some others, the development of potent neuromuscular reflexes and of new bodily shapes capable of rapid and powerful movements ensured survival.

The fossil record reflects a complex sequence of increasingly elaborate structural mechanisms in vertebrate development; this is consistent with increasingly enriched behavior. Morphological vertebrate phylogenesis represents confirmation of an adaptive significance for increasing evolutive complexity (Simpson, 1953). It may be easily inferred that, during the several hundred million years required for proto-living precursors to evolve to metazoans, the characteristics of the surrounding environment have been profoundly changed, requiring new behavioral patterns to survive. In this continuously changing frame, the pervasiveness of adaptation in non-nervous vertebrate structures suggests that the evolution of the nervous system may be even more directly adaptive than is generally accepted (Hotton, 1976).

It is obviously not the purpose of this book to review the morphology and history of vertebrates; nevertheless, some data from this field appear relevant to an understanding of the primitive roots of behavioral evolution. In this general context, as schematically summarized in Fig. 1, the earliest record of vertebrates

FIG. 1. The life evolution. General schema outlining life evolution on earth. (Except for birds, molluscs, insects, plants and others.)

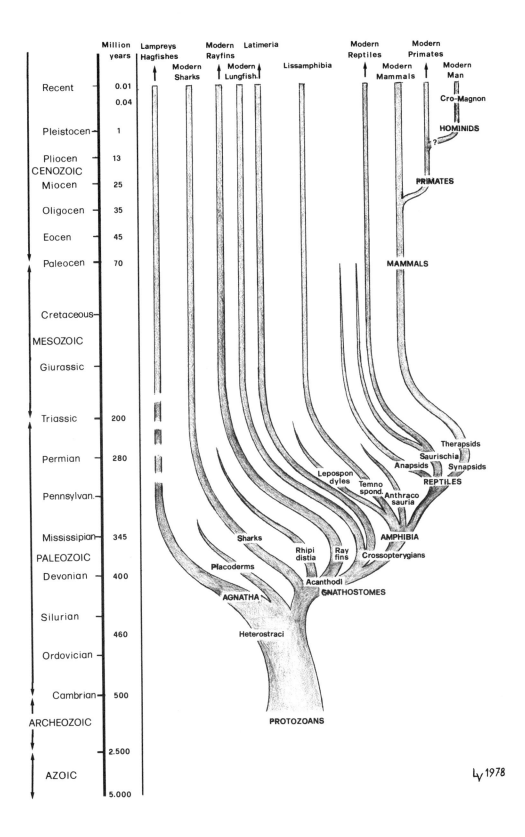

Million
years

Recent 0.01
0.04
Pleistocen 1
Pliocen 13
CENOZOIC
Miocen 25
Oligocen 35
Eocen 45
Paleocen 70

Cretaceous
MESOZOIC
Giurassic

Triassic 200
Permian 280
Pennsylvan.
Mississipian 345
PALEOZOIC
Devonian 400
Silurian
460
Ordovician
Cambrian 500
ARCHEOZOIC
2.500
AZOIC
5.000

Lampreys
Hagfishes
Modern
Sharks
Modern
Rayfins
Modern
Lungfish.
Latimeria
Lissamphibia
Modern
Reptiles
Modern
Mammals
Modern
Primates
Modern
Man
Cro-Magnon
HOMINIDS
?
PRIMATES
MAMMALS
Therapsids
Saurischia
Anapsids
Synapsids
Lepospon
dyles
Temno
spond.
Anthraco
sauria
REPTILES
AMPHIBIA
Sharks
Rhipi
distia
Ray
fins
Crossopterygians
Placoderms
Acanthodi
AGNATHA
GNATHOSTOMES
Heterostraci
PROTOZOANS

Lv 1978

is provided by *Heterostraci,* pertaining to the class of *Agnatha* (Fig. 2), strange frog-tadpole-shaped animals living some 460 million years ago; these measured from 10 to 100 cm in length, were covered by a heavy armour of dermal bone, and lacked grasping jaws. The mouth was represented by a small ventral opening with nearly immobile margins. The lack of paired fins, characteristic of the later jawed vertebrates, and the short tail suggest that the swimming motions of Agnatha must have been mainly accomplished by wriggling from point to point; they probably fed simply by ingesting bottom mud, rich in living and dead organic materials. The fossils Agnatha were directly dependent on the substrate as the main source of food, being capable of ingesting only relatively small particles and small naked molluscs. The adaptative significance of the class may be conceived of as bottom dwelling, preeminently of sedentary habit, and detritus or filter feeding (Hotton, 1976).

In these primitive animals, the initial differentiation of elementary brain structures begins to be recognizable since, at least in the order *Heterostraci,* a pineal

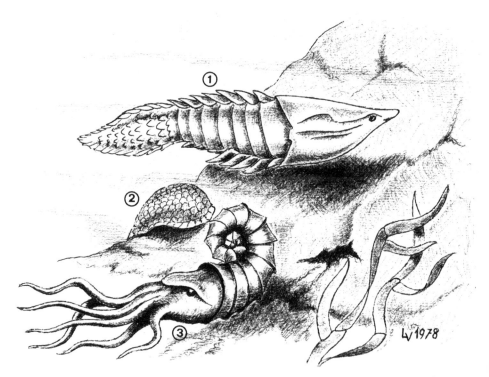

FIG. 2. The first vertebrate.
1. *Heterostracus (Agnata)*
2. *Cystoid Gasteropod*
3. *Spiral Nautiloid*
4. *Alga*

body and semicircular canals were partially preserved as impressions on the internal surfaces of dorsal plates (White, 1935). To Agnatha *Anaspids* also pertained and was characterized by a fusiform-shaped body which antecedes the more efficient swimmers of the fish-like type.

At present, even though important differences from their more primitive ancestors exist, lampreys and hagfishes are considered as the only survivors of this group (Stensio, 1968).

The need for new sources of food, different from mud and the detritus feeding of Agnatha, has probably contributed to the evolution of new animal lines. Nearly 60 million years elapsed before a new and formidable apparatus for predation appeared: the jaws. This fundamental modification first appeared on earth with *Acanthodian* fishes, a subclass of *Gnathostomes,* characterized by a fusiform and fish-like body and a large terminal mouth, supported by jaws armed with efficient teeth. In addition, paired terminal nostrils and large eyes indicate a level of exteroception consistent with the requirements for active hunting.

With the appearance of these actively motile animals, ranging from 2 to 8 cm to about 2 meters in length, the "sedentary predation" of Agnatha was replaced by "active predation" (Hotton, 1976). This form of feeding later evolved into behavior seen in classes of bony fish and sharks, although their precise predecessors remain in question (Nelson, 1968; Stensio, 1963), involving Placoderms, Acanthodians, and others. These species are known to have appeared in the class of *Chondrichthyes* (fishes with a cartilaginous skeleton; Fig. 3) some 340 million years ago, and have since been characterized by a fusiform body and large mouth, with strong jaws and sharp teeth; thus, since their origin, these have always been active and insatiable predators.

The climatic conditions of the surface of the earth had also been evolving with the course of stable seasonal cycles, and some local concurrent situations which may have provided for the further adaptive transformations of water-adapted living organisms of that period. It is possible to determine the alternation of extremely wet and dry seasonal cycles, as still persist in African and South African environments, which influence cyclic restoration and drying of freshwater pools, together with seasonal overpopulation of these nearly rigidly circumscribed basins; these conditions forced the inhabitant animals to develop new anatomical and behavioral strategies for survival.

In the previously cited class of Gnathostomes, comparable with primitive sharks, the *Osteichthyes* of bony fishes appeared; these were characterized by a well-ossified internal skeleton with a neurocranium covered by thin plates of dermal bone, the body coated by small scales, a terminal and large mouth with long jaws armed with teeth. In addition, these animals had two pairs of flexible fins, allowing a high degree of swimming maneuverability and active predative behavior. Bony fishes at that time were essentially represented by two subclasses: the *Actinopterygii,* or rayfins, which over the past 200 million years evolved into a large variety of fish that dominated almost all aqueous

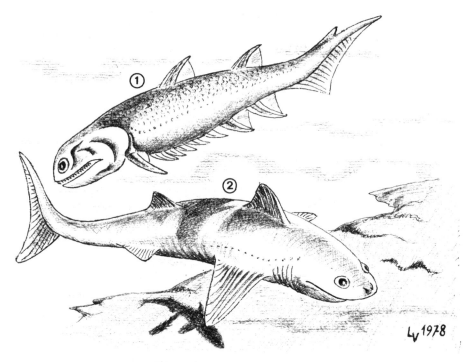

FIG. 3. Gnatostomes.
1. Acanthodian
2. Late Devonian shark

environments; and the *Sarcopterygii,* or lobefins, which included the precursors of tetrapods (Hotton, 1976).

Lobefins derived their name from the fact that the initial part of the fin skeleton, covered by the muscles that operate in the fin, protrudes from the body wall as a rudimental and narrow limb from which short and slim bony rays that support the fin originate (Fig. 4).

These animals, ranging in length from 25 to 150 cm, included the orders *Crossopterygians* and *Dipnoi* or lungfishes; interesting differences between these were the jaw suspension and dental armament. The crossopterygian jaws, the upper of which were movable on the braincase, as in most bony fishes, were also long and armed with large conical teeth, revealing the highly predatory capability of these aqueous animals, resembling the pike or fresh-water dogfish. The upper jaws of lungfishes instead were firmly fused to the braincase, shorter than those of crossopterygians, and with flattened and strongly ridged teeth. Such a tooth morphology suggests a multiple shearing capacity, and the aggressive disposition of living lungfishes indicates that their ancestors would hardly have been less predatory than crossopterygians. Lungfish would have utilized the lobate fins in "walking" on the bottom in search of food, whereas the lungs,

FIG. 4. Crossopterygian.

by which lobefins can breath air, enabled them to survive when waters became contaminated.

It may be inferred that, when water pollution became severe, some lobefins solved the problem by walking from a fouled and crowded pool to others large enough to permit survival until the rainy season came again. These animals thereby became exposed to some of the selective effects of a land environment, and the survivors of this experience eventually evolved as tetrapods (Hotton, 1976).

From fossil records, it may be estimated that the time required for the first vertebrate to reach land from his original aqueous environment through adaptive evolution was approximately 120 million years; this corresponds to 340 million years ago.

At that time, an amphibian of the order *Ichthyostega* had priority of terrestrial adaptation. The representatives of this order ranged from the size of a modern salamander to more than 5 meters in length, had a short trunk and stout and massive limbs; these features reflect the requirements of a land environment, where the effect of gravity was no longer counteracted by the buoyancy of water (Fig. 5).

Most of *Labyrinthodont* amphibians, from which the class *Reptilia* later

FIG. 5. Amphibia.
    1. *Ichthyostega*
    2. *Temnospondylous*

evolved, retained an aquatic habit throughout their life cycle, which indicates their terrestrial adaptation. The opportunity for adaptative evolution in a terrestrial environment has been related to new sources of food. As a matter of fact, when some crossopterygians began to feed on land, the terrestrial surface supported a diversified flora on which some insects had already been living (Hotton, 1976). These two forms of life were insignificant competitors with vertebrates, thereby facilitating the ultimate settlement of amphibians on the earth's surface. Terrestrial adaptation, however, required the elimination of the aquatic larval stage, and this was accomplished in reptiles through the origin of amniotic eggs, which represents the characteristic of the reptilian organization available in some tetrapods of 300 million years ago (Carroll, 1969).

Even though the early origin of dinosaurs remains unclear (Cox, 1976), the adaptative theme of reptiles seems to have coincided with an enhanced trend toward the increasing terrestrial propensity shown by earliest amphibians. Reptiles then became the dominant land faunas, in a large variety of types, over some 200 million years. Among the different kinds of reptiles, the *Synapsida* and *Anapsida* are perhaps the two best documented reptilian subclasses (Carroll and Baird, 1972; Reisz, 1972), and Synapsid, from which mammals ultimately spring, dominated the terrestrial surface for about 85 million years. The primary descendants of the Synapsids are represented by the order of *Pelycosauria,* highly predatory animals ranging in length from about 30 cm to more than 4 meters.

FIG. 6. Reptiles.
1. *Therapsid Cynodont*
2. *Pelycosaurian Dimetrodon*
3. *Tyrannosaurus Rex*

During the second half of the period during which they were dominant, Synapsids underwent a secondary adaptative descent with the order *Therapsida* which produced a variety of large and small herbivores, the first of their kind among tetrapods, insectivorous and large and small voracious predators. Among the therapsidian predators some features, such as endothermy (Seymour, 1976), large heads accommodating long dental batteries, differentiation of teeth, appearance of secondary palate and nasal turbinals became evident; these features anticipate the advent of mammals. Finally, in *Cynodont* reptiles (Fig. 6) the skull and general features became mammal-like in most details, and mammals eventually derived from some of these animals.

Based on the above briefly outlined scheme of vertebrate evolution, adaptation and competition appear important in stimulating developmental processes. There are arguments against this assumption which reject the innovative properties of adaptation, but, according to Hotton (1976), when fossil records and functional role are properly taken into account, it becomes evident that the differences between vertebrates express the result of adaptive processes directed toward survival. Vertebrate evolution suggests that in a new environment living organisms multiply rapidly and soon attain environmental limits. When this occurs, the multipotentiality of structural components of the organisms permits some animal forms to survive just at these limits.

This trend may be recognized in the sequence of events by which acanthodians derived from agnathans, crossopterygians from acanthodians, amphibia from crossopterygians, and reptiles from amphibia. In contrast, a similar trend does not apply to mammals because of their origin, which is still not as clearly defined as the origin of earlier classes. A similar consideration applies to man, who does not show any consistent adaptive improvement or advantage over his presumed ancestral class.

What is of fundamental importance is that no adaptive evolution can occur if there is no suitable receptive system to respond to the demands of environmental change. Therefore, in this context the concept of life must coincide with the presence of nervous activity or of a nervous system, and also, adaptive evolution can exclusively operate in the presence of a nervous system.

BRAIN EVOLUTION

The sequence of brain evolution remains at best a matter of inductive reconstruction. Such a reconstruction suffers from a bias of reliable data since nervous system structure remains poorly represented in fossil records. Only a few primitive vertebrates which have survived in current fauna can provide some information about the brain organization of their more remote ancestors. The broad similarity between the nervous structures of living animals today means to some that the central nervous system is evolutionarily conservative, whereas to others the limitations of established phylogenies suggest that brain evolution is by definition inaccessible.

The adaptative plasticity of the skeletomuscular system and of the specialized sense organs clearly results as the dominant theme of vertebrate evolution, and the brain must reflect this plasticity because of its close correlation with these structures (Hotton, 1976). From another viewpoint, brain plasticity is such that changes in environmental characteristics consistently provoke biochemical and anatomical variations of the brain of young animals, which also reflects upon their behavioral patterns (Bennett et al., 1964; Rosenzweig and Bennett, 1969). It might also be possible to infer that environmental changes that persist for thousands or millions of years can produce brain changes which promote bodily modifications required for behaviors needed for survival. This obviously does not mean that the brain has been capable of creating new organs that did not previously exist, but that it may have been capable of modifying an existing organic mechanism for more proper functional adjustment to altered environmental demands.

The fields of neuromorphology and comparative neuroanatomy have established that the cerebellum was not present in invertebrates and that it became a highly specialized component of the central nervous system exclusively in vertebrates. Cerebellar evolution, as partially summarized in Fig. 7, agrees well with the above considerations of the reciprocal influences of central nervous system development and environmental demands.

The cerebellum is evidently utilized as a neuronal machine (Eccles et al., 1967) in computing complex information about the relationship of the living organism to the external world; its development must bear a relationship with the control of movement, as reflected in the struggle for survival (Eccles, 1969).

Recalling the "sedentary predation" of the agnathans and "active predation" of the gnathostomes, it is not surprising to observe that in the indirect progeny of the first, the lampreys, the cerebellum is rudimentary (Nieuwenhuys, 1967), whereas it is remarkably well developed in sharks which are probably the most active sea predators.

Similar differences may be observed in those modern fishes which display reduced sea bed activity, and in which the cerebellar volume is from 3.2 to 6.6% of the total brain volume; highly active fishes that are active predators possess cerebellums which comprise 12.6 to 15.2% of the total brain volume (Blinkov and Glazer, 1968). In modern amphibians and reptiles, the cerebellum is less developed than in fishes, and barely developed, whereas in mammalian brain it becomes a highly specialized structure.

Due to the increased degree of specialization of several brain regions and structures, in mammals the percent ratio between brain and cerebellum volume varies from 14% for lowest mammals to 12% for some primates and 11% in humans (Blinkov and Glazer, 1968).

The line traced by evolution represents a continued need for meeting the demands of life. Competition and predatory behavior probably represent the major motivated states for several hundred million years, as suggested by fossil records. This implies that primitive brains, and in particular those of the various

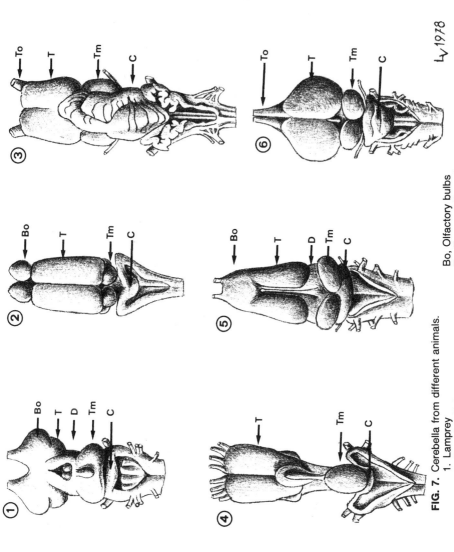

FIG. 7. Cerebella from different animals.

1. Lamprey
2. Bonyfish Polypterus
3. Shark Carcarias
4. Lungfish Protopterus annecteus
5. Frog
6. Alligator Mississipiensis

Bo, Olfactory bulbs
To, Tractus olfactorious
T, Telencephalon
D, Diencephalon
Tm, Tectum mesencephali
C, Cerebellum

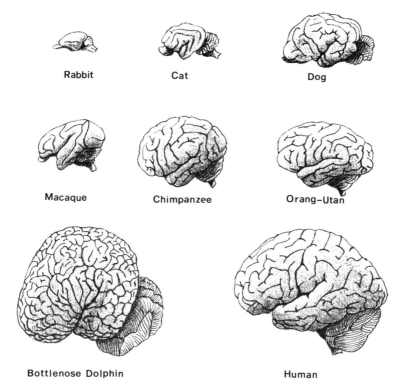

Rabbit   Cat   Dog

Macaque   Chimpanzee   Orang-Utan

Bottlenose Dolphin   Human

FIG. 8. Various Mammalian Brains.

animal forms from Gnathostomes to Reptiles included, must have been especially settled and adjusted for continuous hunting, aggression, and predatory behavior. This also implies that the means for locating and detecting prey would have been as efficient as the motor apparatuses devised to catch and to kill it. It may be further inferred that because of nonexistent feeding alternatives, cannibalic predatory behavior may have been performed frequently; those brains were without efficient mechanisms to limit their innate destructiveness.

A brain formed from an assemblage of coarse neuronal components consists of efficient and enlarged visual and auditory centers, thalamic and hypothalamic structures, motor ganglia and a cerebellum, whereas olfactory structures reach their greatest development in vertebrates which have left the water for land. This broad reconstruction does not appreciably deviate from the general structure of the brains of modern descendants of those ancient animals (Fig. 9).

A relatively simple brain, with minor generic modifications, has been adequate for the maintenance of animal life, since it permitted stable survival and continuous evolution for hundreds of millions of years. The brain of the largest dinosaurs, such as *Tyrannosaurus* or *Brontosaurus* (measuring from 14 to more than 22

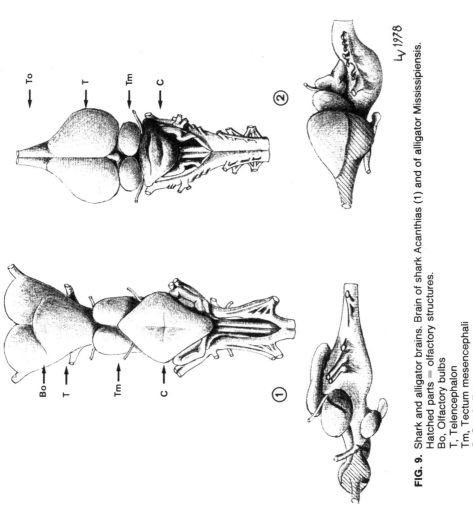

FIG. 9. Shark and alligator brains. Brain of shark Acanthias (1) and of alligator Mississipiensis.
Hatched parts = olfactory structures.
Bo, Olfactory bulbs
T, Telencephalon
Tm, Tectum mesencephali
C, Cerebellum

Lv 1978

meters in length) reached a proportionally increased complexity with a very small ratio between brain volume and body size; these data have been provided in fossil records (Fig. 10).

Primitive brains should have allowed the owners to function mainly as cybernetic machines, programmed for feeding, reproduction, and self-preservation. Behaviors, however, should have mostly been confined to almost automatized reactions to specific stimuli, and then mainly rigidly repetitive and highly codified responses, as may be observed in birds and in other reptilian descendants.

The limited flexibility of such a brain, in view of increasing environmental complexities, deriving from the multiplication of other animal forms or reduced size, gives rise to new manifestations of competition; this could constitute one of the several and controversial reasons for dinosaurs' disappearance. According to Carroll (1969), the significance of size lies in the fact that problems of support, movement, and fresh-water conservation in a terrestrial environment are less difficult for small animals than for large ones. In small reptilian representatives wherein the ratio between brain mass and body size was more favorable than in dinosaurs, the mammalian forms derived in which a proliferation and refinement of brain structures permitted new behavioral resources and a more suitable environmental adaptation.

A newly evolved brain structure consisted of new structures and integrated formations; these coincide with the limbic system of the brain of modern mam-

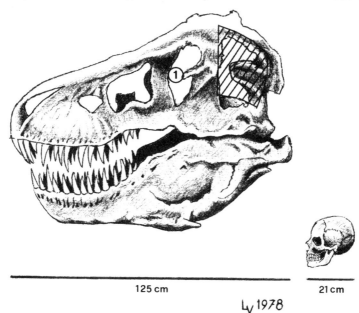

125 cm   21 cm

Lv 1978

FIG. 10. Skull of Tyrannosaurus Rex compared to that of Homo Sapiens.
1) Orbital opening.
Hatched part = brain case.

mals. This nervous system is known to derive information from its environment in terms of affective and emotional feelings (Fulton, 1951; MacLean, 1955, 1958; Papez, 1937); through these new properties, the brain structures evolving from reptiles (Therapsid Cynodonts) to mammals acquired functions for self- and species-preservation with a wide range of adaptive flexibility. This new brain, which represents a common denominator for the nervous system of the various mammals, including man, has served animal evolution well for about the last 70 million years.

Coinciding with the appearance on earth of the earliest primates some 20 million years ago, the brain underwent its last transformation with the development of a sophisticated neuronal mantle, the neocortex (Sanides, 1969; Stephan and Andy, 1969), which has increased the flexible capacity and response complexity of man. Fossil evidence for the evolution of the brain and torso in mammals indicates that there has been a progressive increase in relative brain size (Jerison, 1970). A similar trend is also present in the progression of fossil hominids to man (McHenry, 1975*a*); paleontological evidence suggests that bipedal walking should have evolved before that brain started to expand (McHenry, 1975*b*). The hypothesis outlining human evolution as derived from arboreal quadrumanous animals descending to the ground, adapting to bipedalism, and eventually evolving into large-brained *Homo sapiens* was first proposed by Lamarck (1809) and further developed and refined by Darwin (1859, 1872*a*) and Haeckel (1868). Nevertheless, the lineage of human evolution has not yet been clearly elucidated (Libassi, 1975; Oxnard, 1975). Furthermore, it is most likely that, from the biological standpoint, the brain of modern man does not substantially differ from that of his direct ancestor of 40,000 years ago, the Cro-Magnon; the differences are mainly cultural and in the applications of brain function (Klemm, 1972).

A further point is that man does not represent a consistent evolutionary improvement on Primate descent. As a matter of fact, from an evolutionary standpoint man not only did not acquire greater biological efficiency, but he has lost several animal skills; thus he runs, climbs, swims, and digs less efficiently than other species, and certainly does not survive well in pure animal conditions (Cockrun and McCauley, 1969). Man's highly developed neocortical brain allows him, for the first time in the history of the earth, to reason, abstract, and to deduce, and to force the environment to adapt to him.

2

Brain and General Behavior

THE NEURON AND ITS FUNCTIONING

The neuron is a highly specialized cell, structurally organized to select information, store it, and then transmit it throughout the living organism. The neuron is, in effect, the communication unit of the body, and the fundamental working unit of every brain or nervous system; its activity is performed through chemical and electrical functions.

The neuron is the most variable cell of the organism as far as size and volume are concerned, so that, within the limits of the nervous system as a whole, the largest neurons may be even 1,000 times larger than the smallest (Blinkov and Glezer, 1968). For example, according to Sholl (1953), in the motor cortex of the cat, neuronal surfaces may vary from 31 to 2,520 μm^2, and the corresponding neuronal volumes from 540 to 11,550 μm^3.

Classically, the neuron is divided into four anatomically distinct regions, each one subserving particular functions.

1. A *receiving* section, corresponding to dendrites, transduces the incoming chemical messages from adjacent presynaptic neurons by means of neurotransmitter molecules into electrical potentials; the dendrites maintain the highest enzyme activity of the nerve cell (Bishop and Clare, 1955; Koelle, 1951, 1954; Lowry, 1953; Pope et al., 1957).

2. The *cell body* contains the nucleus; the tigroid, formed by clusters of osmiophilic granules (Palade, 1955; Shultz et al., 1957) containing enzymes (Sjöstrand, 1956); the Golgi apparatus or agranular reticulum, which plays an important role in the biosynthesis of neurosecretory molecules (Baker, 1954; Nath, 1957); the neurofibrils (Palay and Palade, 1955; Shultz et al., 1957); the mitochondria, in which the metabolic processes for energy production take place (Hydén, 1960); and all of the nonmitochondrial neuronal DNA and RNA; this section is the *protein-synthesizing* site of the cell (Wurtman and Fernstrom, 1974).

3. A *conducting* section is the axon, which contains longitudinally oriented fibrils (Hartmann, 1953), particles of fibrous proteins (Hydén, 1960), and neurotubules (De Robertis and Schmitt, 1948); it carries electrical impulses and maintains an axoplasmic flow of amino acids (Karlsson, 1977), proteins (Grafstein et al., 1970), enzymes synthesizing and metabolizing neurotransmitters (Wooten and Coyle, 1973), and neurotransmitter-containing vescicles to the nerve terminals (Dahlström, 1971; De Robertis, 1964; Geffen and Livett, 1971; Salmoiraghi et al., 1965).

4. A *transmitting* section, formed by the nerve terminals or axonal endings, synthesizes, stores, and releases the neurotransmitting molecules (Carlini and Green, 1963; De Robertis, 1964, 1967; Potter and Axelrod, 1963; Ryall, 1964; Whittaker, 1964).

This general outline of the neuron is depicted in Figure 11.

In mammals the length of the dendrites of neurons in the cerebral cortex is consistently related to the size of the neuronal cell body, which, in turn, is proportional to the body weight of the specific animal (Bok, 1959). For instance, the dendrites of a mouse cortical neuron are 900 μm, whereas in rat, guinea pig, and rabbit they are, respectively, 1,200, 1,675, and 2,100 μm. With the dendrite length and the body weight of the mouse taken as a reference unit, the dendrite length in the rat is 1.33, in the guinea pig 1.86, and in the rabbit 2.33; the corresponding body weight ratios, respectively, are 1.4, 1.9, and 2.2 (Blinkov and Glezer, 1968).

An increase in body weight as proportional to an increased dendritic length may be related to the reflex response rate and behavioral variations among different sized animals; this has relevance for the previously cited relationship between body size, survival, and evolution (Chapter 1, p. 4).

The basic features of neuronal organization, both structurally and biochemically, have not significantly changed during the millions of years of evolution. No substantial differences exist between neurons from planaria, snails, squids, sea anemones, molluscs, fishes, reptiles, and mammals; their neurons contain the same neurotransmitting chemicals found in mammalian brains (Dahl et al., 1963, 1966; Garattini and Valzelli, 1965; Giacobini and Holmstedt, 1958; Gosselin et al., 1962; Welsh and Williams, 1970; Whittaker et al., 1972). The species differences in brain and behavioral complexity are most likely due to differences in organization; at the neuronal level, such differences probably reflect the number of synaptic interconnections. This appears to be a crucial point, since at the synaptic level, through a series of biochemical mechanisms, the transmission of nerve impulses and information from one neuron to others takes place. The number of operating synapses represents a sophisticated degree of anatomical interconnections and functional integrations, and this might be further regarded as one of the mechanisms of cerebral evolution.

Nearly a century ago, Ramon y Cajal (1895) said that "cerebral exercise, since it cannot produce new cells (neural cells do not multiply as do muscular cells), carries further than usual the development of protoplasmic expansions

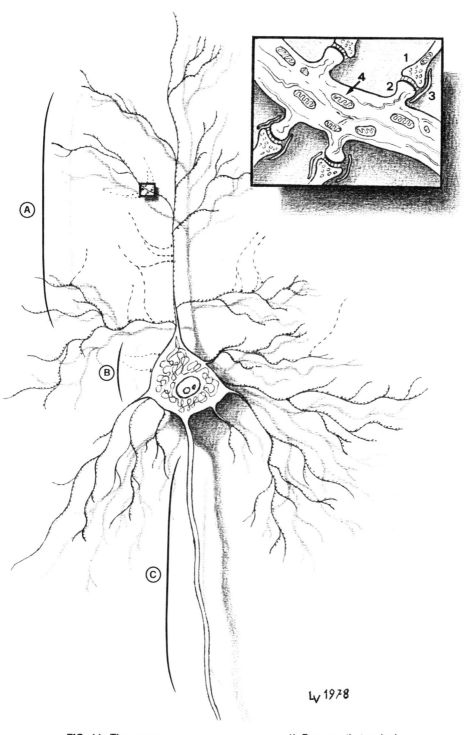

FIG. 11. The neuron.
A) Receiving section
B) Protein-synthesizing section
C) Conducting section

1) Presynaptic terminal
2) Postsynaptic terminal
3) Glia processes
4) Mitochondrium

and neural collaterals, forcing the establishment of new and more extended intercortical connections." Thus, environmental characteristics and the adaptive and multipotential plasticity of the brain are once again represented in the history of evolution. Recent findings have shown that animals reared in complex and enriched environments, which provide high levels of sensory stimulation and socioenvironmental interaction, show increased cortical thickness, with an increment in neuron size, and an increased number of branched dendrites and synapses, which account for a total increase in brain size and weight (Bennett et al., 1964; Diamond et al., 1964, 1966; Greenough, 1975; Rosenzweig and Bennett, 1969). These neuronal changes, together with others involving neurochemical transmitters (Bennett et al., 1964, 1973; Riege and Morimoto, 1970), amino acid incorporation in brain (Levitan et al., 1972*a, b*), and brain RNA content (Ferchmin et al., 1970), subserve marked behavioral changes which include increased emotional stability, facilitation of learning processes and of problem-solving ability, and an efficient memory storage (Forgays and Forgays, 1952; Greenough, 1975; Greenough et al., 1972; Krech et al., 1962; Manosevitz, 1970; Manosevitz and Joel, 1973; Rosenzweig, 1964, 1966).

To thoroughly understand the meaning of these observations, and the biological relevance of synapses, it may be useful to consider that, normally, from 56 to 69% of neuron surface is covered by synapses, which form a characteristic mosaic (Illis, 1964); in addition to these "somatic" synapses, "axodendritic" and "axoaxonic" synapses (Blinkov and Glezer, 1968; Stevens, 1966) provide further neuronal interconnections. In the human brain, which contains at least 30 billion neurons (Harth et al., 1970), each of which may have 1,000 or more synapses and since some neurons of the cerebral cortex have as many as 200,000 synapses, such a setting has been calculated to allow for a potential range of possible operational stages that can number as much as $1.5 \times 10^{3,000,000,000}$.

From a functional standpoint, an individual neuron continuously monitors and summates all external excitatory and inhibitory influences. When the net excitatory input is great enough to reach the threshold of neuronal excitation, it depolarizes the cell, and an *action potential* is generated, subsequently evoking the release of its own neurotransmitter molecules at the nerve endings (Fig. 12).

Each neuron is in fact thought to synthesize and release only a single type of chemical neurotransmitter, which may be excitatory or inhibitory depending on its biochemical effect on postsynaptic neurons; this effect remains a function of the particular ion, the inward or outward flux of which is mediated by a neurochemical transmitter (De Robertis, 1967; Wurtman and Fernstrom, 1974). However, *Dale's principle*, which states that each nerve cell produces and releases only one neurotransmitter (Eccles, 1976), is now likely to be re-examined. There is evidence that some neurons can release more than one transmitter (Burnstock, 1976), while it has been shown that serotonergic neurons can incorporate and concentrate catecholamines under given circumstances (Barrett and Balch, 1971).

Since Dale's principle was formulated some 45 years ago, coinciding with

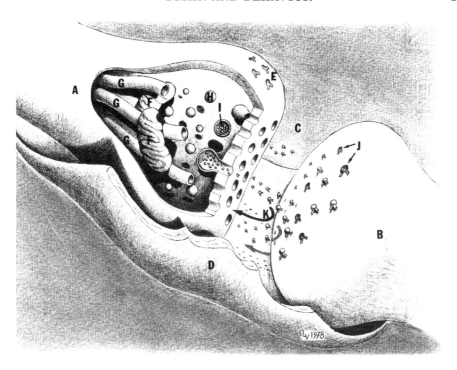

FIG. 12. The synapsis.

A) Presynaptic terminal
B) Postsynaptic terminal
C) Synaptic cleft
D) Glial process
E) Auto-receptor
F) Mitochondria

G) Neurotubules
H) Vescicles
I) Neurochemical transmitter in granular form (granules)
J) Receptors
K) Reuptake

the discovery of norepinephrine and acetylcholine, a variety of other substances have been established or suggested as putative neurotransmitters. These include glutamate, glycine, aspartate, carnosine, glutamine, proline, taurine, octopamine, γ-aminobutyric acid (GABA), dopamine, epinephrine, histamine, serotonin or 5-hydroxytryptamine (5-HT), piperidine, adenosine triphosphate (ATP), and peptides such as substance P, somatostatin, bradykinin, endorphins, and enkephalins (Agranoff, 1975; Burnstock, 1972; Gerschenfeld, 1973; Hebb, 1970; Iversen, 1970; Krnjevíc, 1974; McLennan, 1970; Otsuka and Takahashi, 1977). Another important advance in the field of neurotransmission is the discovery that other substances may be released at nerve endings together with the principal transmitting chemical. In adrenergic neurons, for example, norepinephrine is released together with chromogranin A, dopamine β-hydroxylase, and probably ATP; the role of these accompanying substances and their fate after release remain unclear (Burnstock and Costa, 1975; Smith and Winkler, 1972).

The postsynaptic membrane is believed to carry specialized chemical groups called *receptors,* with which the neurochemical transmitter combines, and which are conceived of as molecular "cavities" on the membrane surface, shaped in such a way that transmitting molecules will fit exactly into them (De Robertis, 1971; Durell et al., 1969; Goldstein et al., 1969; see Fig. 12). As soon as contact takes place, transmission occurs, and the *postsynaptic potential* develops and spreads through the postsynaptic neuron.

GLIAL CELLS

Glial cells embryologically derive from the same group of ectodermal cells that gives rise to neurons. It has been calculated that about half of the volume of the vertebrate brain consists of glial cells, which in humans outnumber neurons roughly 10 to 1. Although glial cells were described more than a century ago (Virchow, 1859), little is understood about their function and role. It has usually been argued that because neurons are closely enveloped by glial cells, the latter serve to "cement" neurons together (Virchow, 1859), to repair or regenerate them (Ramon y Cajal, 1913; Weigert, 1895), to have undetermined secretory functions (Nageotte, 1910), or to serve a nutritive function for neurons (Burns, 1956). It has recently been suggested that glial cells may provide a protein vital for the growth and differentiation of the neuron, based on the discovery that a tumor derived from glial cells contains large amounts of a protein known as *nerve growth factor* (Longo and Penhoet, 1974). Although the hypothesis is legitimate on theoretical grounds, it is weakened by the fact that other neoplastic cell lines from subcutaneous and adipose tissues, embryonic tissues, and several other sources also release nerve growth factor (Bueker et al., 1960; Levi-Montalcini, 1975; Oger et al., 1974).

Interestingly, as early as 1886, Nansen suggested that glia represent a basis for intelligence and are correlated with it, since the volume of these cells increases from lower organisms to the higher species. In addition, glial population has also been described as increasing as a function of the maturation processes of the brain and the age of the animal (Brizzee et al., 1964; Johnson and Sellinger, 1971). Analyzing glial function on this basis, which directly involves behavioral production, Galambos (1961) hypothesized that glial cells collaborate with neurons in determining animal behavior. Glial cells are thought to "plan" neuronal activity to such an extent that the latter presumably act only in response to the instructions sent to it by glia.

A functional and biochemical re-evaluation of glial cells has been made also by Hydén (1960) and Giacobini (1961), and the observation of glial processes interposed in the intersynaptic cleft (De Robertis, 1956; De Robertis and Bennett, 1954; Gray, 1959; Palade and Palay, 1956) could consistently suggest that these cells interfere in the mediation of nerve impulses, somehow modulating transmission. Recently binding sites for dopamine and dopamine-sensitive receptors have been found in glial cells, which suggests the presence of metabolic responses by these cells to the neuronal release of dopamine (Henn et al., 1977). Glia also play a specific role in brain ammonia detoxification, and almost exclusively contain the brain enzyme glutamine synthetase; this enzyme regulates the production in the brain of two putative neurotransmitters, glutamic acid and GABA (Martinez-Hernandez et al., 1977). Since glutamic acid is believed to be an excitatory neurotransmitter, whereas GABA is thought to be inhibitory (Davidson, 1976; Krnjević, 1974), glial cells appear to be involved in the regulation of *brain excitability* (Henn and Hamberger, 1971; Krnjević, 1974; Roberts, 1976) and *epileptogenesis,* in which glial abnormalities (Pollen and Trachtenberg, 1970), and irregularities in glutamic acid, GABA, and ammonia metabolism (Davidson, 1976; Meldrum, 1975; Stone, 1969; Wiechert and Herbst, 1966) have been associated with seizures.

Another glial-neuronal interaction is that neuronal discharge evokes the depolarization of glial cells, which is mediated by the release of potassium ions (K^+) from the neuron (Futamachi and Pedley, 1976; Orkand et al., 1966). During depolarization, glia take up K^+ to maintain the constancy of the neuronal environment by the process of *spatial buffering* (Kuffler, 1967; Trachtenberg and Pollen, 1970). Alterations of glial K^+ uptake can contribute to the generation of neuronal epileptic discharges (Pollen and Trachtenberg, 1970), and, conversely, an experimentally induced overloading of glial cells with sodium ions impairs the glial K^+ uptake and spatial buffering, and induces both glial swelling and epileptiform firing of adjacent neurons (Grossman and Seregin, 1977).

These recent findings should be considered as an indication of the importance of glia in regulating brain functions, and consequently in behavioral production; it seems no longer possible to disregard this brain component in the evaluation of central nervous system activities.

BEHAVIOR AS BRAIN PRODUCT

As can be easily deduced from the analysis of behavioral studies, most behaviors produced by any living organism arise from and subserve the various aspects of the demand of survival. Different neural substrates of the brain are alerted and mobilized in several ways to allow the organism to satisfy needs, to resist enemies, and to neutralize threats.

Subcortical brain structures are involved in *motivation* and *emotion,* which may be considered as the oldest and the major mechanisms subserving survival (McCleary and Moore, 1965; Smythies, 1970).

Motivation, the first of these two mechanisms, involves awareness of the physiochemical balance of the body, in such a way that any condition that arouses hunger or any other basic needs results in that behavior or sequence of behaviors which satisfies the need and maintains the integrity of the body. This arrangement, which operates overtly in driving animal and human behavior, from the phylogenetic standpoint must have almost certainly been the primitive way of behaving. This means, in turn, that primitive nervous systems were organized and adequate to transform basic *drives*—such as hunger, thirst, sex, and self-preservation—into their corresponding pattern of behavior: feeding, drinking, reproductive and, according to circumstances, defensive, hostile, and aggressive behaviors.

We might also speculate that innate behavioral responses were not sharply limited by a relative overabundance of feeding resources, selection, and survival of the most viable forms of life. Some conception of this may be given by the insatiable and voracious predation displayed by the modern representatives and descendants of ancient reptiles, the brain and behavioral organization of which do not allow for a fine modulation of feeding or for a clear separation between it and aggressive reactions.

Interestingly, parental or maternal drives, which sustain the most strongly motivated behavior of mammals (Jenkins et al., 1926; Nissen, 1930a, b; Stone, 1942), do not exist in reptiles, and one of the behavioral observations which distinguishes the latter from the former, is based essentially on the absence of parental concern (Cockrum and McCauley, 1965).

Within the context of motivation, the general activity of the central nervous system, as expressed by several behavioral patterns, can be directed toward maintenance of the *internal homeostasis* that results from the balance between basic needs and their satisfaction. This suggests that even the most elementary pattern of behavior may be represented by bipolar forces of *reward* (satisfaction, gratification, pleasure) and *punishment* (dissatisfaction, frustration, pain), with the obvious aim of avoiding punishment and achieving satisfaction. This setting must be considered the most primitive in the phylogenesis of behavior, since it is evident not only in mammals but also in fishes and reptiles (Campbell, 1972).

Because reward cannot exist without its unpleasant counterpart, the excess of one of these two polar forces at the expense or without the counterbalancing effect of the other immobilizes behavioral output and reduces brain activity. Evidence of this can be found in experiments of *intracranial self-stimulation* (Milner, 1970; Olds, 1958, 1962; Olds and Milner, 1954), in which animals with electrodes implanted in the septal "pleasure" area learned to press a lever to stimulate their own brain at rates of up to 7,000 responses per hour. The animals never achieved satiation, and disregarded food and water until they became exhausted and died.

Animal behavior becomes as complex and rich as the complexity of brain development; this indicates that internal homeostasis develops on an enriched variety of multiple gratifications which, in turn, reflect a multiplication of needs, which may include emotional, affective, learned, and cultured components. The balance of internal homeostasis coincides with that of adaptation (Chapter 1, p. 14), so that a disturbance of homeostasis is a basis for disease, whereas its complete destruction causes death (Valzelli, 1979a). Reward and punishment have well-defined anatomical substrates within the brain (Fig. 13) of both animals and man (Delgado et al., 1954; Heath, 1963, 1964, 1972, 1975; Nauta, 1963; Milner, 1970; Olds et al., 1964; Poschel and Ninteman, 1963; Wright and Craggs, 1977).

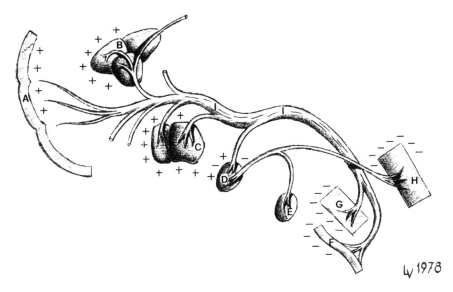

FIG. 13. Reward (+) and punishment (−) system.

A) Frontal cortex	F) Ventral tegmentum
B) Septal nuclei	G) Lateral tegmentum
C) Hypothalamus	H) Periaqueductal gray matter
D) Mammillary body	I) Medial forebrain bundle
E) Interpeduncular nucleus	

It has also been suggested that the neural system regulating reward and punishment might be involved in human mental illness (Stein and Wise, 1971); it might provide a physiological explanation for Freudian theories of Eros and Death instincts (Freud, 1920), as well as for other psychoanalytic issues (Glusman, 1975; Kelly, 1975).

As noted previously, emotion is the second mechanism subserving and promoting survival (McCleary and Moore, 1965; Smythies, 1970). Emotion greatly enlarges the range of behavioral responses, making them more specific and better suited to environmental demands. Darwin (1872b) first analyzed the survival value of emotions in terms of their motivating properties, and he concluded that fear, by motivating the animal to caution, subserves species preservation—as do anger, by motivating it to remove obstacles to survival, and friendliness, by promoting socialization.

Feelings such as joy, anxiety, sorrow, anger, and fear; and responses such as activity level, vocalization, hostility, and wildness or timidity refer to emotions. Emotions are the precursors of affect, and the influence of emotions on behavior results in *emotional* or *emotional-affective behavior,* which has also been called "temperament" (Hall, 1941). Considerable evidence has suggested that differences in emotionality are inherited (Broadhurst, 1957, 1960; Hall, 1941; Tryon et al., 1941a, b; Valle, 1970; Wilcock, 1968), and that a number of factors such as handling, sex, testing, social interaction, environmental characteristics, and probably several others can interact and interfere with and modify the underlying emotional state (Archer, 1974; Bennett et al., 1964; Denenberg and Morton, 1962a; Denenberg et al., 1964; Essman, 1968; Henderson, 1973; King and Appelbaum, 1973; Manosevitz and Montemayor, 1972; Masur, 1972; Russell and Williams, 1973; Wilcock, 1968; Williams and Russell, 1972).

Emotions and affect produce new drives in mammals and several motivated behaviors not observed in reptiles; these include a maternal (or parental) drive, hoarding, and social behavior. Hoarding is a behavior pattern by which many mammals, including man, store food in excess of their immediate needs (Morgan, 1947; Stellar and Morgan, 1943; Wolfe, 1939). This behavior may be seen in animals as a rudiment of what may become *cognition* in man, in that by hoarding the animal seems to anticipate possible future difficulties in obtaining sufficient food.

Behavior enhanced and elaborated through evolution reached its greatest complexity in man; consequently, the maintenance of a suitable internal homeostasis requires an increase in the sources of reward, with which there is a corresponding increase in frustration and punishment. In this increasingly intricate framework, frustration and reward play an important role, especially during infancy, when frustration becomes of primary importance in shaping adult behavior (Hunt, 1941; Hunt et al., 1947; Turner et al., 1969; Widdowson, 1951; Yarrow, 1961). This occurs as a function of the degree to which reward reinforces and consolidates a given behavioral response, whereas frustration thwarts the expected gratification (Hunt and Willoughby, 1939; Scott, 1958a, b).

To education, it is important that punishment and frustration can attenuate or abolish those behavioral patterns that are partially or completely replaced by socially appropriate or environmentally adequate responses. In this connection there is the often debated question of what behaviors are normal or abnormal. This argument represents fertile ground for bizarre philosophical, sociological and political lucubrations, which occasionally demonstrate the exact opposite of what was claimed before. However, from a biological standpoint, normal behavior must be considered as a "continuum," with regard to both spontaneous and learned environmental responses formed by a harmonic transition of response patterns one into the other according to environmental demands; and in this framework, abnormal may be considered as any behavioral response that dominates the others, regardless of the demands of the environment, and impairs the level of behavioral productivity.

In studying behavior, Hoge and Stocking (1912) first demonstrated that discriminative learning is facilitated more by shock for errors than by food for correct responses, and that a combination of shock and food, that is, of punishment and reward, produced the most efficient results. This relative value of punishment and reward in learning has been confirmed by others (Diserens and Vaughn, 1931; Warden and Aylesworth, 1927). It has been concluded that punishing the incorrect responses while rewarding the correct ones leads to more efficient learning than does punishment or reward alone (Munn, 1950).

The neuroanatomical substates of emotional-affective behavior are best represented by the *limbic system* of the brain (Valzelli, 1978*a*), which was first described by Broca in 1878. This brain system constitutes an important advance in the phylogeny of brain evolution and represents a region found in the oldest mammals (Papez, 1958); it is an anatomical common denominator in mammals and man (Anand, 1957; MacLean, 1958; see Fig. 14), which mediates emotional feelings and produces the behavioral patterns required for self- and species preservation (Fulton, 1951; MacLean, 1955, 1958; Papez, 1937).

In general terms, the primitive limbic cortex is structurally less complex and complicated than the neocortex and, although the limbic system was once believed to receive information principally from the olfactory and visceral systems, such that it was named the "visceral brain" (MacLean, 1949), it was later shown that signals reach it from oral, visual, auditory, and somesthetic receptors (MacLean, 1955; MacLean and Delgado, 1953).

A third mechanism of survival, and probably the most subtle, is *consciousness*. Consciousness, originally defined as self-awareness of one's knowledge and thought, is now more largely intended as awareness of all experiences, including sensations, perceptions, emotions, affects, desires, and volition. Aside from the initial negation of consciousness by behaviorism (Watson, 1913) and the subse-

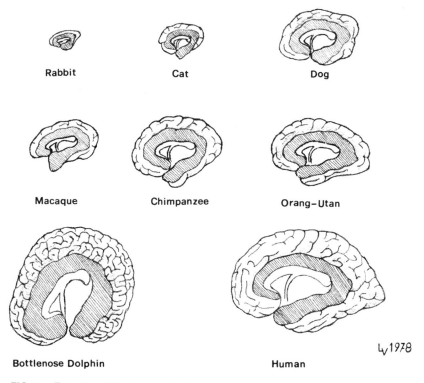

Rabbit   **Cat**   **Dog**

Macaque   **Chimpanzee**   **Orang–Utan**

Bottlenose Dolphin   **Human**

FIG. 14. Extension of limbic lobe (hatched area) in various mammalian brains.

quent restoration of the concept (Burt, 1962; Eccles, 1970; Sperry, 1969), as well as the psychoanalytical theories concerned with the three levels of consciousness, subconsciousness, and unconsciousness (Freud, 1961), consciousness biologically deals with awareness and attention, which involve selection and interpretation of incoming stimuli. Both awareness and attention support the relationship between body and mind (Barrett, 1968; Brown, 1976; Bunge, 1977; Zangwill, 1976).

Yet the precise identification of the physiological basis of consciousness remains undefined (Ernst, 1976). The cerebral hemispheres initially were thought to be the seat of consciousness, although not all parts of them were conceived of as equally important for mental life (Fechner, 1860). Later on, some subcortical structures including reticular formation, pons, thalamus, and hypothalamus were shown to be responsible for the activation of cortical activity (Allen, 1932; Hess, 1954, 1967; Morison and Dempsey, 1942; Moruzzi, 1954; Moruzzi and Magoun, 1949), eventually resulting in a conscious state. Further, more recent studies have confirmed that, within the visual system, distinct neocortical areas are activated, and subserve operations of different cognitive complexities (Buser, 1976).

Since these observations have been made almost exclusively in experimental animals (Mountcastle, 1968; Rose, 1973), there has been some debate about whether consciousness is exclusively a human attribute. Evolutionary data offer no great assistance in this issue, mainly because it is not clear at which point in the supposed ascent of man consciousness arose. It has been suggested that there is a hierarchy of consciousness which might be correlated with the number of cells (and probably of their interconnections) in the brain, and the size of the neocortex and of association areas. A formula that relates the level of consciousness with these variables would place man at the top of the list, far from the next species in line (Rose, 1973). Even though a brain similar to that of mammals has been claimed to be unnecessary for consciousness as defined by its activity (Ernst, 1976), consciousness must be regarded as a question not of "quantity" but of "quality." In this context a brain like that of man or, probably better, a neocortex similar to that of man is required to achieve comparable depth of consciousness and a similar range of application.

Consciousness must be intended as the final neocortical product of a complex series of multiple integration processes of the brain. Based on a large number of clinical observations of brain-bisected or *split-brain* patients, Sperry and his co-workers (1969) have been able to demonstrate beyond reasonable doubt that each brain hemisphere may operate, under appropriate circumstances, independently of the other. This suggests that each hemisphere may have its own sphere of consciousness for feelings, perceptions, and other mental activities.

HUMAN TRIUNE BRAIN

A certain mystical connotation has been attributed to the number three, especially in view of the sacred meaning that Pythagoras attributed to it in his ancient philosophy. The concept of trinity is also fundamental to various religions, having widely influenced thinking and beliefs. The unity of man was viewed as a triad of physical, psychological, and spiritual components by Hermes Trismegistus some 15 centuries ago. More recently Yakovlev (1970) said that the structural and functional trinity of the body consisted of three systems of: (1) organs (visual, somatic, and nervous); (2) the brain, as formed by three systems of regulation of body motility (autonomic, extrapyramidal, and pyramidal); and (3) behavior, as divided into three spheres of motility (visceral, expressive-emotional, and manipulative-transectional).

These empirically derived representations have been replicated in other three-item theories concerned with brain functions. We have rapidly reviewed the survival meaning of behavior as founded on motivation, emotion, and consciousness, which may grossly find their corresponding features in Freudian theories of Id, Ego, and Super-Ego, and of unconscious, subconscious, and conscious.

Recently MacLean (1970, 1973, 1976) elaborated on human brain functions as reflecting the evolutionary derivation of their structures; this led the author to view the human central nervous system as a *triune brain.* According to MacLean, the brain of primates has evolved in a hierarchical manner along three basic patterns which mark the main steps of evolution, and which have been accordingly labeled as *reptilian, paleomammalian,* and *neomammalian.* Each of these cerebrotypes has its own kind of intelligence, its own specialized memory, its own sense of time and space, its own motor skills and special functions, and its own neuroanatomical architecture and preferential neurochemical setting. These three kinds of brains are extensively interconnected, and obviously strictly dependent on one another functionally, although there is evidence that each of them is capable of operating somewhat independently under certain circumstances (MacLean, 1976).

In anatomical terms, the human reptilian brain is represented primarily by the assemblage of caudate nucleus and putamen with globus pallidus and peripallidal structures such as the amygdala, the substantia innominata, the basal nucleus of Meynert, the nucleus of the ansa peduncularis, and the entopeducular nucleus; and with thalamohypothalamic structures, corpora quadrigemina, substantia nigra, and the phylogenetically oldest part of the cerebellum (Fig. 15).

This ancient cerebrotype is centered primarily on the corpora striata or *striatal complex,* the functions of which have not been clearly understood, despite more than 150 years of experiments. This is a reflection of the traditional position viewing these structures as merely involved in motor functions, which is disproved by the finding that large bilateral lesions of striatal complexes in mammals result in no apparent motor deficit (Kennard, 1944; Ranson and Berry, 1941). The striatal complex has instead been shown to subserve several basic and genetically transmitted behaviors, such as home site selection, establishment and defense of territorial areas, and hunting, feeding, mating, competition, dominance, aggression, and imitative patterns (MacLean, 1962, 1964, 1972*a;* Ploog and MacLean, 1963; Rosvold, 1968; Stevens et al., 1961). These behavioral patterns coincide primarily with motivation (this chapter, p. 27), thus subserving the

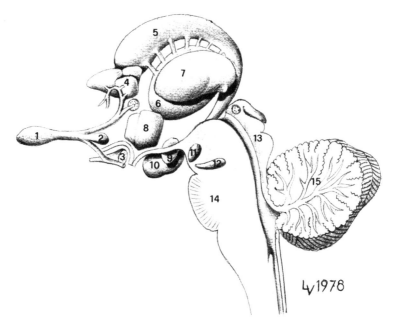

FIG. 15. Reptilian brain.

1) Olfactory bulb
2) Olfactory tubercle
3) Optic chiasma
4) Septal nuclei
5) Caudate
6) Putamen + globus pallidus
7) Talamus
8) Hypothalamus
9) Mammilary body
10) Amygdala
11) Interpeduncular nucleus
12) Substantia nigra
13) Quadrigeminal bodies (Tectum mesencephali)
14) Pons
15) Cerebellum

elementary demands of survival. Equally fundamental to survival is imitative behavior, which allows for differentiating "like" individuals in a group from the "unlike" and potentially dangerous ones; this also usually includes such features as dress, attitudes, and habits, which vary widely in man.

Imitative behaviors are generally repetitive, compulsive, and highly ritualized and automatized, so that even partial presentation of the usual stimuli can elicit replicative behavioral responses. Of the many examples of this, one of the best known is that of babies responding to rudimentary, incomplete representations of the human face. In reptiles, birds, and lower vertebrates, dummies or even parts of dummies elicit courtship or aggressive displays and other replicative behaviors. This mechanism is present even in mammals: the mirror reflection of a single monkey eye can elicit the dominance ritual ("display behavior") in squirrel monkeys (MacLean, 1964). Phantom-like and shadowy forms or partial representations are known to evoke fearful and paranoid-like reactions in animals and man.

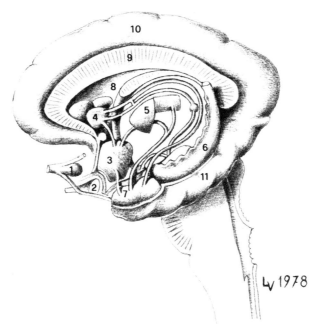

FIG. 16. Paleomammalian brain.

1) Olfactory tubercle	7) Amygdala
2) Optic chiasma	8) Septum
3) Hypothalamus	9) Corpus callosum
4) Septal nuclei	10) Cingulate gyrus
5) Thalamic nuclei	11) Hippocampal gyrus (Temporal
6) Hippocampus	lobe)

Reptiles have only a rudimentary cortex, and it is presumed that in those animals, in which the cortex is represented by an evolutionary transition between reptiles and mammals (Chapter 1, p. 20), this cortical rudiment increased in size, becoming further differentiated. In modern mammals, most of this phylogenetically old cortex corresponds to the large brain convolution that Broca (1878) called the "great limbic lobe," since it surrounds the brainstem structures; more recently and more precisely, MacLean (1952) defined this area as the limbic system.

This lobe resembles, like a mold, the ring-like configuration of corpus striatum; it may be regarded as the paleomammalian cerebrotype (Fig. 16).

The complex anatomical web forming the limbic system, and described in detail elsewhere (Valzelli, 1979a), also includes cerebellar components; these involve ventral tegmental area, interpeduncular area, periaqueductal gray areas, locus ceruleus, hippocampus, amygdala, and septal nuclei (Costin et al., 1970; Snider and Maiti, 1976). Electrical stimulation of the anterior cerebellum elicits arousal, predatory attacks, and feeding, mimicking the responses elicited by stimulation of the amygdala (Anand et al., 1959; Ball et al., 1974; Berntson et al., 1973; Reis et al., 1973).

As stated previously (this chapter, p. 31), the limbic system is considered the seat of animal and human emotional-affective life, and the functional significance of this system mainly resides in the high level of integration of visceral and exteroceptive inputs, sensations, perceptions, and affects. In 1949 MacLean elaborated Papez's theory of emotion (1937) and suggested that impulses from intero- and exteroceptive systems reach the hippocampus through the hippocampal gyrus; the hippocampal structure was viewed as capable of combining internal and external information into affective feelings. The latter was further elaborated and expressed through connections with amygdala, septal area, basal ganglia, and hypothalamus; they then reentered the limbic lobe and completed the "Papez circuit." Thus the hippocampus can be considered the key structure of the entire system, and the amygdala-hippocampal complex the functional key unit responsible for normal or abnormal feelings and behavior (MacLean, 1976; Mishkin, 1978; Valzelli, 1973a, 1978a).

Aside from data from animal experiments, the best evidence for the role of the limbic system in emotions and affect derives from clinical observations in which neuronal discharges of the limbic cortex of the temporal lobe trigger a series of vivid affective feelings. Basic and general affects evoked by surgical stimulation or pathological foci are usually associated with threats to self-preservation (MacLean, 1958). More specifically, basic affects include feelings related to hunger, thirst, and visceral messages; specific affects relate to unpleasant tastes, odors, sounds, and somesthesic sensations such as pain, tingling, and genital sensations; affects are associated with feelings of anxiety, fear, terror, sadness, wanting to be alone, familiarity, unfamiliarity, distorted perception, depersonalization, hallucination, and, rarely, anger (MacLean, 1952, 1973; Penfield and Erickson, 1941; Penfield and Jasper, 1954).

Thus several types of human psychopathology may stem from disturbed functions of the limbic system. Some structures of this system also relate to the neural mechanism subserving reward and punishment (this chapter, p. 28, and Fig. 13, p. 29); this is the case, for instance, of the septal nuclei (reward) and of periaqueductal gray structures (punishment).

Abnormal neuronal discharges originating in or near the limbic cortex spread in and remain largely confined to the limbic system. The hippocampus is almost always involved with the discharge, which either originates in it or spreads into it from related structures, whereas the simultaneous recording from the neocortex may show only minor changes during this kind of limbic seizure, except for a diffuse desynchronization. This led MacLean (1954) to postulate *schizophysiology* of limbic and neocortical systems, suggesting this condition as partly responsible for conflicts between what is affectively "felt" and what is "known." Further, since evidence indicates that somatic, auditory, olfactory, gustatory, and visual information is channeled to the hippocampus (Benjamin and Burton, 1968; Jones and Powell, 1970; MacLean et al., 1968; Van Hoesen

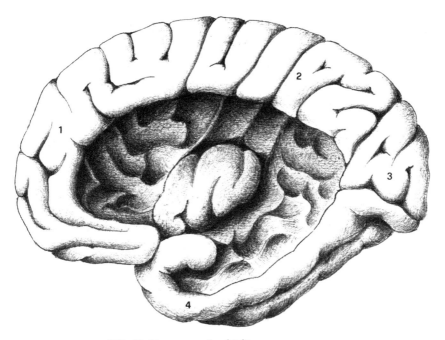

FIG. 17. Neomammalian brain.
1) Frontal lobe   3) Occipital lobe
2) Parietal lobe   4) Temporal lobe

et al., 1972), feelings triggered by limbic epileptogenic foci may involve any of the sensory systems; the ictal disruption of limbic functions possibly results also in changes in the sense of reality of the self and of the environment, changes of mood, distorted perception, depersonalization, and paranoid delusions (Mac-Lean, 1973).

Finally, the third cerebrotype is represented by the neocortex, which is characteristic in mammals more recent in evolution. This neomammalian brain (Fig. 17) has rapidly proliferated, covering entirely the two preceding brains, and reaching its ultimate expansion and development in man.

Because of its numerous infoldings, the human neocortex has a surface area of more than 1.5 m² and a total volume of 561 cm³; in comparison, neocortical volume is approximately 29.9 mm³ in mice, 102 mm³ in rabbit, and 41.96 cm³ in chimpanzee (Blinkov and Glezer, 1968).

The neocortex communicates widely in both input and output with the two cerebrotypes lying below, but interestingly, it receives no direct information from the outside and inside spheres. This means that all stimuli reaching the neocortex, to be rationally evaluated and logically elaborated are more or less emotionally colored, as are the signals emitted in response to such stimuli. Differences among people, which reflect the wide degree of variation in these emotionally colored cortical processes and responses, might account for the debated and unclear concept of *personality,* including such extremes as the psychopathological personality. A distorted cortical input brings about a distorted elaboration, which results in a distorted output. As anatomical evidence of this, many neocortical areas project to the hippocampus via the entorhinal cortex (Van Hoesen et al., 1972).

In relatively modern or recent mammals, the neomammalian brain seems to be unessential for basic life, since decorticated rats retain most of the patterns of locomotion, climbing, grooming, feeding, and fighting (Vanderwolf et al., 1978), but show disruption of mating, nest building, maternal behavior, and learning complex tasks (Beach, 1937, 1943, 1967; Lashley, 1929). When the hippocampus is removed and the neocortex is ablated, nest building, grooming, and social behavior are also lost to a great extent (Kolb and Nonneman, 1974; Shipley and Kolb, 1977; Vanderwolf et al., 1978); a reptile-like behavioral regression is produced, most likely due to the functional prevalence of a reptilian brain. These observations indicate that in mammals the nerve structures in the spinal cord, brainstem, and cerebellum are responsible for the complex spatiotemporal patterns involved in behaviors such as walking, rearing, climbing, and biting (Berntson and Micco, 1976).

In man, the neocortical mantle is thought to be the seat of logical and mathematical reasoning, knowledge and understanding, analytical and synthetic processes, invention and fantasy, philosophy and religion, meditation and intuition. However, in man, too, some behaviors and aspects of mental diseases suggest the regression of brain functioning to a predominantly paleomammalian (limbic) or reptilian level. In this last instance, as has been observed in animal experiments, the breakdown of social, familial, parental behavior, and personal care is often accompanied by the emergence of asocial, hostile, and aggressive behaviors, and "reptilian man" eventually emerges.

Even though the human triune brain is an elegant expression of evolution, such a complex and subtle system carries the threat of dissociation and malfunction that may rout the intrinsic biological controls ensuring normal functioning of the brain of higher animals.

HUMAN BRAIN PREROGATIVES

The human brain is unquestionably the most complex structure in our presently known universe: it performs in relation to memory, consciousness, language, and culture in a way that uniquely distinguishes it from even the most highly developed cerebral structure of any other animal. Nevertheless, we still do not know how these subtle properties came to be associated with a neural structure that does not materially differ from those highly organized brains of animals such as primates and dolphins.

It should also be asked what (or Who?), some two million years ago, altered the brain functions of an ape-like *Homo habilis* into that of *Homo sapiens;*

and why, if the natural history of human evolution is correct, did only some of man's believed ancestors become *Homo sapiens;* and also, what suddenly triggered humans to create civilizations like those of the Egyptians or the Mesopotamians while other men still behave primitively in an almost neolithic manner.

Such issues as consciousness, intelligence, and personality, once believed to be exclusive appanage of man, elude precise definition, and are also part of the brain functioning of several mammals; thus these properties do not differentiate man from animals. As already stated (this chapter, p. 32), the differences between the brain activities of animals and those of man are probably qualitative rather than merely differences of degree and quantity. If this is so, this situation has probably reflected upon the degrees of behavioral choices and freedom, which have in turn resulted in the amazingly creative adaptability of man.

In this framework, utmost importance should be attributed to a new way of utilizing information. Animals perform primarily in response to information directly relevant to their needs and survival, whereas man stores information—even that which seems irrelevant or minor—until it can be usefully retrieved, coordinated with other information, and utilized. Such an information-collecting process is further consciously elaborated in culture, meditation, philosophy, and intuition, which subserve an incredibly wide range of free choices, new behaviors, and creative activities. In this resides, unlike in animals, the freedom of man to be evil or good.

Relevant to this issue is the characteristic of language, which is peculiar to man. Interestingly, some three centuries ago, Descartes said that "none is so depraved and stupid, without even excepting idiots, that they cannot arrange different words together forming of them a statement by which they make known their thoughts while, on the other hand, there is not other animal, however perfect and fortunately circumstanced it may be, which can do the same." (Chomsky, 1966). The development of language in man is the most significant milestone in all of evolutionary history and in cognitive annals of man. Mathematics, engineering, literature, theater, music, as well as all other typically human productions developed, disseminated and stored because of language; these cultural advances have been made possible by detailed and written documentation, which allows unlimited dissemination of propositions, deductions, concepts, ideas, objects, and events. Every advance in these fields represents an important addition to individual and collective consciousness, perception, and valuation, also increasing cultural integrity.

Evidence from cave painting and tool making, implicative of the cognitive processes underlying symbolization and language formation, indicates the emergence of spoken language somewhere between 50,000 and 500,000 years ago (Critchley, 1960). The uncertainty of establishing an exact date most probably reflects the time span during which it might be inferred that anthropoid vocalizations, presumably originating in the brainstem, have been transformed into verbal symbolizations, according to phonetic and syntactical rules produced by neocortical representations (Lenneberg, 1974). More exact estimates of the origin of written language date its beginnings approximately 50,000 years ago (Barnes, 1935); the development of practical rules, however, can be traced to only about 7,000 years ago in Sumerian or Indo-Hittite notations (Wescott, 1976).

Certainly human civilization had spread all over the earth soon after the invention of writing. With it came, just as rapidly, trends, preferences, similarities, and differences, which were reflected in living, education, ideology, and beliefs; all of these coincide with the history of aggression, which in its multiple expressions, has become an almost accepted facet of human life.

3

Affective Behavior

A PREMISE

Among the important issues related to studies of brain and behavior, one of the most fascinating deals with the anatomicophysiological and neurochemical correlates of cognition and affect. Generally, this argument is a modern restatement of the old problem of mind-body relationship (Money, 1956; Power, 1965; Rupp, 1968). This issue, which can be traced back two and a half millennia to the philosophical and medical speculations of Hippocrates, Aristotle, Galen, Aretheus, and others, developed into contemporary psychiatric and psychological sciences. However, modern psychiatry and psychology are still caught on the horns of a treacherous dilemma with regard to mind and body. While these disciplines obviously deal with the study of the mind, they are also biological disciplines; and to biology, concepts such as cognition, emotion, affect, personality, and intelligence are too vague and intangible to define with scientific vigor.

Consequently, divided in their allegiance, investigators of the mind are prone to search for concepts and terms—such as psychophysiology, psychobiology, psychosomatic, and biological psychiatry—that do not, at least overtly, flout biology. A similar uneasiness seems to have troubled Freud. After he tried to theorize the integration of psychological and physiological functioning of the brain, he wrote in 1898 to his friend Fliess: "I have no desire at all to leave the psychology hanging in the air, with no organic basis. But, beyond the feeling of conviction [that there must be a physiological basis], I have nothing either theoretical or therapeutic, to work on, and so I must behave as if confronted by just psychological factors only" (Freedman et al., 1975).

The need to find an organic basis for brain activities and assign them to specific cerebral structures has resulted in several curious propositions. In the sixteenth century, Leonardo da Vinci (1452–1519) postulated that some fundamental faculties of human mind were localized in cerebral ventricles. Accordingly, the first ventricle, or *cellula phantastica,* was considered the seat of sensation and imagination; the second ventricle, or *cellula rationalis,* the center of thoughts; and the third ventricle, or *cellula memorialis,* the seat of memory.

Nearly a century later, René Descartes (1596–1650) in his treatise *Les Passions de l'Âme,* issued in 1649, claimed that mind, or *res cogitans,* is bound to body, or *res extensa,* by means of a specialized brain structure, the pineal, saying that "the seat of passions is not the heart . . . it exists in the brain a little gland, in which the mind performs its functions" (Descartes, 1649). Later, La Peyronie, Royal Surgeon at the Court of Louis XV of France, declared that corpus callosum was the seat of the mind (La Peyronie, 1709, 1744).

More recently, Franz Joseph Gall, a Viennese anatomist, attempted to compartmentalize moral and intellectual faculties in different areas of the brain, maintaining that lumps of the skull corresponded to the shape of the brain which, in turn, was molded by the cerebral qualities and faculties. In 1800 he was joined by Johan Kasper Spurzheim, who attempted to improve Gall's idea by calling it *phrenology* and by devising a series of phrenological charts which illustrated topography of brain function (Fig. 18).

New hypotheses stemmed from the discovery of *animal electricity* in 1780 by the Italian anatomist Luigi Galvani (1737–1799). In the middle of the nineteenth century, Ernst Brücke, Theodor Meynert, and Sigmund Exner represented a "triumvirate" of German scientists dominating scientific thought in Europe. They believed that the brain operated by "transmitting" a quantitatively variable excitation from afferent to efferent nerve endings. The nature of this neural impulse was not agreed on, but Brücke believed that it was electrical in nature. Shortly thereafter, Alexander Bain claimed that the mind is nothing more than a breadboard of electrical circuitry, and concisely stated: "No nerve currents, no mind." (Bain, 1885).

An active proliferation of studies concerned with the anatomicophysiological and neurochemical correlates of behavior originated from this fertile ground. Differences between current and prior approaches have been that instead of searching for specific brain loci responsible for specific behaviors, investigators now accept the concept of different but extensively interconnected neural circuits, the activity of which occurs through and is regulated by the availability of putative neurochemical transmitters at the receptor sites. Such a regulation is modulated by functional needs, thus reflecting, from a neurochemical standpoint as well, brain plasticity (Chapter 2, p. 24). Consequently, it is considered possible to correlate various behavioral responses with changes in the neurochemical activity of specific brain circuits. It is also believed possible to relate brain neurochemical changes, induced by drugs or physical means, to behavioral modifications.

Yet these advances, which as never before promise a better future for mankind—based on increased understanding of brain capacities and the possible elimination of mental diseases—have aroused sociopolitical rumination that threatens to hamper the progression of brain studies. The term "brain control" has been bandied about in modern rhetoric; its connotation of a robotized society ruled by a perverse dictator and supported by equally depraved scientists is a striking example of an irrational fear, not too different from superstitions and

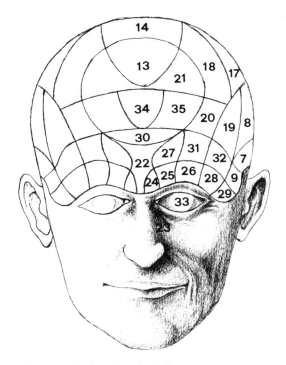

FIG. 18. Example of a phrenological map.

7. Secrecy	24. Dimensions
8. Acquisitiveness	25. Weight and resistance
9. Constructiveness	26. Color
13. Kindness	27. Location
14. Respectfulness	28. Order
17. Expectancy	29. Calculation
18. Wonderfulness	30. Contingency
19. Ideality	31. Time
20. Mirthfulness	32. Harmony
21. Imitation	33. Language
22. Individuality	34. Comparison
23. Configuration	35. Causality

witchcraft which, born in the Middle Ages, flourish even in the shadows of the skyscrapers. In this contrasting atmosphere, it is paradoxical that man, who has developed a technology allowing exploration of the moon and planets, cannot unravel the mystery of his brain, which is the source of his technology, his behavior, and his existence.

Several factors complicate a study of the neurochemical and neuroanatomical correlates of behavior. Behavior production is normally a "continuum" with regard to both spontaneous or learned responses to environmental demands and neurophysiological changes involved therein. According to Fentress (1973), in studying behavior "categories must be formed, but the investigator must not believe them." The same is true for any neurochemical investigation of a given behavioral variable. Hence, it should be considered inappropriate to select a single behavioral variable as an index of differences and similarities, and to attribute it to possible metabolic changes influenced by a single neurochemical constituent of the brain. In this framework, only those spontaneously determined or experimentally induced situations in which a single behavioral variable dominates others allow reliable investigation of the neurochemical and neuroanatomical correlates of that variable, especially when selected brain areas or circuits can be assumed as functionally related to the behavior being considered.

There is also the question of the extent to which a single neurochemical transmitter involved in the activity of selected brain structures may directly govern specific behaviors. Changes in the neural activity and metabolism of any neurotransmitter (Chapter 2, p. 25) are likely to affect the function and metabolism of other neurochemicals (Blondaux et al., 1973; Burnstock and Costa, 1975; Hery et al., 1973; Johnson et al., 1972; Jouvet, 1973; Kostowski et al., 1974; Lichtensteiger et al., 1967). Consequently, it is inappropriate to attribute the production of a specific behavior to the activity of a single neurotransmitter (Valzelli, 1977*b*, 1978*a*).

The challenge that psychological sciences pose to neurophysiology, concerning development of a logical explanation of how neurochemical principles and reactions could produce either normal or abnormal behavior, seems therefore to remain open.

Nevertheless, new approaches to behavioral neurophysiology and neuroanatomy are now available. Recent evidence has shown that neurotransmitters are localized within specific neural pathways in the brain, leading to the concept of *neuroanatomical biochemistry*. The role of these neuroanatomical sites, which are biochemically specific, in producing normal or abnormal behavior and mental illness, is of great importance (Breese et al., 1974*a, b;* Lilly and Miller, 1962; Stein and Wise, 1971; Stevens, 1973; Valzelli and Sarteschi, 1977; Wide et al., 1974). Maps of serotonergic, noradrenergic, dopaminergic (Fuxe et al., 1971; Ungerstedt, 1971), cholinergic (Eng et al., 1974; MeGeer et al., 1974; Rossier et al., 1973), histaminergic (Garbarg et al., 1974), GABAergic (Coyle and Enna, 1976; Fonnum et al., 1977; Gottesfeld and Jacobowitz, 1978), and substance P (Cuello and Kanazava, 1978; Hong et al., 1977; Paxinos et al., 1978*a, b*) neurons and transmitter-specific neural projections and pathways in brain are available (Figs. 19–21).

These advances are undoubtedly valuable, from both speculative and practical standpoints. The course and distribution of these pathways link the oldest brain structures to the limbic system and neocortex, creating a series of neurochemically specific systems of integration, the distribution and general topography of which do not differ substantially between experimental animals and man (Falk et al., 1962; Nobin and Björklund, 1973; Nyström et al., 1972; Olson, 1974; Olson et al., 1973*a, b, c*). This finding, probably as never before, allows animal experiments to be relevant for human studies, interpretation, and theories.

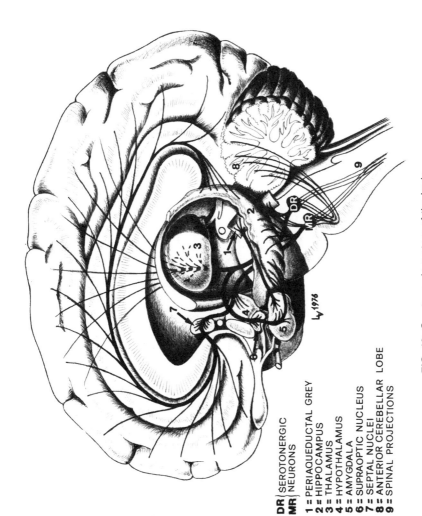

DR } SEROTONERGIC
MR } NEURONS

1 = PERIAQUEDUCTAL GREY
2 = HIPPOCAMPUS
3 = THALAMUS
4 = HYPOTHALAMUS
5 = AMYGDALA
6 = SUPRAOPTIC NUCLEUS
7 = SEPTAL NUCLEI
8 = ANTERIOR CEREBELLAR LOBE
9 = SPINAL PROJECTIONS

FIG. 19. Serotonergic system of the brain.

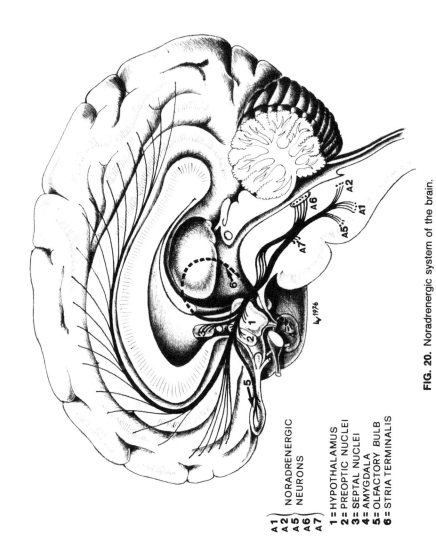

A1
A2
A5 } NORADRENERGIC
A6 } NEURONS
A7

1 = HYPOTHALAMUS
2 = PREOPTIC NUCLEI
3 = SEPTAL NUCLEI
4 = AMYGDALA
5 = OLFACTORY BULB
6 = STRIA TERMINALIS

FIG. 20. Noradrenergic system of the brain.

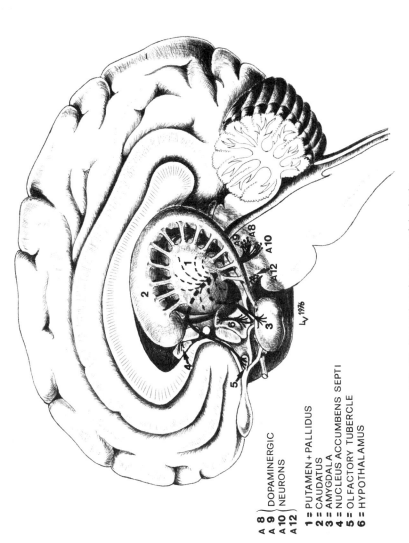

A 8
A 9 } DOPAMINERGIC
A 10 } NEURONS
A 12

1 = PUTAMEN + PALLIDUS
2 = CAUDATUS
3 = AMYGDALA
4 = NUCLEUS ACCUMBENS SEPTI
5 = OLFACTORY TUBERCLE
6 = HYPOTHALAMUS

FIG. 21. Dopaminergic system of the brain.

COMPONENTS OF AFFECTIVE BEHAVIOR

As stated earlier, affective behavior largely regulates activities which subserve self- and species-preservation. Thus animal behavior is almost entirely affective, with response varying according to position on the evolutionary scale. Only humans show a highly personalized modulation, and sometimes unpredictable outcome, of their emotional-affective reactions as a consequence of ethnic, cultural, logical, and cognitive influences and differences.

In a behavioral context, it is important to distinguish *motives* from *drives.* When behavioral activities are directed toward *incentives,* motives rather than drives arise, since drives coincide mainly with *instincts,* and as such are usually "blind," in that they initiate random behavior; this does not evolve into the corresponding and appropriate behavior pattern until learning has occurred. Motives, instead, activate specialized channels of behavior aimed at satisfying needs. Thus hunger, thirst, and sex are drives, and they initiate motivated patterns of behavior such as feeding, drinking, and sexual activity.

Motives differ in intensity, and motivational strength can be measured by the extent to which "resistance" or "blockage" can be overcome (Morgan, 1923). Motivational strength is tested by imposing an aversive stimulus block—such as an electrified grid—variable in intensity, between a deprived animal and the related stimulus-objects, such as food, water, pups, or another animal of the opposite sex. The persistence and the percentage of animals crossing the grid to reach the goal measure motivational strength (Jenkins et al., 1926; Morgan et al., 1943; Moss, 1924; Nissen, 1930a, b; Stolurow, 1948; Stone, 1937, 1942). In this way it has been possible to devise a sequence of motivational strengths which, in decreasing order, include maternal, drinking, feeding, hoarding, sexual, exploratory, and social behavior.

Other emotional-affective components, such as wildness, timidity, anxiety, fear, and possibly other related variables, elude definition and measurement, since frequently they either interact or stimulate opposite patterns of behavior. For example, fear can lead to freezing, flight, hostility, rage, or overt aggression, depending on environmental factors that may or may not permit escape from the threat. Further, previous experience, involving a similar situation that produced frustrating or rewarding results, has a role in determining the choice of behavior to be displayed (Hunt et al., 1947; Scott, 1958a, b). In addition, the already mentioned genetic predeterminants of emotional traits (Chapter 2, p. 29) and learned or culturally transmitted restraints may further modify the behavior induced by frightening circumstances.

Thus experiments dealing with fear or similar elusive emotional issues remain difficult to perform reliably and interpret correctly. To simplify the puzzle, some investigators have suggested that fear is the conditioned response to pain (Miller, 1948; Mowrer, 1950, Mowrer and Lamoreaux, 1946), but fear is aroused not only by painful stimuli but also by visual, olfactory, and auditory cues (Curti, 1935, 1942; Kunkel, 1919; Small, 1899). In addition, fear and pain may elicit contrasting responses (Baenninger, 1967; Myer and Baenninger, 1966). In rats the choice between freezing or avoidance behaviors, which are the two main defensive responses to fear, is determined by the features of the threatening situation (Blanchard and Blanchard, 1969*a, b,* 1970*a, b*). The rapidity of development and the specificity of reactions elicited by a frightening situation also suggest that fear associated with such stimuli is acquired before the experiment (Blanchard et al., 1970); thus it represents a *species-specific defensive reaction* of adaptive significance for the preservation and evolution of that animal species (Blanchard and Blanchard, 1971; Bolles, 1970).

It is even more difficult to provide consistent definitions of and experiments for anxiety, a term which has been used inexactly to indicate fear (Mowrer and Lamoreaux, 1946), thus falling again within the context of the above mentioned considerations. Anxiety is more appropriately considered as that emotional feeling which—correctly or incorrectly—forewarns the possibility of experiencing fear, pain, or other unpleasant sensations. Within these limits, anxiety mainly arises from uncertainty and insecurity. Nevertheless, it is still difficult to overcome the lack of reliable experimental data on this affective component.

Other factors can significantly modify the sequence of emotional-affective behaviors, both during animal or human development and in adulthood. These factors may be summarized by the concept of the *mean level of environmental stimulation* (Welch, 1965), where the environment is considered not only in terms of the physical or structural complexity of the territory in which the organism lives, but mostly in terms of emotional, affective, and social production and interactions of the subjects inhabiting that area, which constitute a complex web of reciprocal influences (Valzelli, 1977*a*). In this context, "environment" coincides with the French word *milieu* preferred by some clinicians (Fisher, 1970; Klerman, 1963; Linn, 1959; Malitz, 1963), since it implies the concept of *sociability,* and the meaning of social relations and social experiences. Environmental stimulation and social interaction increase in intensity and complexity as the number of inhabitants of a given territory increases. There is an optimal ratio between the number of inhabitants and the territorial area, and the limits of this ratio vary according to the physical characteristics and resources that the environment offers. At the extreme limits of such a ratio, the devastating effects of isolation (Hatch et al., 1965; Valzelli, 1973*a*) and of overcrowding (Calhoun, 1973; Freeman, 1978) include behavioral disruption of the individual or disintegration of social relations and integrity of a group or population.

EFFECT OF SOCIOENVIRONMENTAL TEXTURE

Socioenvironmental variables profoundly affect the developing brain. In addition to the previously mentioned environment-dependent changes in brain structure and chemistry (Chapter 2, p. 24), handling and gentling of animals during weaning result in a more stable emotional baseline in adulthood. Such treatment also increases exploration and adaptation to new tactile experiences (Bernstein, 1952; DeNelsky and Denenberg, 1967a, b; Denenberg et al., 1964), as well as strengthening resistance to stressful events, infant death (Weininger, 1953, 1956), and deprivations later in development (Levine and Otis, 1958). Social interaction between emotionally stable and emotionally labile animals results in the eventual improvement of emotional characteristics of the latter (Denenberg et al., 1964). Socioenvironmental enrichment also induces more stable emotional-affective behavior (Denenberg and Morton, 1962a, b; Denenberg et al., 1969; Manosevitz, 1970), even improving some genotype-dependent emotional characteristics (Manosevitz and Joel, 1973; Manosevitz and Montemayor, 1972), and enhances learning and problem-solving capability (Forgays and Forgays, 1952; Hymovitch, 1952; Krech et al., 1962; Rosenzweig, 1964) and efficiency of memory storage (Greenough et al., 1972).

Imitative learning, by which one animal observing another performing a specific behavioral task rapidly learns to accomplish the same behavior (Pallaud, 1969a, b), is important. Imitation originated in the reptilian brain structures and functions (Chapter 2, p. 34), and such learning is a prominent behavioral feature of childhood and adolescence. It exerts a potent effect in shaping adult behavior, which becomes to a great extent "bad" or "good" according to the imitative schemes formerly provided. Behavioral interaction and imitation are so influential that even retarded children show considerable improvement in their impaired behaviors, learning capacity, intelligence quotient, and mind range when the quality and intensity of socioenvironmental stimulation are appropriately increased (Butterfield and Zigler, 1965; Cheyne and Jahoda, 1971; Clarke and Clarke, 1954; Crissey, 1937; Das, 1971; Kephart, 1940; Miyamoto, 1960; Vogel et al., 1967; Zigler and Williams, 1963).

In contrast, low-level socioenvironmental stimulation induces an emotional-affective dearth which results in consistent alterations of several anatomical and neurochemical parameters of the brain, such as a decrease in brain weight and cortical thickness, decreased cholinesterase activity in visual and somesthetic areas (Bennett et al., 1969; Krech et al., 1962, 1966; Rosenzweig, 1966; Quay et al., 1969), pineal hypertrophy (Quay et al., 1969), alteration of arousal and behavioral inhibition (Campbell and Raskin, 1978; Morgan et al., 1975), altered sexual behavior (Gerall et al., 1967; Gruendell and Arnold, 1969; Spevak et al., 1973), impairment of learning and memory (Bennett and Rosenzweig, 1968; Bennett et al., 1970), and increased susceptibility to neoplastic disease (LaBarba and White, 1971; LaBarba et al., 1972). These alterations, although primarily described in mice and rats, disrupt behavior in other animal species (Cheal, 1975; Geissler and Melvin, 1977; Hull et al., 1973; Scott, 1974; Spencer et al., 1973), including monkeys and primates (Cross and Harlow, 1965; Davenport et al., 1961, 1966; McKinney, 1974; McKinney et al., 1972a, b; Missakian, 1969; Newman and Symmes, 1974; Prescott, 1971; Turner et al., 1969; Young et al., 1973), in which *isolation,* or separation distress, induces decreased sexual behavior, reduction of play activity, social withdrawal, loss of grooming, stereo-typed behavior, hyperirritability, and early mortality.

Moreover, emotional-affective and social interactions in general, and maternal care in particular, are also essential in shaping appropriate behavioral stability in children (Bender and Yarnell, 1941; Bowlby, 1952; Casler, 1963; Freud and Burligame, 1944; O'Connor, 1971; Widdowson, 1951; Yarrow, 1961). Maternal and emotional-affective deprivation in children has been reported to induce verbal retardation, slow physical growth, poor locomotor control with restless-ness, emotional shallowness, mood instability, distractibility, low intelligence quotient, impaired capacity for abstract thinking, hyperreactivity, hyperirritabil-ity, outbursts of hostility, violence, and aggression and, later, apathy, self-seclu-sion, and organic decline (Bowlby, 1952, 1953; Clancy and McBride, 1975; Prescott and Essman, 1969).

Prolonged socioenvironmental deprivation or isolation also has profoundly negative effects in adulthood, so that adult male mice singly housed in individual cages show persistent, compulsive, and apparently aimless aggressive behavior (Al Maliki and Brain, 1977; Brain, 1975; Brain and Benton, 1977; Crawley et al., 1975; Essman, 1969, 1971a, b; Karczmar and Scudder, 1969; Kršiak and Janků, 1969; Lagerspetz, 1969; Valzelli, 1967a, b, 1971a; Yen et al., 1959). This abnormal behavior is sex and strain dependent (Valzelli, 1971a; Valzelli and Bernasconi, 1978; Valzelli et al., 1974), and exhibits circadian and seasonal components (Sofia and Salama, 1970; Valzelli et al., 1977). In addition to aggres-sion, prolonged isolation also induces several changes in brain chemistry, includ-ing neurochemical transmitter metabolism; changes in brain enzymes, amino acids, polypeptides, proteins, and nucleic acid distribution (Conde and DeFeudis, 1977; DeFeudis, 1972a, b, c; DeFeudis et al., 1976a; Essman, 1968, 1969, 1971a; Essman et al., 1972; Garattini et al., 1969; Marcucci et al., 1968; Miller et al., 1979; Modigh, 1973; Nishikava et al., 1976; Segal et al., 1973; Welch and Welch, 1971; Welch et al., 1974); changes in blood pressure (Henry et al., 1967, 1971); gastric ulceration (Essman and Frisone, 1966); deviant and/or decreased sexual activity (Charpentier, 1969; Hautojärvi and Lagerspetz, 1969); unexpected convulsive attacks (Takemoto et al., 1975); and impaired exploration, learning, and memory consolidation (Essman, 1970, 1971c, d; Valzelli, 1969, 1971b, 1973a; Valzelli and Pawłovski, 1979; Valzelli et al., 1974). These changes have been described as an experimental *isolation syndrome* (Valzelli, 1973a). Prolonged isolation has been also described to induce "timidity" in about 45% of male Albino Swiss mice (Kršiak, 1975).

A similar "isolation syndrome" can be induced in rats (Hatch et al., 1956); it also leads, depending on strain and sex, to either neurochemical or behavioral alterations (Gerall, 1963; Gerall et al., 1967; Goldberg and Salama, 1969; Salama and Goldberg, 1970; Spevak et al., 1973; Valzelli, 1971a; Valzelli and Garattini, 1972; Welch et al., 1974). The latter include muricide or other killing behavior (Chapter 7, p. 126); changes in adrenal, thyroid, and pituitary weight, and in liver glycogen, adrenal cholesterol, and plasma corticoids (Hatch et al., 1965); altered ontogeny of arousal (Campbell and Raskin, 1978); modified production of sex steroids (Dessi-Fulgheri et al., 1975); and changes in sexual and social behavior (Spevak et al., 1973). "Neophobia," behavioral inhibition, and hyperphagia (Morgan and Einon, 1975; Morgan et al., 1975), changes of heart rate and of cardioacceleratory response to novel stimuli (Leigh and Hofer, 1975), decreased behavioral reaction to intense foot-shock (Nishikawa and Tanaka, 1978), and, as in isolated mice, changes in the effects of several drugs (DeFeudis et al., 1976*b*; Kršiak, 1975; Nishikawa and Tanaka,1978; Valzelli, 1977*a*; Valzelli and Bernasconi, 1976, 1979) and in morphine self-administration (Alexander et al., 1978) have been also described.

In humans, socioenvironmental deprivation induces anxiety states, depressive ideations, sleep disturbances, regression, and consciousness and personality alterations (Adams et al., 1966; Kellerman et al., 1977; Persky et al., 1966; Wilkinson, 1975; Zuckerman et al., 1966). These symptoms vary in intensity according to the psychological baseline of the volunteers subjected to the experiment, and also account for the behavioral changes shown by prisoners, in whom partial or complete confinement frequently produces headache, loss of appetite, nervousness, irritability, and outbursts of self- or extero-directed aggression (Drtil, 1969*a*, *b*). Total sensory deprivation may even induce schizophrenic-like syndromes in man, with visual, acoustic, and somesthetic hallucinations, emotional instability, dissociation symptoms, hostile behavior, and unmotivated aggression (Heron et al., 1956; Kracke, 1967; Leff, 1968; Miller, 1962; Pollard et al., 1963; Rosenzweig and Gardner, 1966; Ziskind and Augsburg, 1967). Such experimentally induced syndromes have been described as being often more intense than those produced by hallucinogenic drugs such as mescaline (Rosenzweig, 1959). In contrast, isolation has been reported to be without effect in patients with organic brain damage or mental deterioration (Boman, 1964), or even to ameliorate stutterers' speech (Šváb et al., 1972).

Environment, with its social connotations and physical characteristics, is a potent factor in shaping animal and human behavior. However, despite this, and contrary to certain pseudosocial and pseudoscientific claims, environment is neither exclusively responsible for nor the sole culprit of perverted or aberrant human behavior and mental disease. Rather, environment together with imitation, learning, and culture interact with the individual's genetically predetermined emotional-affective profile. The problem of human consciousness may serve as a basis for psychobiological approaches to its explanation (Davidson and Davidson, 1978), and to such fascinating issues as the brain knowing about itself and God being in the brain (Mandell, 1978). Mandell explains "religious ecstasy" and "hallucinations of God" as deriving from prolonged afterdischarge hypersynchronous seizures restricted to the hippocampal-septal-reticular-raphe circuit (without amygdala involvement); this is supported by several findings with associate ecstatic states, religious conversions and experiences, and compulsive metaphysical writing and preaching with epilepsy (Allison, 1967; Christensen, 1963; Dewhurst and Beard, 1970; Glaser, 1964; Howden, 1872; Mabille, 1899; Sargant, 1969; Sedman and Hopkinson, 1966). Nevertheless, just as there clearly exists a mental disease of God, so does the concept of God reside in man's brain.

Hence, having God in our mind, men have free will (Chapter 2, p. 39) to accept Him for behaving as men or to refuse Him and revert to primates.

NEUROANATOMY OF AFFECTIVE BEHAVIOR

As stated earlier, the neuroanatomy of emotions and affect corresponds with the limbic structures, primarily the paleomammalian brain, which modulates and tunes behavior of both animals and man (Anand, 1957; MacLean, 1958) depending on emotional responses associated with self- and species preservation (Fulton, 1951; MacLean, 1955; Papez, 1937, 1958).

Historically, there have been debates about which structures of the inner brain belong to the limbic system; initially, rhinencephalon, hypothalamus, amygdala, hippocampus, and septal nuclei were almost exclusively included. A more recent, and still approximate, inventory includes the olfactory bulbs and bandelets, prepyriform cortex, insula, uncus, amygdala, hippocampus, cingulate gyrus, central gray matter, hypothalamus, medial thalamus, and superior quadrigeminal bodies. Inferior quadrigeminal bodies, preoptic areas, mammillary bodies, habenula, caudate-putamen complex, tip of temporal lobes, and temporal, prefrontal, and frontal cortex also significantly contribute to limbic activities (Adey, 1959; Anand and Dua, 1955, 1956; Blanchard and Blanchard, 1968; Delgado, 1962, 1963a, 1966a, 1969a; Egger and Flynn, 1967; Fonberg, 1969; Gumulka et al., 1970; Heath, 1976; Horel and Misantone, 1974; Horvath, 1963; Kaada et al., 1954; Kling, 1965, 1968; Klüver and Bucy, 1938, 1939; MacLean, 1955, 1972b; MacLean and Delgado, 1953; Moyer, 1968; Ojemann, 1966; Reis and Gunne, 1965; Robinson, 1963; Ursin and Kaada, 1960; Vergnes and Karli, 1963a, 1969a, b; Weiskrantz, 1956; Wood, 1958). Further, other brain structures such as labyrinths and cerebellum, formerly considered to have no role in regulating emotions and affect, have been recently shown to contribute to the limbic system in both animals and man (Babb et al., 1974; Ball et al., 1974; Berman et al., 1974; Costin et al., 1967, 1970; Heath and Harper, 1974; Piggott et al., 1976; Riklan et al., 1974; Smith et al., 1974; Snider and Maiti, 1976; Snider and Snider, 1977). The cerebellum also has functional connections with basal ganglia and sensorimotor areas of cerebral cortex (Mehler and Nauta, 1974; Oka et al., 1976).

The limbic system also must be regarded in its entirety, as a "continuum"; its significance is primarily in its manifold associative and integrative functions. The activation or inhibition of a component of the system initiates a sequence of events that are distributed to most brain structures. This may explain why even the most elementary pattern of behavior becomes complex and refined in relation to the evolutionary level of the brain of the species considered. However, considering the good advice of Fentress (p. 43), a set of behavioral responses can be attributed to the activities of some experimentally identifiable brain structures and circuits.

The hypothalamus has an important role in *feeding behavior,* which also implies specific forms of aggression. Stimulation of the lateral hypothalamus leads to feeding, increased food intake (Delgado and Anand, 1953; Miller, 1957, 1961; Oomura et al., 1967*b;* Smith, 1956), and predatory behavior when the prey appropriate for the animal under experiment is available (Wasman and Flynn, 1962). Electrical stimulation of the ventromedial nucleus of the hypothalamus, in contrast, stops feeding behavior (Oomura et al., 1967*a*), as does stimulation of the basolateral nuclei of the amygdala (Fonberg and Delgado, 1961*a*). Hyperphagia with consequent obesity results from lesions of the ventromedial hypothalamic nucleus (Anand and Brobec, 1951; Ehrlich, 1964), basolateral amygdala (Fuller et al., 1957; Graff and Stellar, 1962; Morgane and Kosman, 1960) pallido-hypothalamic tract, immediately dorsal to the fornical columns (Morgane, 1961), dorsolateral tegmentum (Sprague et al., 1961), and periaqueductal gray matter beneath the superior collicula (Skultety and Gray, 1962). Feeding and predatory behavior are also induced by stimulation of the anterior cerebellum and superior cerebellar peduncle (Ball et al., 1974; Snider and Maiti, 1976).

Eating, irritative, hostile, or overtly aggressive behaviors, together with submission, mimicry, grooming, vocalization, threat, dominance, and sexual activity, are included in the more general framework of *social behavior;* and it is important to understand the brain structures which regulate social interaction. Delgado (1961, 1963*a, b,* 1964, 1965, 1966*a, b,* 1967*a, b,* 1969*a, b, c,* 1974, 1979; Delgado et al., 1979) has performed several interesting and relevant experiments. These species-specific behavioral patterns are relatively fixed and predictably ordered in animals, whereas they are culturally influenced and rapidly changing in human societies.

An important observation Delgado gathered from his experiments (1969*b*) is that electrical stimulation of the brain produces no strange or distorted behavior, but acts only as a trigger or modulator of cerebral functions by influencing the processing of sensory inputs and the release of previously acquired behavior patterns. According to studies in monkeys, most head and facial motility, threatening mimicry, and vocalization, which are never followed by actual aggression (menacing or threatening behavior), are evoked by stimulating the sylvian fissure, facial nucleus, superior olivaris nucleus, lateral lemniscus reticular mesencephalic nucleus, superior colliculus, and ventromedial hypothalamus (Lipp and Hunsperger, 1978). Stimulation of the rostromedial part of the red nucleus evokes immediate motor activity, low-toned (menacing) vocalization, and overt threats against subordinate monkeys, following the pattern of dominant behavior. Stimulation of the ventrolateral-posterior thalamic nucleus or of the periaqueductal gray matter evokes behavior patterns modulated by environmental circumstances. For example, if the monkey is alone in its colony cage, increased walking, circling, and low-toned menacing vocalization may be seen, whereas when the stimulated monkey is in its own colony, hostile and aggressive behavior, indistinguishable from spontaneous hostile activity, is directed against other subjects which may compete for dominance and social rank in the colony (Delgado, 1967a). Dominance in rats is decreased by septal lesions (Gage et al., 1978), but an intermittently programmed electrical stimulation of the head of the caudate nucleus of a dominant monkey abolishes its dominance, thus changing the hierarchical structure of the entire social group (Delgado, 1974). The results of this experiment may represent a source of concern for several political leaders.

Stimulation of the posterior hypothalamic nucleus produces a complex behavioral sequence that activates curiosity and *exploration,* whereas crouching, inhibition of spontaneous motor activity, and increased sleeping time are induced by stimulating the basolateral nuclei of the amygdala. Low-level stimulation of the basolateral amygdala evokes exploration in rats (Weingarten and White, 1978). Drowsiness and *sleep* are also induced by cerebellar fastigium stimulation (Snider and Maiti, 1976). Among other effects, stimulation of the anterior hypothalamus or of the supraoptic nucleus produces penile erection, but no further sexual activity (Delgado, 1967b; MacLean and Ploog, 1962), thus most likely pertaining to the already mentioned pattern of display behavior (Chapter 2, p. 34) and dominance. The fimbria of the fornix is involved in the offensive-defensive system; together with the posterior hippocampus, this constitutes one of the major links in the central integration system for nociceptive responses (Delgado, 1955). In monkeys, as in other experimental animals, stimulation of the dorsolateral pontine tegmentum, of the cerebellar dentate and fastigial nuclei, and of the cerebellar peduncle induces *grooming* activity; whereas stimulation of the ventrolateral pontine tegmentum results in threatening behavior (Ball et al., 1974; Berntson, 1973; Reis et al., 1973).

Bilateral ablation of the prefrontal or anterior temporal cortex in monkeys leads to a critical decrease of grooming, facial motility, vocalization, maternal and sexual behaviors, and other social interactions (Brody and Rosvold, 1952; Bucher et al., 1970; Franzen and Myers, 1973; Myers et al., 1973; Snyder, 1970; Suomi et al., 1970). The hypothalamus, amygdala, hippocampus, and temporal lobe are consistently involved in the regulation of sexual behavior, and the functional integration between the hippocampus and the medial amygdala is critically involved in the feedback system controlling progesterone production in female animals. In this context, the hippocampal-hypothalamic-pituitary-gonadal axis represents the loop of positive feedback regulation, and the medial amygdaloid nuclei with stria terminalis and habenula constitute the loop of the negative feedback of the system regulating progesterone production (Elwers and Critchlow, 1961; Kawakami et al., 1967). Stimulation of the nucleus accumbens septi in female animals induces both prolactin secretion and increased milk production (Smith and Holland, 1975).

A complex syndrome involving several behavioral changes, including alterations of sexual activity, has been described in monkeys by Klüver and Bucy

(1938, 1939) as a consequence of bilateral temporal lobectomy. This surgical procedure had been primarily performed to study the connections between the visual area of the brain and the temporal cortex; the failure of surgically ablated animals to correctly adapt to visual stimuli was termed "psychic blindness" (Horel and Misantone, 1974). However, temporal lobectomy also induces a reduction of emotional responses, and a peculiar docility, which seems to be based on an apparent loss of fear, since such animals repeatedly expose themselves to the same previously known harmful situation (Pribram and Bagshaw, 1953). Operated animals can not eat properly or select appropriate food, and they display indiscriminate and exaggerated sexual behavior (Klüver and Bucy, 1938; MacLean, 1958; Pribram and Bagshaw, 1953). Male cats submitted to a fronto-temporal lobectomy mount other male cats or other male and female animals such as dogs, monkeys, and even chickens (Schreiner and Kling, 1953).

An additional neural system involving part of the hippocampus, cingulate cortex, and septal area has been identified as regulating pleasure, grooming, and sexual behavior. One hour after the cessation of seizures induced in cats by either chemical or electrical stimulation of the intermediate part of the hippocampus, pleasure responses and grooming behaviors are exhibited; there is an unusual receptivity to genital stimulation, often accompanied by sustained penile erection (MacLean, 1957a, b; 1958). Similar responses are also observed in rats following hippocampal afterdischarge (MacLean, 1955, 1957b). Stimulation of the medial nuclei of the septum, which provide afferents to the hippocampus, also induces pleasure reactions (Chapter 2, p. 28) and occasionally penile erection. This latter effect is also induced by stimulation of either the anterior portion of the supracallosal cingulate gyrus or the cortex above the posterior cingulate gyrus (MacLean, 1954, 1958), whereas in monkeys bilateral ablation of the former structure, probably also involving a portion of the underlying bundle of the cingulum, induces a loss of mimicry and grooming activity and a hypoemotional status (Ward, 1948).

Evidence that pleasure and aversion have their own specific mechanisms within the brain has been already presented (Chapter 2, p. 28–29, and Fig. 13) as primarily suggested by the experiments of Olds and Milner (1954). The anatomical substrate of pleasurable and rewarding feelings is a telencephalic-diencephalic sequence which includes the medial forebrain bundle, prefrontal cortex, subcallosal part of the cingulate gyrus, septal nuclei, preoptic nuclei, olfactory tubercles, Broca's diagonal band, centromedial amygdala, caudate nucleus, supramammillary area, and the region from the mammillary bodies to the interpeduncular nucleus, subfornical area, substantia nigra, the area beneath the medial lemniscus, locus ceruleus, raphe nuclei, and cerebellar fastigial and dentate nuclei (Clavier and Corcoran, 1976; Clavier and Fibiger, 1977; Clavier et al., 1976; Crow, 1972; Crow et al., 1972; Heath, 1975; Kant, 1969; Markowitsch and Pritzel, 1976; Olds, 1974; Olds and Olds, 1962, 1969a, b; Stiglick and White, 1977; Van der Kooy and Phillip, 1977; Wright and Craggs, 1977). Potent aversive and punishing sensations are elicited by stimulation of the mesodiencephalicperiventricular area (Olds, 1962; Stein, 1968), essentially corresponding to the "mesodiencephalic limbic area" of Nauta (1960, 1963, 1964). This anatomical model for a pleasure and aversion regulating system has been found to be common to various mammals including cats (Brady, 1955; Brown and Cohen, 1959; Roberts, 1966), dogs (Sadowsky, 1972; Stark et al., 1962), monkeys (Burnsten and Delgado, 1958; Lilly, 1958; Porter et al., 1959), bottlenose dolphins (Lilly and Milner, 1962), and men (Delgado and Hamlin, 1960; Heath, 1975, 1976; Heath and Mickle, 1960; Higgins et al., 1956). Pleasurable feelings are also induced by stimulation of some peripheral tactile receptors in fish, reptiles, and mammals (Campbell, 1972), and this observation emphasizes the evolutionary importance of the pleasure-aversion system, and the presence of direct associations between peripheral sensory circuits and the central mechanisms regulating reward and punishment.

In humans, hypersexual behavior was classically reported by Erickson (1945) in a 55-year-old woman who had a tumor of the paracentral cortical lobule, which lies just above the cingulate gyrus. Clear hypersexual behavior has also been described in the presence of degenerative diseases and tumors of the temporal lobes (Poek and Pilleri, 1965; Van Reeth et al., 1958). Inappropriate or deviant and abnormally increased sexual behavior has been observed after bilateral partial or total surgical ablation of the temporal lobes in man (Hill et al., 1957; Terzian, 1958; Terzian and Dalle Ore, 1955). Epilepsy, instead, has been associated predominantly with hyposexual activity, but abnormal or even increased sexual behavior has also been reported (Blumer, 1969, 1970; Gastaut and Colomb, 1954; Hierons and Saunders, 1966; Hoenig and Hamilton, 1960; Johnson, 1965; Maeder, 1909; Taylor, 1969). Temporal lobectomy in epileptic patients has been shown to be capable of reversing previous hypersexuality or changing the aberrant sexual behavior (Blumer, 1970; Hill et al., 1957; Mitchell et al., 1954; Taylor, 1969).

Stereotaxic lesions of the ventromedial nucleus of the hypothalamus (Cajal's nucleus or the tuber cinereum) have been claimed to eliminate sexual deviation without impairing normal sexual activity (Roder, 1966), and septal stimulation has been successfully used in a male homosexual patient to initiate heterosexual behavior (Moan and Heath, 1972). In addition, erotic fantasies reportedly have been induced by stimulation of the medial geniculate body, whereas anxiety follows stimulation of the cingulate gyrus (Heath, 1976). Stimulation through electrodes implanted in those brain structures described as relevant for the "pleasure system," such as the septal region, medial amygdala, head of the caudate nucleus, and anterior cerebellum, performed in patients with various neurological or psychiatric disorders has been consistently observed to decrease anxiety, hostility, and aversive feelings in general, while inducing confidence, friendliness, relaxation, euphoria, pleasure, and sexual well-being, sometimes resulting in orgasmic climax (Bishop et al., 1963; Delgado, 1969d; Delgado and Hamlin, 1956; Delgado et al., 1973; Heath, 1963, 1964, 1972, 1976; Heath et al., 1978; Moan and Heath, 1972; Riklan et al., 1974; Sem-Jacobson, 1964).

Bechtereva and her associates from Leningrad (Bechtereva et al., 1975) have suggested that repeated electrostimulation of deep brain structures in men chronically implanted with gold electrodes for various mental diseases may induce a "mobilization" of the wide brain resources, leading to the brain's relearning and/or acquiring normal or new schemes of functioning. This would cause the formerly acquired pathological structures to disintegrate and allow the restitution of normal conditions (Bechtereva and Bondartchuk, 1968; Bechtereva et al., 1972.)

A summary of the proposed anatomical network regulating emotional-affective behavior is reported in Table 1; this scheme will rapidly become outdated by the extensive proliferation of studies in this field.

Nevertheless, even a brief overview of the table is sufficient to appreciate that most of the identified anatomical loci participate in the regulation of more than a single behavior, and that some of them, while triggering one or more behavioral responses, also inhibit other behaviors. These facts can be taken as convincing evidence for the concepts of either anatomical or functional brain integrative activities. Moreover, the brain areas governing "positive" behaviors such as grooming, sexual and maternal behavior coincide mainly with those pertaining to pleasurable sensations in both animals and man. In the context of the various interactive functions of brain areas, which ultimately result in the different behavioral patterns, it may be inferred that the level of intensity of a given stimulus, on the one hand, and a flexible adjustment or a pathological change of the excitability threshold of some brain structures, on the other hand, can result either in differences in behavior or in behavioral abnormalities. For example, stimulation of peripheral tactile receptors, which are classified as pleasure triggers (Table 1; Campbell, 1972), can elicit pleasant or unpleasant and painful sensations depending on stimulus intensity. Such stimuli may also contribute to sado-masochistic tendencies resulting in such extremes as self-mutilation and the infliction of life-threatening lesions (Akhtar and Hastings, 1978); this has been observed in cases of mental retardation (Bach-Rita, 1974; Lester, 1972; Talkington et al., 1971), schizophrenia (Kohut, 1972), severe factitious dermatoses (Blacker and Wong, 1963; Krupp, 1977), Cornelia de Lange syndrome (Bryson et al., 1971; Shear et al., 1971), and some psychosexual disturbances (Roy, 1978).

The relevance of neurochemical mediators to behavioral production may be related to regulation of the above cited flexible adjustability of the neuronal firing threshold, and thereby to the excitability of various brain structures. In this way neurochemical mediators are likely to participate in regulating behavioral patterns produced by identifiable neural circuits. The previously mentioned concept of "neuroanatomical biochemistry" as springing from identified anatomical pathways operating by means of a preferential neurotransmitter molecule (this Chapter, p. 44) assumes further prominence, especially since those neurochemically specific projections impinge on such subcortical associative pathways as the medial forebrain bundle, stria terminalis, and stria medullaris, the primary

TABLE 1. *Anatomical correlates of some emotional-affective behavioral items*

Feelings and behavioral patterns	Brain structures involved as:	
	Triggers	Suppressors
Feeding	Lateral nuclei of hypothalamus Anterior cerebellum Superior cerebellar peduncle	Globus pallidus Pallidohypothalamic tract Ventromedial nuclei of hypothalamus Basolateral amygdala Periaqueductal gray matter Dorsolateral pontine tegmentum
Anxiety	Cingulate gyrus	Anterior cerebellar cortex
Aversion	Periaqueductal gray matter Nauta's limbic area	(see Pleasure triggers)
Rage	Lateral hypothalamus Dorsomedial hypothalamus Centromedial amygdala Temporal lobes	Septal nuclei Medial amygdala Head of caudate nucleus
Threatening	Sylvian fissure Ventromedial nucleus of hypothalamus Superior colliculus Reticular mesencephalic nucleus Ventrolateral pontine tegmentum Facial nerve nucleus Superior olivaris nucleus Lateral lemniscus	Head of caudate nucleus
Dominance	Anterior nuclei of hypothalamus Ventromedial nucleus of hypothalamus Supraoptic nucleus Septal nuclei Ventrolateral posterior thalamic nucleus Rostromedial red nucleus Periaqueductal gray matter	Head of caudate nucleus
Grooming	Prefrontal cortex Anterior cingulate gyrus Cortex above posterior cingulate gyrus Anterior temporal cortex Hippocampus Dorsolateral pontine tegmentum Superior cerebellar peduncle Cerebellar fastigial nucleus Cerebellar dentate nucleus	
Pleasure	Prefrontal cortex Subcallosal cingulate gyrus Septal nuclei Olfactory tubercles Preoptic nuclei Head of caudate nucleus	(see Aversion triggers)

TABLE 1. *(Continued)*

Feelings and behavioral patterns	Brain structures involved as:	
	Triggers	Suppressors
	Medial geniculate body	
	Medial hippocampus	
	Medial amygdala	
	Medial forebrain bundle	
	Supramammillary area	
	Subfornical area	
	Substantia nigra	
	Raphe nuclei	
	Locus ceruleus	
	Cerebellar fastigial nucleus	
	Cerebellar dentate nucleus	
	Peripheral tactile receptors	
Sexual behavior	Supracallosal anterior cingulate gyrus	
	Cortex above posterior cingulate gyrus	
	Paracentral lobule	
	Medial septal nuclei	
	Ventromedial nucleus of hypothalamus	
	Hippocampus	
	Medial amygdala	
Maternal behavior	Prefrontal cortex	Medial amygdala
	Nucleus accumbens septi	Habenula
	Hippocampus	
	Anterior temporal cortex	
Exploration	Posterior hypothalamic nuclei	
	Basolateral amygdala	
Sleep	Medial thalamus	
	Basolateral amygdala	
	Raphe nuclei	
	Cerebellar fastigium	

and secondary afferents and efferents of which connect most of the structures involved in emotional-affective behavior (Valzelli, 1979*a*).

In any consideration of behavior and human conduct, there remains an inadequately understood issue. In order to assemble the relevant heterogenous components of this problem, one must determine the contribution of physical, parental, social, educational, and cultural influences, all of which can modify neural activity. In turn, brain activity springs from the integrative functioning of cerebral structures and from their neurophysiological operations; these are genetically predetermined for their perfection or defects, but still remain poorly understood. Any "one-way" approach—exclusively sociological, psychological, or neurochemical—or, even worse, any deliberate omission of the multiple components which make up complex human conduct, will certainly lead to more vague and less well understood behavioral theories and conclusions.

It is in such a context of caution that the concept of aggression is considered.

4

Is Aggression in Our Nature?

INTRODUCING THE ARGUMENT

Such terms and concepts as *aggression, aggressiveness, aggressive behavior, predation,* and other related issues have been purposely used sparingly in the preceding chapters, although it is difficult to deal with the issues at hand without some use of the vocabulary. In preliminary considerations it might be viewed as completely senseless to speak about evolution, brain, mind, and behavior without taking aggression into account; conversely, even any simple approach to these arguments, and specifically to human nature, which disregards aggression is deceptive or pedantic. Nevertheless, examples of both these approaches appear in the incredible number of theories, propositions, and essays that are delivered by a number of self-appointed "experts."

Aggression, like many other commonly used words, is an apparently easy and self-defining concept, capable of immediately arousing a series of contrasting emotional responses with religious, philosophical, sociological, and political connotations that confuse the meaning of the term. As a matter of fact, the word aggression covers so many situations, and is used so broadly, that even confining the problem to a biological viewpoint, it becomes virtually impossible to formulate a single and comprehensive definition. This essentially is the result of the fact that aggression is not a biologically unitary concept (Jacobsen, 1961; Moyer, 1968*a;* Valzelli, 1976*b*) but is articulated in different types of aggressive behaviors which have been accurately described by Moyer (1968*a,* 1969). In turn, the "fluidity" of the argument has allowed the growth of a large and diverse literature in which ethology, biology, genetics, physiology, neurochemistry, and psychiatry are mixed, and often conflict with psychology, philosophy, sociology, and political views. Such confusion is by no means clarified by claims that "human diversity" does not allow the animal studies to be relevant for understanding human behavior and aggression (Barnett, 1972); or that studies on aggression and the various definitions of this behavior depend merely on value judgments of the experimenter rather than on the character of the behavioral responses or effects being studied (Tedeschi et al., 1974). The suggestion that abandoning the term "aggressive" as deriving from the "stubborn belief" that this term is most useful for classifying those behaviors which can legitimate coercive power and actions (Tedeschi et al., 1974), is not of greater assistance to the comprehension and significance of aggression. Conversely, such a proposition appears to stem more from personal and ideological convictions than from new scientific information. This perhaps unwitting but implicit ambiguity contributes to those "eight deadly sins" of our civilized society, so clearly described and analyzed by the Nobel laureate Lorenz (1973), which are greatly responsible for the continuously increasing uneasiness of mankind.

It is clear that aggression operates consistently and is widespread in the animal kingdom. But man, among the thousands of species that fight, alone fights destructively, with cruelty and malice, and is capable of becoming a mass murderer (Carthy and Ebling, 1964; Dillon, 1896; Freeman, 1964; Lorenz, 1967, 1973, Tinbergen, 1973). Since man has a poor memory for the cruelty in his own history, many past and present atrocities may serve as an appropriate reminder. For example, in Walker's *History of the Law of Nations* (1899), it is reported that, during the Middle Ages, Basil the Second (1014) blinded 15,000 Bulgarians, leaving an eye to the leader of every hundred; Saracen marauders, 30 years later, were impaled by Byzantine officials; the Greeks of Adramyttium, in the time of Malek Shah (1106–1116), drowned Turkish children in boiling water; the Emperor Nicephorus (961) cast from catapults into a Cretan city the heads of Saracens killed in an attempt to raise the siege; and the Crusader Prince of Antioch (1097) cooked human bodies on spits to provide his warriors with the terrifying reputation of cannibals.

It may also be illuminating to know that the aforementioned mass-murdering tendency of man has resulted in the death of over 59 million human beings from wars, or other disputes, between 1820 and 1945 (Richardson, 1960). In addition, since the end of the Second World War in 1945, there have been more than 150 wars, scrimmages, *coup d'états,* and revolutions. During this period of deceitful peace, the annual mean is 12 acts of war occurring simultaneously in different countries, with only 26 days of actual peace. The total number of human beings killed during the last 35 years amounts to some 25 million people, which even exceeds the total number of soldiers killed during the two World Wars (Cafiero, 1979). Further, a simple but broad calculation indicates that between 1820 and 1945 (125 years), some 472,000 people have been killed yearly from acts of war. This frightening figure has been nearly doubled (735,000 people per year) in the last few years. An additional statistic shows that in the last 3,427 years of recorded history, there were only 268 years of peace (Burke, 1975), which corresponds to 7.8% of the entire life of "civilized" man.

For such a history man has achieved the distinction of being the only misfit in his own society (Tinbergen, 1973), and the wicked seed from which his own destruction will likely stem.

The last statement need not be assumed as an inescapable destiny. Man still retains his freedom of choice (Chapter 2, p. 39). But no more time can be lost in pseudo-democratic dissertations or in domesticated scientific theories. In this context, Lorenz (1973)—even though completely aware that today, in reference to human behavior, one cannot use the words "inferior" or "valuable" without arousing suspicion that he is advocating the gas chambers—maintains that, because of our "liberal" permissiveness, something has gone wrong with human society. This allowed wrong behaviors to be transformed in defective mutation and genetic decay, which spread through society, so that the man's natural sense of justice now tells us to take action against antisocial behaviors (Lorenz, 1973). Similar conclusions have also been reached by Tinbergen (1973), who further said that the effects of human behavior endanger the survival not only of man but also of the entire planet.

We now face an almost incredible paradox: aggressive behavior, which was established in the behavioral repertoire of both animals and man for subserving self- and species preservation through billions of years of evolution, in man is now heading toward self- and species annihilation. This reflection may provide some indication of how unnatural the nature of modern man has become. In 1974 I publicly expressed my conviction that our ignorance of brain function greatly impairs our understanding of behavior, thus leading to intolerance and conflict. This is in complete agreement with Moyer's suggestion (1968b) that brain research can greatly contribute to a peaceful world by permitting physiological control of man's destructive violence. This task demands considerable wisdom and a sense of responsibility toward mankind, which I feel those of us engaged in brain research have achieved; we are at a critical threshold, as were the atomic scientists 35 years ago (Valzelli, 1975).

Since my statement in 1974, nothing has happened to disprove my theories or change my mind and, I suppose, that of most of my colleagues. Rather, I feel a much more urgent need for proper national and international measures to control individual and group violence. An important point is the concept of "control," since it would be foolish to believe that the aggressive component of human behavior could be forcibly abolished. This would correspond to a true mutilation of the natural human prerogatives subserving self- and species preservation. What is instead to be controlled is abnormally destructive, and arbitrarily violent and criminal behaviors, which increasingly endanger our society. Such control, which must be completely respectful of human mind and personality, is already present and available, even though most political representatives are still debating its "democratic" or "anti-democratic" connotation, completely unaware that no effective socialization is possible without the regulation of aggression (Bernstein and Gordon, 1974; Hartup and De Wit, 1974). It may be instructive for some of them to learn that such regulation or control of aggressive behavior has already been achieved and operates efficiently in the animal kingdom.

Among the numerous definitions of aggressive behavior available in the pertinent literature, I thought it useful to focus on a few biological definitions that are well integrated among themselves. According to Timbergen (1973), aggression is that behavior which tends to remove an opponent or make him change his behavior in such a way that he no longer interferes with the attacker. According to Kahn and Kirk (1968), aggressiveness is a goal-directed behavior having deep biological roots, and released by frustration or by any other impulse linked to self- and species-preservation. Within these limits aggressiveness supports and assures the success of these impulses or needs through a series of graduated behavioral patterns, ranging from the simple consolidation of supremacy to destructive outbursts. A third definition of aggresiveness is that of a specifically oriented behavior directed toward removing or overcoming whatever threatens the physical and/or the psychological integrity of a living organism (Valzelli, 1967*b*, 1978*d*). All of these definitions more or less overtly include two main biological principles—maintenance of internal homeostasis and self- and species-preservation—but none of them clearly reflects the existence of different kinds of aggression as identified by Moyer (1968*a*, 1969), nor the "altruistic" value associated with some aggressive reactions displayed by animals in defense of a fellow against a predator (Lorenz, 1967, 1973).

A biologically more precise definition might be the following: *aggressiveness is that component of normal behavior which, under different stimulus-bound and goal-directed forms, is released for satisfying vital needs and for removing or overcoming any threat to the physical and/or psychological integrity subserving the self- and species-preservation of a living organism, and never, except for predatory activity, initiating the destruction of the opponent.*

The last point, which actually differentiates animal from human aggression, is of utmost importance since it represents a potent factor harnessing the harmful effects of aggressive behavior. This is manifested by the conquered opponent by display of a highly ritualized code of *submission,* colloquially expressed as "I surrender"; this immediately stops the aggressive activity of the victor, allowing the loser to flee. *Territorialism,* by which animals of several species divide among themselves, and according to their rank in the colony, the available living space by establishing territorial limits, plays an important part in facilitating both avoidance and attack. The *attack-avoidance system* is activated when neighboring territory owners meet near their common boundary in such a way that both attack and withdrawal tendencies are equally provoked in the potential opponents. In this case a ritualized threatening code of behavior, colloquially signaling "keep out," is displayed, and instead of intruding, which would require the use of force and an actual fight, most potential trespassers withdraw.

In addition to this "individual" territorialism, there is also a "group" territorialism, by which animals of some species defend the territories belonging to a group or clan (Kruuk, 1966). An important aspect of group territorialism is that the members of a colony solidly unite when in hostile confrontation with another group approaching or crossing the boundaries of their feeding territory. This implies that groups confront groups as units, in which the attack-avoidance system starts to work as a result of the interindividual fellow members' attitude in the group; most potential group clashes can thereby be avoided. When a concrete possibility of flight is open, "precautional" and reciprocal fear provides an efficient agent for control of aggressiveness in the animal kingdom.

Does man still carry the animal heritage of individual and group territorialism? Such a question implies speculation about human evolution. Biologists use comparitive methods which allow the study of contemporary living organisms to provide insights into the evolutionary history of various species, with a degree of probability closely approaching certainty (Tinbergen, 1973). Unfortunately, this is unknown to most psychologists and behavior scientists, who tend to warn their students against the so-called zoomorphic interpretations of human behavior (Barnett, 1972). Such a warning could in fact be an important point of caution in speaking about single individuals, but it may be deceiving in studying the behavior of man as a species or mass human behavior. This alleged unreliability of the analogies between animals and man (Montagu, 1968) had been particularly emphasized in the case of territorial aggression and defense (Barnett, 1972), ignoring that man, in evolutionary terms, has started his career as a "social Ape who has turned carnivore" (Freeman, 1964; Morris, 1967). About this last point, man has already been described as the most formidable of all the predators, and the only one that systematically preys on its own species (James, 1911). Incidentally, *cannibalism,* which distinguishes humans from other primates (James, 1911), is only one expression on man's carnivorous nature which has been reported for almost all parts of the world; based on data from paleo-anthropology, cannibalism was probably once a universal practice (Blanc, 1961). Therefore, as a social, hunting primate, man must originally have been organized on the principle of group territories (Tinbergen, 1973), showing most of the aggressive repertoire and strategies displayed by chimpanzees and baboons in their natural habitats (Hamburg, 1971; Goodall, 1964; Rioch, 1967). This means that group territorialism is one of man's ancestral characteristics, which is reflected in his cultural pattern of parceling out his living space on earth along the lines of tribes, towns, countries, nations, federations, and now even "bloc" areas. Such grouping has not united mankind; instead, it has permanently divided it in a potentially explosive way.

The continuous perfection of technology directed toward the development of new and more powerful weapons, from clubs to atomic bombs and missiles, has considerably weakened the efficacy of the attack-avoidance system as a regulator of human aggression. Moreover, a subtle mechanism, exclusive of man, has greatly enhanced his destructiveness by driving him to the *depersonalization of killing.* To correctly understand this definition, one must remember that most aggression in the animal kingdom is controlled through the codified patterns of threatening and submissive behavior (p. 64). Animals are equiped with clear-cut and unequivocal social signals and responses which enable them to avoid injurious conflicts. Barnett (1972) pointed out that it is remarkable to a zoologist that, with no repertoire of universally recognized signals, man lives most of his life at peace with his fellows. But this statement seems to be somewhat unguarded for at least two reasons; first, except for interrelationships between single individuals or limited groups of individuals, when the word "man" is used to indicate his whole gender, probably no year of his life on earth has been spent without local or extended bloody quarrels and conflicts; second, man too has threatening or submissive gestures and facial expressions, which are used as menacing, reassuring, or appeasing signals, such as a friendly smile or a cordial handshake. These signals, unlike speech, are unequivocal and universally understood, so they have cross-cultural and species-specific value.

The unfortunate history of human aggression probably began as soon as man discovered how to hurt his opponent at a distance by means of slings or bows and arrows; the distance rendered the threatening, appeasing or submissive signals less clearly seen, and less efficient than the directed arrow. The control of aggression through the attack-avoidance system thereby became considerably weakened. This situation became more serious when firearms were invented, and as their range increased, until aggressors were provided with such pseudo-alibis as "We did not see them surrender in time." But the trend has greatly worsened, reaching the point of the previously mentioned depersonalization of killing with the adoption of long-range weapons. Today a battleship shooting from miles to another ship, or an aircraft dropping bombs from thousands of feet on sleeping towns no longer involves men killing other men, but only expensive "devices" purposely conceived for destroying "targets."

At present, the same principle applies, to an even greater degree, to remote-control weapons, since the man who may press the release button is completely shielded from any direct emotional involvement with the terrible consequence of his action. According to Lorenz (1967) and Tinbergen (1973), this situation, which induces a false and dangerous feeling of non-responsibility, can understandably drive even a perfectly good-natured man to lay carpets of incendiary bombs, and to release atomic missiles on "enemy targets," thereby apparently unaware of killing hundreds, thousands, or millions of children and men in a few seconds.

This is a perfect depersonalization of our schizoid civilization.

Once again we see the paradox of human aggressive behavior which, accepted as subserving self- and species preservation, has malignantly turned against human self- and species preservation. There is, also, the frightening absurdity of the miracle of the human brain that, equipped to reach the highest summit of consciousness, refuses to control itself, instead searching for more and more efficient devices for self-annihilation.

The advent of atomic power for destruction has, however, brought about a significant return to the primitive biological control of aggressive displays of the world "bloc" areas; through the alternating balance of threatening and submission signals, the attack-avoidance system has been restored. The system now operates under more sophisticated and up-to-date labels, such as Strategic Arms Limitation Talks (SALT) and Mutual Assured Destruction (MAD). These agreements are fraught with other disquieting initials, such as ICBM (Intercontinental

Ballistic Missiles), MIRV (Multiple Independently Targetable Reentry Vehicles), and ALCM (Air-Launched Cruise Missiles). But such systems are by no means able to guarantee the sustained survival of mankind.

The ancient precept engraved about 3,000 years ago on the face of the temple dedicated to Apollo in Delphi "γνῶθι σεαυτόν," know thyself, is still the best defense against ourselves. In this exhoration resides the already mentioned importance of extending the study of brain function and, specifically, of aggression.

We have now made a full circle, and in our presently unclear perspective of the future, we might also recall that "the darkest hour is that preceding sunrise."

The question is: will man be mindful enough to enjoy it?

IMPULSES TO AGGRESSION

During the past 20 years, an increased amount of research related to aggression has yielded over 3,800 scientific contributions on the topic (Crabtree and Moyer, 1977). These have resulted in the identification of: specific functions of some brain mechanisms implicated in aggressive behavior (Moyer, 1971*a*, 1976; Valzelli, 1967*b*, 1973*a*, 1978*d;* Wasman and Flynn, 1962, 1966); the role of stimuli acting as releasers (Berkowitz and Le Page, 1967; Scott, 1958*a;* Tinbergen, 1953, 1973); the acquistion of aggression through modeling (Bandura, 1973; Eibl-Ebesfeldt, 1974*a*) or direct reinforcement (Lovaas, 1961; Walters and Brown, 1963); and the different goals subserved by the various kinds of aggressive behaviors (Feshbach, 1970; Moyer, 1968*a*).

However, still unresolved is the issue of whether or not there is an innate drive for aggression in both animals and man. Some investigators maintain that there is no drive for aggression, at least in the sense that there are drives for eating, drinking, or sex (Altman, 1965; Hinde, 1967; Marler and Hamilton, 1966; Scott, 1958*a*, 1962, 1965, 1966*a*, *b*).They assume that there are no pysiological mechanisms for spontaneous internal stimuli for fighting, and that the physiological mechanisms associated with fighting are different from those subserving eating and sexual behavior (Scott, 1965). Further, Scott and Fredericson (1951) have maintained that no spontaneously arising internal stimulation capable of causing a need for fighting *per se* exists, and that aggresiveness is enhanced only in response to predictable external stimulation. Other scientists in the same field of study have reached the opposite conclusion, and affirm that in both animals and man there is indeed a drive for aggressive behavior (Feshbach, 1964; Lagerspetz, 1964; Lorenz, 1967).

Like Tinbergen (1973), I surmise that most of such disagreement lies mainly in the differences with which the various authors view the internal and external variables. These differences may well be a matter of semantics. According to Moyer (1969), much of the controversy stems from variations in defining "drive," so that the concept has assumed different meanings for different people. Drive is frequently, and confusingly, referred to as the status of a genetically predetermined intervening variable that produces a state of tension when it functions. This tension, or psychic excitation, which to most contemporary psychoanalysts corresponds to Freud's concept of instinct, motivates the organism into action to alleviate such tension (Freedman et al., 1975). Obviously, this definition which may be rather generally inclusive, reflects little concern with the internal environment of the organism, whereas the term "drive," to be useful, should have a sound physiological basis.

In the preceding chapter (p. 48), we discussed drives, considering the concept that they may be viewed as "blind" multipotent energy initiating random behavior, which is converted into a motivated and directed pattern of behavior when both appropriate targets and suitable learning occur. In this framework, aggression cannot be considered as an instinct or a drive, but simply as a series of graded "behavioral means" for properly reaching precise and specific goals, such as the prey, the self, group and territory defense, and other purposes related to self- and species preservation.

The previous statement by Scott and Fredericson that no internal need for fighting exists, that aggression is merely a response to predictable external stimulation, applies to all basic patterns of behavior since, as Moyer (1969), Eibl-Ebesfeldt (1974*a*), and others observed, no behavior takes place *in vacuo*. For example, a starved animal, regardless of the intensity of its hunger, does not chew randomly or eat inedible objects. The same applies to aggression, by which normal animals respond to a limited number of stimulus objects. The statement that the physiological mechanisms associated with fighting are different from those underlying sexual behavior and eating, or that there is no physiological mechanism by which spontaneous internal stimulation for aggression arises (Scott, 1965) is also inadequate. It has been shown that stimulation of the lateral hypothalamus induces eating even in satiated rats (Akert, 1961), which eat laboratory chow or lap milk, but do not drink pure water (Miller, 1957, 1961); this suggests that the response to the experimental manipulation is clearly stimulus dependent. In a similar way, a cat stimulated in the lateral hypothalamus will display predatory aggression when its usual prey is available (Wasman and Flynn, 1962), although in the absence of the stimulus-object, the animal may explore restlessly without displaying any aggressive behavior (Moyer, 1969). Further, although there are obvious marked differences between sexual and aggressive behaviors, the anatomical data already reported (Chapter 3, p. 53; Adams, 1968; MacLean, 1965), coupled with physiological results (Bronson and Desjardins, 1971; Ervin and Mark, 1969; Moyer, 1968*a;* Roberts and Kiess, 1964; Young et al., 1964), suggest specific connections between the functional mechanisms governing sexual and aggressive behaviors (Feshbach, 1974*b*). In man, experimentally induced inhibition of anger and aggression has been shown to produce a significant decrement on a subsequent measure of sexual arousal induced by erotic stimuli (Fesnbach and Jaffe, 1970), whereas sexual arousal, in both male and female volunteers, produced a significant increase of the aggressive tendency (Jaffe et al., 1973; Zillman, 1971).

In summary, when such terms as "drive," "instinct," "internal need," and others are avoided, most of the controversy with such views as those of Lorenz and Scott disappears. Instead, it is necessary to emphasize, that as we have already seen for the neural substrates of emotions and affect (Chapter 3, p. 53), there are both brain circuits and physiological mechanisms whose activation leads the organism to display the different kinds of normal aggression, as will be shown next. It should be recalled that a distinction has been made between "normal" and "abnormal" (Chapter 2, p. 30).

I would also like to clarify what is intended by attributing to normal aggression the significance of a behavioral mean (p. 68). For instance, when we study the feeding behavior of a hungry laboratory animal in its home cage, the animal simply has: (1) to approach the food container, and (2) to eat, directly displaying its feeding behavior; in this instance such behavior almost entirely corresponds to consummatory behavior, with the laboratory pellets constituting an exclusive "one-way" situation without any chance of feeding choice. In contrast, an equally hungry but wild animal searching for food in a natural and free environment must display a much more complex sequence of behaviors that principally includes: (1) exploratory behavior, to become acquainted with favorable or potentially dangerous environmental characteristics, (2) orienting behavior, for establishing fixed points of reference allowing the recognition of its own spatial position within the environment, (3) goal-directed searching behavior, for detecting appropriate food objects; further, once the feeding source has been identified, according to its characteristics and the presence or absence of other animals, related or not to the hunting subject, (4) instrumental, competitive, or defensive aggressions could become necessary intermediate and graded behavioral means before the display of (5) predatory aggression, which allows the animal (6) to eat. Again, unlike the earlier example, in this case feeding behavior does not necessarily correspond to consummatory behavior; the latter mostly reflects the extent of food intake dependent on taste preference when such free choices are available.

In general terms, the above examples indicate that environment, through its demands and characteristics, modulates aggressive displays externally to a great extent, as has already been shown for other behaviors (Chapter 3, p. 49). Nevertheless, the facilitation, reduction, or inhibition of the different kinds of aggression through experimental lesions and stimulation, as well as the uncontrolled outbursts of pathological aggressive behaviors as a consequence of the internal stimulation of certain brain areas (Jonas, 1965), indicate that both animal and human brains are "technically" equipped for the internal activation to aggression. It is wrong to emphasize exclusively the external "social" components that may promote or inhibit aggression, completely disregarding the internal brain mechanisms that govern aggressiveness; and the opposite approach is equally wrong. Interestingly, the first position is held mostly by sociologists, psychologists, and psychiatrists, whereas the second is primarily maintained by neurophysiologists, neurochemists, and most biologists. Consequently, the previously cited dilemma (Chapter 3, p. 41) dividing the bulk of the investigators of mind and behavior into two "fighting" camps arises again in connection with the issue of aggression, which is often burdened with uncontrolled emotion that transforms a potentially fruitful scientific debate into a largely unproductive ideological or, at best, philosophical quarrel. As a few examples of this, evidence for the influence of economic and political ideologies on developmental psychology has already been given (Riegel, 1972), and currently available psycho-chemical and psycho-environmental measures (Chapter 3, p. 49), potentially capable of increasing human intelligence, depend more on the trends of different "social" politics than on scientific reliability (Krech, 1967).

The assumption that man is "anthropologically" free from aggression, the denial of any human aggressive propensity, and of any evolutionary value of aggression for man (Montagu, 1974) reflect an unconscious desire to attribute an "innate" saintliness to human nature. The proposed lack of correlation between aggression in animals and man (Barnett, 1972; Lion and Penna, 1974; Montagu, 1974) depends mostly on a lack of distinction between aggression and violence, which is, once more, the difference between normal and abnormal aggression. In fact, as for other aspects and characteristics of human brain production (Rose, 1973), human aggression differs from that of animals almost exclusively in terms of quality rather than quantity, also being dependent on man's often cited freedom of choice in selecting his behavior. Even though this does not comply with such pseudo-democratic doctrines which claim that all men are equal by nature, and that all human behavior is learned and structured by conditioning, thereby being susceptible to change and correction to an unlimited extent, men still differ according to genetic predisposition, neurophysiological setting, brain use, and cognitive production. More precisely, interindividual differences (Pitkänen, 1973), genetic constitutional factors (McClearn, 1974; Sheldon, 1942), ego development and sociocultural differences (Ferracuti and Wolfgang, 1963; Pinderhughes, 1974), peer modeling and teacher influences (Cohen, 1971), imitative, reinforcement, and model learning (Bandura and Walters, 1963; Berkovitz, 1962; Kahn and Kirk, 1968), tribal, ethnical, and national influences (Bolton, 1973; Bolton and Vadheim, 1973; Lynn, 1971; Montagu, 1974), and positive or negative social influences (Borden and Taylor, 1973) are all factors which, despite any sociological dream, differentiate to a large extent between human beings, as well as their tendency to behave more or less aggressively in response to the same stimulus-object or situation.

Another factor that affects the occurrence of aggression is population density (Chapter 3, p. 49, Calhoun, 1973; Stokes and Cox, 1970). It has been observed that, within a population, aggression increases six times faster than the rate of increase in density (Burns, 1968; Myers, 1966; Southwick, 1955). This may result from the breakdown of perhaps the most rudimentary territorial system:

individual distance (Hediger, 1950). Many animals will not tolerate the approach of another animal beyond some minimum distance. When this distance or minimal surface area is infringed on, the intruded individual reacts either by attacking or by moving away. This kind of personal territorialism, or *body-buffer zone,* like the previously described group territorialism (p. 64), is also present in man. The spatial distance that is tolerated by man varies inversely with culture and education (Hall, 1959, 1966), mental equilibrium (Horowitz et al., 1964), and propensity to violence (Kinzel, 1970). The extended areas available to some primitive human ethnic groups or tribes, which were reported to peacefully display their activities and interactions to the point of being assumed to represent the pacific nature of man (Montagu, 1974; Thomas, 1959), may merely represent a potent factor limiting aggression. Obviously the opposite is true in our overcrowded cities.

This consideration represents another point of serious concern to our politicians.

With regard to the influence of learning, it is true that aggression can be increased or decreased by a directed education, and that like many other behavioral patterns, it can be modified to some extent through experience. Nevertheless, such curbing of aggression occurs within the limits imposed by genetic bounds, which underlie the aforementioned inter-human differences. A large amount of literature (Baron, 1977; Crabtree and Moyer, 1977) demonstrates that learning processes leading to the facilitation or reduction of aggressive behavior are not substantially different in animals and man. These processes are such that *instrumental conditioning,* with various kinds of positive or negative reinforcements can spontaneously induce docile animals to fight, and aggressive animals to avoid fighting (Azrin and Hutchinson, 1967; Baenninger, 1970; Corum and Thurmond, 1977; Farris et al., 1970; Flory et al., 1977; Hutchinson et al., 1971; Leitenberg et al., 1970; Melvin and Ervey, 1973; Tondat and Daly, 1972).

The avoidance of fighting does not, however, imply the absence of aggression and the already cited genetic components (Lagerspetz and Lagerspetz, 1971) together with environmental variables (Ader, 1975; Galef, 1970a; Heiligenberg and Kramer, 1972; Lagerspetz and Hyvärinen, 1971; Powell and Creer, 1969) affect the outcome of instrumental learning. Human beings also are recognized as being capable of acquiring aggression through instrumental conditioning (Bandura, 1973, 1977; Feshbach, 1974b; Zillman, 1971, 1978), and are influenced by a socioenvironmental component (Geen and Stonner, 1971; Gentry, 1970). Obviously, no animal will fight for such insignificant objects as toys, money, decorations, or Rolls-Royces, but men do so, as these are taken as material incentives (Buss, 1971; Gaebelein, 1973; Walters and Brown, 1963), for which the attribution of value represents a means of enhancing self-worth and esteem. This trend is typical of the childlike behavior of our society, which demands instant gratification while rejecting any sense of responsibility and showing no consideration for the feelings of others (Lorenz, 1973). Nevertheless, I do not believe that we are victims of society, at least in the sense that Montagu (1977), as many others, has suggested. Because society is a product of each one of us rather than an outside entity, its errors and immaturity reflect our dangerous mistakes and infantilism. Thus, if we are victims of society, we are victims of ourselves, of our errors, and of our lack of knowledge concerning the functions of our brain, as previously pointed out (p. 63).

The foregoing observations also seem to agree with the finding that, exclusively in man, the signs of pain and suffering in the victim serve as a form of reinforcement to many individuals (Baron, 1974, 1977a; Feshbach et al., 1967; Hartmann, 1969), thus strengthening the tendency to engage in such behavior subsequently. Berkowitz (1974) has also shown that human aggression is reinforced when man learns that he has injured his "enemy." This is especially true of homicide, an expression of the pathology of aggression or violence, which can also represent the basis from which originated the widespread and aberrant practices of persecutions, concentration camps, and torture.

What is not exclusively limited to human behavior is *xenophobia*, which also facilitates aggression among nonhuman primates (Holloway, 1974; Southwick et al., 1974), probably as an extension of territorialism. Xenophobia in man is further elaborated into various forms and destructive expressions of ideological, social, and racial prejudice, which clearly reinforce or enhance uncontrolled aggressive displays (Buss, 1961; Kaufmann, 1970). These can result in assassination and genocide, as we have all learned. Experiments have shown that highly prejudiced subjects are significantly more aggressive than unprejudiced individuals (Genthner and Taylor, 1973).

Another factor influencing aggression is *sex*. Among animals, as we will see later, females fight only under limited and specific circumstances. Among humans, it is commonly believed that men are almost always more physically aggressive than women, and that women show indirect or displaced forms of aggressiveness. This has led some to the playful suggestion that the so-called emancipation of women may act as a prophylactic against tyranny (Barnett, 1972). But this is not true. There is evidence that women are often as openly hostile and as directly aggressive as men, and occasionally even more so (Frodi et al., 1977). Some researchers have shown that women are more likely than men to be child-abusers (Gelles, 1973; Gil, 1968, 1970), that violent crime rates for younger women are constantly rising (Noblit and Burcart, 1976), that female aggression may be reinforced by the overt suffering of the victim (Brock and Buss, 1962; Deaux, 1971; Jaffe et al., 1974; Levitt and Viney, 1973), and that, even if men and women tend to react differently to external aggressive cues and provocations, when aggression is perceived as ideologically justified or "prosocial," women act as aggressively as men (Frodi et al., 1977).

Much of the socio-psychological literature on aggression has also been based on Dollard's assumption that any interference in a goal-directed behavior, resulting in *frustration*, is a necessary and sufficient cause for aggression (Dollard et al., 1939). Conversely, in Dollard's view, every act of aggression must have originated in previous frustrations. In psychoanalytical terms, it has also been said that aggression comes into being when libidinal and ego forces are frustrated (Kuiper, 1972). According to such "frustration-aggression theory," frustration becomes a persistent internal instigation toward aggression; but aggression can be blocked or inhibited by fear of punishment. Nevertheless, when aggression is punished, Dollard and his colleagues maintain that instigation to aggression remains and may lead to assault against targets other than the frustrating ones, and associated with less inhibition. This corresponds to the general notion of *displaced aggression* (Miller, 1941), to which conditions such as physical limitations, delays in the completion of some activities, omission or reduction of some assumed customary rewards, and the need of choosing between two or more equally desirable targets have been ascribed (Berkowitz, 1962, 1965; Brown and Farber, 1951).

According to Bandura (1973), Dollard's theory is highly appealing, for both its boldness and simplicity. These characteristics made this assumption a readily acceptable and widely held doctrine, which, certainly beyond the intention of its authors, has been adopted by several schools to advocate the abolition of any behavioral limitation, education, and control, based on the premise that such "repressive" measures cause frustration. Obviously, this position does not consider that the prerequisite of all civilized communal life is that people learn to properly control their impulses. Consequently, from such a background, which fails to consider the "frustrative effect" of the improper behavior of some toward others, has emerged the doctrine which maintains that all moral and behavioral defects of criminals depend directly on defects in their environment and education. This has led, especially in recent years, to an overpermissiveness with an increasing rejection of a sense of responsibility toward our offspring, and to a present-day caricature of liberal democracies, which greatly thwart the human sense of justice (Lorenz, 1973).

Public opinion is inert, and reacts to new influences only after protracted "dead time." Further, it favors gross simplifications which reflect exaggerations of the facts. This explains why the effects of the misused Dollard theory are still at work 40 years after its formulation, despite the evidence that frustration only sometimes facilitates aggression (Berkowitz, 1969; Geen and O'Neal, 1976), often is incapable of enhancing aggression (Buss, 1963, 1966; Kuhn et al., 1967; Taylor and Pisano, 1971), and may even serve to reduce aggression (Gentry, 1970; Lange, 1972; Rule and Hevitt, 1971). Such scientific reports will not have the same successful dissemination as Dollard's theory, since they do not appeal to those demagogic and hiddenly proaggressive and pseudo-social philosophies which can survive only by pushing men against men. This should also be considered by many politicians.

5

Cerebral Representation of Aggression

DIFFERENT KINDS OF AGGRESSION

For some years aggression has been recognized as being not a unitary behavioral concept (Bevan et al., 1960; Jacobsen, 1961; Scott, 1958a; Valzelli, 1967b), but rather a series of different specialized behaviors, which can be operationally identified as was first shown by Moyer (1968a, 1969, 1971a). Moyer's studies are relevant since they indicate that various aspects of the aggressive repertoire represent a graded sequence of biological tools that animals have available for achieving different goals. The plurality of aggression (Brain and Al-Maliki, 1978) has also been proven by experiments in which a specific manipulation of proper neurophysiological substrates may facilitate one kind of aggression, suppress another, and have no effect on a third.

By applying the already quoted principle of Fentress (p. 43) to studies of aggression, the apparently uninterrupted "transposition" of one form of aggression into another, according to environmental demands, as hypothetically outlined on page 69 of the preceding chapter, may make it difficult to identify correctly different types of aggressive displays. Another source of confusion is the trend of some researchers to develop and use "personal" and diverse terminology to indicate the same, previously defined responses, as will be later shown. Further, according to Moyer (1968) the proposed classification of aggressive behavior was intended, and probably still is intended, as a tentative scheme which flexibly allows the insertion of new, consistently documentable items.

According to the view, the classes of aggression which may be identified are listed below.

Predatory Aggression

Predatory aggression is characteristically evoked by the concurrence of hunger and the presence of an appropriate object of prey, and deals with most interspecific aggressiveness. Predatory behavior is clearly stimulus-specific, with various preferences shown for target choice (Bandler and Moyer, 1970; Barnett, 1975;

Brain and Al-Maliki, 1978), and a general lack of relevance of the particular test environment (Moyer, 1968). The size of the prey, its mobility or immobility, previous experiences, and previous satiation or starvation may influence the features of a predatory attack (Adamec and Himes, 1978; Bandler and Moyer, 1970; Beddington et al., 1975; Driver and Humphries, 1970; Katz and Thomas, 1977; Paul, 1972; Polski, 1978). Predatory behavior has been experimentally shown to differ from other forms of aggression (Barr et al., 1975; Carthy and Ebling, 1964; Knutson and Hynan, 1973), and it has also been shown to consist of two behaviorally or neuroanatomically separate components: *killing* and *feeding* (Adamec, 1975*a, b;* Adamec and Himes, 1978; Berg and Baenninger, 1974; Desisto and Zweig, 1974; Karli et al., 1969; Krames et al., 1973; Paul and Posner, 1973; Polsky, 1975).

Predatory display has also been suggested as being a class of behavior that does not come within the scope of aggressive activities (Carthy and Ebling, 1964), rather being a normal and unavoidable behavioral means of achieving an appropriate foodstuff. In man, the most direct relationship to animal predatory aggression may be found in his various hunting activities, even though they have almost completely lost any actual hunger motivation. Nevertheless, separating predatory behavior into the two aforementioned components of killing and feeding makes it possible that one of them will prevail over the other, as a consequence of behavioral or neurophysiological alterations capable of leading to an abnormally repetitive, and almost automatically evokable killing behavior (see Chapter 7, p. 126), or to a similarly abnormal way of feeding. In this framework, it might be hypothesized that abnormalities of these two activities may have similarities, respectively, to some violent human conduct and human killing behavior, as well as to some obsessional forms of neurosis such as *anorexia nervosa* (Barry and Klavans, 1976; Story, 1976; Young, 1975); in some cases of the latter, alternating anorectic and bulimic episodes occur, associated with oscillations of more or less overtly hostile behavior. Disorders involving the neurochemical regulation of the integrated activity of the ventromedial hypothalamic satiety center and of the lateral hypothalamic hunger center have been suggested as a neurophysiological basis for anorexia (Cole, 1973); the lateral hypothalamus has been known to be involved in regulating predatory aggression (Valzelli, 1978*d*).

Competitive Aggression

Competitive aggression may also be defined as inter-male aggression or *agonistic behavior,* and is identical to *intrastrain, intraspecific,* or *conspecific* aggression, conflict, and fighting (Brain and Poole, 1974; Clark and Schein, 1966; Price et al., 1976). The most potent stimulus to competitive aggression is the presence of a strange male of the same species, which may be attacked even in the absence of any apparent provocation (Blanchard and Blanchard, 1977; Blanchard et al., 1975; Luciano and Lore, 1975; Moyer, 1968*a*). However, competition for social rank and dominance, for food and water, and for the control of their sources, for the choice of females, and for reproduction are potent sources of motivating animals to vigorous fighting (Clark and Schein, 1966; Levine et al., 1965, Kahn, 1961; Price et al., 1976; Van Kreveld, 1970).

Once hierarchical dominance has been achieved within a group, the fighting brought on by competitive aggression is to a great extent substituted by such gestures as mimicries and vocalizations relating to the previously cited threatening or submissive ritualized behaviors, which will constantly deter further fighting. It is likely that *straight aggression* and *appeal aggression,* two recently described categories of aggression found in Java monkeys (De Waal, 1976), are only highly specialized and species-specific patterns relating to the dominance-submission system of competitive aggression. According to De Waal, the straight aggressor gives exclusive attention to his opponent, whereas the appeal aggressor alternates his attention to the opponent with nonagonistic elements of behavior toward a third group of bystander members; these are almost invariably dominant in respect to both the actor and reactor.

In the opinion of Van Kreveld (1970), the social relevance of hierarchical dominance is found in three main functions. The first is that an integrated group is capable of reacting as a unit to outside threats, and to better defend the territory; dominance, at least in nonhuman primates, does not exclusively represent a priority of rights, but also implies a priority in fulfilling duties towards subordinates and protecting them (Bernstein, 1964; Carpenter, 1963, Crawford, 1942; Maslow, 1940). The second function is the general regulation and limitation of aggressive displays within the group, with the strongest dominant animal also being the most likely survivor when insufficient food is available for the entire group of subordinates; there is, further, the opportunity to reproduce and breed strong offspring (Maslow, 1936; Van Kreveld, 1970). A similarly greater reproductive fitness of dominant males has also been described in rodents (Nyby et al., 1976). Finally, the third function might be development of effective birth control mechanisms by excluding from food, reproductive behavior, and from the group any aberrant, parasitic, or weak individuals (Wynne-Edwards, 1963, 1965). Aggression, and more specificially, competitive aggression and the dominance hierarchy, may be considered vital for the establishment and regulation of primate society and social behavior (Bernstein and Gordon, 1974).

Several factors, other than the threat-submission system, regulate both the frequency and the intensity of competitive aggression. Early social experiences, social rearing, and attachment or affiliation between some not too alien mammalian species, and the young female intruder are all factors that greatly limit aggressive displays (Cairns and Werboff, 1967; Galef, 1970*b;* Poole and Morgan, 1975; Whelton and O'Boyle, 1977). Olfactory cues, as directly related to the home-cage, the subordinate male, female, and their urines, or the smell of a dominant male and of its urine, can accordingly inhibit or reinforce aggression when dominance is in question (Alberts and Galef, 1973; Dixon and Mackintosh, 1971; Fortuna, 1977; Jones and Nowell, 1973*a;* Kimelman and Lubow, 1973;

Mugford, 1973; Zook and Adams, 1975). Sexual experience, cohabitation with and the presence of female animals, and imitative learning enhance fighting between male animals (Flannelly and Lore, 1977; Goyens and Noirot, 1974; De Ghett, 1975; Flandera and Nováková, 1974), whereas competitive aggression between females is mostly, and inversely related to the size of the opponent (Bernstein and Gordon, 1974; White et al., 1969). Hunger, undernutrition in early life, prenatal nutritional deficiency, and an increased intake of l-tyrosine with food are all factors that consistently increase aggression in competition testing (Fraňkova, 1973; Fredericson, 1950; Peters, 1978; Randt et al., 1975; Rohles and Wilson, 1973; Thurmond et al., 1977).

Beyond any doubt, competition has always been and still is one of the most important components of normal everyday human life. From a psychoanalytic standpoint, competition may be assumed to correspond to a large extent with the Freudian principles of Eros and Death instincts, the former having been proposed as subserving self-preservation, and the latter the preservation of the species. These, respectively, underlie sexual and aggressive behaviors (Freud, 1920; Mora, 1975). The limited psychiatric terminology, by which competition is defined as the "struggle for the possession or use of limited goods, concrete or abstract" (Freedman et al., 1975), has necessarily reflected upon an almost general negative connotation (Feshbach, 1974b; Hapkiewicz, 1974; Kagan, 1974) of this potent biological force which, when properly channeled, has always driven man to achievement in diverse areas of the arts and sciences. Then, as previously observed (p. 70), the question is not that of depriving man of his competitive aggression, but rather of channeling it into socially valuable, rewarding goals.

Regrettably, in the absence of such concern, human competitive aggression can easily degenerate into those adverse effects which do not occur in the animal kingdom. Several animal competitive cues or stimuli have maintained their primitive value but under new "exorcistic" denominations, such as social status instead of rank order, self-assertion instead of dominance, or assistant instead of subordinate, immigrant instead of social intruder, and several other verbally euphemistic substitutes. Olfactory cues, even though less important in the biological structure of man, still operate as features of "attractive" smells and perfumes or aversive "bad" smells; the latter are frequently associated with racial differences and prejudices, and with negative feelings such as the "smell of fear" or of blood. The value of olfaction and of display behavior as hallmarks of social rank and hierarchical dominance in animals has merely been replaced by displays of richness, decorations, degrees, and clothes, (Laver, 1964), retaining their original anti-aggressive or pro-aggressive values. Further, just as in animals, hunger, malnutrition, imitation, and female and bystander influences are all elements which interact to enhance or reduce human competition (Baron, 1971, 1977; Borden, 1975; Cumberbatch and Howitt, 1974; Parke, 1974; Shuck, 1974). Competition, through the action of "supporters" or "fans," has a well-known positive or negative influence on several human activities, and those represent forms of redirection for human competitive aggression, as represented by various sporting activities.

Defensive Aggression

Defensive aggression corresponds to *fear-motivated* or *fear-induced* aggression, which is typically displayed against an aggressor when the animal is presented with an inescapable threat. Defensive aggression is then recognizable, based on an attempt at flight, which always precedes the aggressive reaction; this stimulus situation is represented by the obvious presence of a threat and by the extent to which the frightened individual is confined.

The majority of defensive aggression may be observed as part of the aggressive interaction between a predator and its prey when the latter, after having exhausted all escape strategies (Humphries and Driver, 1967), has no further course for survival. Fear is the principal stimulus to defensive aggression (Blanchard et al., 1974, 1975), and it has been shown to be inversely related to the intensity of animals' reactivity and aggressiveness (Svare and Leshner, 1973). Fear may thereby be present, to varying degrees, in other aggressive displays, interacting with competitive, maternal, and irritable aggressive behaviors, while also reflecting primitive analogues of human emotional responses (Blanchard and Blanchard, 1971; Gray, 1971).

In this last respect, little has been systematically reported in the current literature concerning the effect of fear on human aggressive behavior, even though fear may presumably also act in man, by causing him to react violently when physically or psychologically confined, to stop his ongoing behavior, or to take flight, based upon environmental circumstances, previous experience, and his feelings and temperament.

Irritative Aggression

Irritative aggression is evoked by a broad range of attackable animate and inanimate targets, as typically observed in anger or rage reactions. Irritative or *irritable* aggression, in Moyer's opinion, should involve no fear component, although, according to others (Flinn, 1976; Hess and Brugger, 1943; Ranson, 1936), rage is classically described as a mixture of threatening, defense, and attack behaviors. If this is true, the participation of a clearly identifiable fear component may be supposed to also take part in the display of irritative aggression. This may be inferred from the fact that several aversive stimuli, such as a physical blow (Azrin et al., 1965), intense heat (Ulrich and Azrin, 1962), extinction of a previous food reinforcement by food removal (Azrin et al., 1966; Thompson and Bloom, 1966), morphine withdrawal (Boshka et al., 1966), and painful tail- or foot-shock (Azrin et al., 1964; Ulrich and Azrin, 1962) are all capable of eliciting attacks against inanimate objects or other animals (Azrin et al., 1966; Galef, 1970c, d). This reaction has been observed to occur in a wide series of different animal species, ranging from fish and birds to several mammals, including monkeys (Azrin et al., 1966; Flory, 1969; Legrand et al., 1974; Lyon and Ozolins, 1970; Turner et al., 1973; Ulrich et al., 1969).

Since most of the experiments concerned with this type of aggression have been performed with the administration of a painful electric shock, the terms *pain-* or *shock-induced* aggression have been preferred by many. However, the observation that not only the perception of physical pain but also psychologically aversive stimuli and distress can elicit aggressive displays even against inanimate targets suggests that irritation may be considered as a common and general basis for this type of aggressive behavior. This concept is also in good agreement with the general observation that a wounded animal is much more irritable than a healthy one (Scott, 1958a), and is more likely to react with anger and aggression to a wide range of nonspecific stimuli that are ordinarily rather ineffective in arousing aggressiveness in normal animals.

This fact may also account for the involvement of some avoidance and defense components (Blanchard et al., 1978; Dunham and Carr, (1976) in eliciting irritative aggression. Accordingly, this behavior includes either intraspecific or *interspecific* aggressive displays (the latter pertaining more to predatory aggression) when paired conspecific animals or representatives of diverse species are confronted in aversive situations (Reynierse, 1971; Ulrich et al., 1964). Other factors such as hunger and thirst, target location and novelty, previous social habituation to conspecific, sex and age (reflecting mostly the hormonal balance of the subject), previous similar experience, sensory impairment, and the intensity and duration of noxious stimuli may variously increase or decrease, but not abolish, irritative aggression (Creer, 1973; Galef, 1970c, d; Hutchinson et al., 1965; Tondat, 1974; Ulrich, 1966).

Surprisingly, no directly available experiments on human irritability, and irritative aggression, have apparently been performed. A few authors (Berkowitz, 1974; Kagan, 1974; Miller and Miller, 1974; Rule, 1974; Wolff, 1973) have almost incidentally referred to anger in connection with studies on human aggressive behavior. Nevertheless, feelings of irritability capable of precipitating angry reactions mostly develop into verbally hostile explosions; these are not necessarily elicited by the physical presence of a directly "responsible" conspecific antagonist, and often directed against a totally innocent and naive scapegoat. Even inanimate stimuli are sometimes the target of such aggression, and can be more physically injured. Such irritative conduct, which has more commonly been designated as a "bad temper" is a function of the extent to which given individuals are capable of tolerating various aversive stimuli; this is well depicted in the familiar statements, "I'm blowing up" or "I blew my cool," and has obvious similarities with animal irritative aggression.

The significance of having a true bad temper also relates to those "technical" concepts of the participation of ego strength and disturbances in aggressive displays (Mueller and Grater, 1965; Silverman, 1964); these concepts in turn include such components as discomforting feelings, hormonal balance, duration and intensity of the aversive contingency, level of education, and degree of culture. In addition, previous familiarity with the possible animate or inanimate targets can alter human irritative aggression. A typical pain-elicited aggression, during the conduct of behavior therapy, and a classic extinction-induced increase in aggressive responses have been described with humans (Kelly and Hake, 1970; Rachman, 1965). Intense heat has already been mentioned as a stimulus for irritative aggression in animals, with the "common knowledge" that many individuals become more irritable, more likely to exhibit aggressive outbursts, and more negative in their reactions towards others under uncomfortably hot environmental conditions. Serious outbursts of collective violence in major cities occur more frequently during the summer months, when heat-wave or near-heat-wave conditions prevail (Goranson and King, 1970; Griffitt, 1970; Griffitt and Veitch, 1971; U.S. Riot Commission, 1968). In controlled experimental situations, the influence of heat on the aggressive tendency of normal subjects has not been confirmed (Baron, 1977a), even though increased aggression has been noted, even in non-angry subjects; reduced aggression in anger-prone individuals has been shown to be experimentally induced by uncomfortably high temperatures (Baron and Bell, 1975).

Such different effects of intense heat on aggressiveness in man probably reflect only an interaction with different emotional baselines which, in turn, are likely dependent on different physiological and genetic predispositions, as observed in several other instances. Even in the absence of any direct experimental evidence, irritative aggression in normal man may be regarded as the least harmful expression of human aggressiveness. However, explosions of irritative aggression can also drive mentally disturbed and/or drug-addicted subjects to unpredictable and sometimes unintentional homicidal acts.

Territorial Aggression

Territorial aggression is a response of active defense that is triggered by any intruder violating the boundaries of an area in which a subject, or a group of subjects, has already established living activities. As previously described (Chapter 4, pp. 64, 71), both individual and group territorialism occurs widely throughout the animal kingdom (Anderson, 1961; Charles-Dominique, 1974; Crowcroft and Rowe, 1963; Holloway, 1974; Lorenz, 1967; Tinbergen, 1973), also being observable in domesticated and laboratory animals (Flannelly and Thor, 1978; Mackintosh, 1970, 1973; Murphy, 1970a; Thurmond, 1975).

The implicitly defensive and competitive components of territorial aggression do not invalidate its differentiation from other types of aggressive behaviors, due to the basis of the territory as a stimulus on which the presence of an intruder provokes attack; this, in turn, decreases recurrence of intrusion, and pursuit of the intruder forces him further from the invaded territory (Moyer, 1968a). A well-organized system of territories acts as a control of aggressive displays (Valzelli, 1967b), in both animals and humans (Eibl-Eibesfeldt, 1974b;

Hamburg, 1971; Lorenz, 1967; Tinbergen, 1973), through the attack-avoidance system. The value and consequences of either individual or group territorialism in man were discussed in the preceding chapter.

Maternal Protective Aggression

In mammals, this type of aggression is a normal component of parental care which, from a behavioral standpoint, sharply distinguishes mammals from reptiles (Chapter 2, p. 28). Maternal protective aggression is characteristic of female animals for purposes of defending the newborn against any potentially or actually threatening agent (Hafez, 1962; Leblond, 1920; McCabe and Blanchard, 1950; Moyer 1968a). Experiments in female mice have shown that lactating animals display an immediate and intense aggressiveness against both male and female conspecific intruders (Gandelman, 1972a), with the fighting period exclusively covering the lactating phase of each reproductive cycle. Aggressiveness is most intense at the beginning of lactation and decreases, then disappears at the end of the lactation period (Svare and Gandelman, 1973, 1976a). During the first half of the lactation period, when aggressiveness reaches a peak, aggressive displays are directed against both male and female intruders; during the second half, aggression is still displayed against males, but no longer against females. Such differences in behavior could reflect "precaution" against cannibalistic activity which is seldom displayed by adult conspecific males against pups (Paul and Kupferschmidt, 1975). Generally, however, adult conspecific males will not kill pups (Myer and White, 1965), since the pups normally have an odor which inhibits predatory behavior by conspecific members (Myer, 1964); mouse pups also emit an ultrasonic "distress call" which stimulates maternal retrieval (Noirot, 1972).

Lactating mice seldom attack conspecific animals to which they were previously exposed, whereas they vigorously attack a strange intruder, which, however, will not be molested after pup removal (Svare and Gandelman, 1973). Thus the presence of the offspring represents an important component for stimulating maternal protective aggression; this, however, is also displayed during pregnancy, also obviously being affected by female hormonal balance (Gandelman and Svare, 1974; Noirot et al., 1975).

Aside from that exhibited by nonspecific animals, this type of aggression includes intraspecific aggressive components, so that wild female rabbits have been seen to bravely defend their offspring from ravens, brown hawks, and other predator birds (Mykytowycz, 1959; Mykytowycz and Dudziński, 1972).

In humans, even though accounts on the matter are almost exclusively anecdotal, informal, and bookish, maternal protective aggression may be assumed to still actively survive, and to operate at an acceptable level.

Female Social Aggression

This behavior represents another type of animal aggressiveness, which still, however, remains to be more extensively studied and better delineated. Southern

(1948), while describing the behavior of wild rabbits in their natural environment, remarked on the particularly aggressive responses of females toward juvenile members of a different social group. Further observations of rabbit populations under either experimental or natural conditions have shown that the aggression of adults toward kittens is sometimes responsible for the death of a proportion of some immature members of the species (Myers, 1964; Mykytowycz, 1959, 1960). The previously cited studies on the effects of overcrowding on aggression (Chapter 4, p. 70) have shown that social friction and hostility increase at a greater rate than the increase in population density (Burns, 1968; Calhoun, 1973; Southwick, 1955), the highest rate of mortality, and aggression, respectively, being among and directed toward the youngest members (Chitty, 1964; Myers, 1966; Myers et al., 1971) of "other" groups (Mykytowycz and Dudzínski, 1972).

In such conditions adult male rabbits tolerate young males, regardless of their parentage, and show friendly behavior toward them, even protecting and defending them from the attacks of adult females (Mykytowycz and Dudzínski, 1972). Similar observations have been made in other rodents, where this behavior has been suggested as one of the mechanisms regulating population density (Archer, 1970). This issue also includes the aggression that females display against males in overcrowed situations (Blick et al., 1971). Overcrowding also results in decreased female fertility and a lower birth rate; an increased number of spontaneous abortions also occurs, and the intensity of aggression toward males becomes directly proportional to the degree of crowding (Blick et al., 1971).

No direct similarities have been found between female social aggression and the responses of the human aggressive repertoire, with the possible exception of so-called human *prosocial aggression,* discussed by Sears (1961); this refers to a socially approved way of expressing the moral standards of the group that may reflect some unconscious or remote connection with aggression. Since moral standards have been and remain to a large extent subject to contingent political immorality, such issues as the recurrent advocation to genocide, often performed in the name of an overt declaration of "racial purity," injure the youngest; and the strong claims for the unrestricted right to abortion may also have implications for the issue of the population control in animals; the point also reflects a prominent province of human females against their unborn and vulnerable offspring.

Sex-Related Aggression

This type of aggressive behavior is assumed to be released by the same stimuli which produce sexual responses (Barclay and Haber, 1965), with both the aggressive and sexual displays having in common, in either animals or man, a state of increased arousal (Berne, 1964; Bindra, 1959). The experimental studies on animal behavior, although quite scant, appear to indicate that after either sexual or aggressive behavior have been performed, one no longer influences the practice of the other (Lagerspetz and Hautojärvi, 1967).

In contrast with the other types of aggression, and aside from the fact that sex-related aggression is obvious to anyone who reads the newspapers, there is much more information available in humans than in animals. Several data imply that sexual arousal will often, if not always, encourage the occurrence of overt aggressive actions by man, and heightened sexual arousal has been reported to facilitate overt attacks against others even without apparent provocation (Jaffe et al., 1974; Meyer, 1972, Zillman, 1971). Other experiments (Baron, 1974b; Baron and Bell, 1973; Frodi, 1977) indicated instead that heightened sexual arousal may actually reduce later aggression, and the suggestion has also been made that the existing relationship between sexual arousal and aggression is based on the simultaneous arousal and distracting effects of erotic stimuli (Donnerstein et al., 1975). When distracting effects predominate, aggression is inhibited, whereas it is enhanced when arousal predominates.

Thus, as previously observed for heat-induced aggression (p. 81), heightened sexual arousal exerts a bimodal and opposite effect, which is likely to involve differences in physiological, emotional, and genetic elements and predispositions. In addition, this issue must not be considered an exclusively male affair, since heightened sexual arousal may generally have the same consequences on later aggression in females as well (Baron, 1977a,b).

Instrumental Aggression

According to Moyer's definition (1968), any of the above classes of aggression may result in a change of the social environment so as to represent a reinforced response; this would result in an increased probability of an aggressive response in analogous situations. In other words, this type of aggression is "instrumental" in that the learned effect of previous positive experiences motivates a subject to use it more frequently; it is also employed as the most appropriate means of responding to manifold stimuli, even to those normally evoking other kinds of behavioral patterns or other kinds of aggression (Valzelli, 1978d). For example, in a monkey colony with an established hierarchical structure, even in the absence of any overt provocation or intentional competitive approach by subordinates, the boss monkey often displays threatening gestures and mimicry which "instrumentally" serve to reaffirm and consolidate dominance.

Instrumental aggression is largely limited to threatening behavior, differing exclusively from other forms of aggression insofar as the significance of warning behavior preceding an imminent overt competitive display (p. 76). In anthropomorphic terms, instrumental aggression may be considered as being conceptually similar to authoritative command behavior.

The situation in man is much more confusing, largely because of a traditional reluctance in accepting any analogy to animal behavior. It is therefore not surprising that, based on Moyer's sequence of aggressive displays, it is still believed to be impossible to identify a similar number of different types of aggression in man (Hartup and De Wit, 1974).

The major dichotomy utilized to categorize human aggression is a distinction between hostile versus instrumental aggression (Buss, 1961, 1971; Feshbach, 1964, 1970; Rule, 1974). Even though it remains difficult for me to conceptually discern "friendly" aggression as a hypothetical counterpart of hostile aggression, the latter term has been applied to episodes of aggression in which the major goal of the aggressor is that of injuring the victim and causing him pain. In contrast, human instrumental aggression is assumed to not intentionally inflict harm on others as an end in itself, but rather to be used as a technique for obtaining various prizes.

According to Hartup and De Wit (1974), the distinction between hostile and instrumental aggression is far from clear, since (1) both instrumental and hostile elements are often involved in the same situation (Bandura, 1973; Feshbach, 1970); and (2) this distinction incorrectly suggests that there is no instrumental value in hostile aggression (Hartup and De Wit, 1974), whereas both types of aggression are actually directed toward specific goals; in this sense they both can be labeled as instrumental (Bandura, 1973). To obviate such criticism, Zillman (1978) has recently proposed that the terms hostile and instrumental aggression should be replaced by, respectively, *annoyance-motivated* and *incentive-motivated* aggression. The former refers to those aggressive displays undertaken primarily to reduce or terminate noxious stimulation such as intense anger or mistreatment by others; the latter refers to aggressive actions performed in order to attain various extrinsic incentives. Within these broad limits, I see no need for changing labels because no substantial difference appears to exist between those behaviors included in human hostile or annoyance-motivated aggression and those already described for either defensive or irritative aggressive displays (p. 79). Similarly, no substantive distinction between human instrumental or incentive-motivated aggression and the manifold expressions of competitive aggressiveness (p. 76) appears obvious. Rather, it seems to me that instrumental aggression, as observed and described in animals, has obvious similarities with several aspects of authoritarianism (Elms and Milgram, 1966) or "executive" behavior in man; the latter is also largely a learned and reinforced behavior that is never directly harmful.

NEUROANATOMY OF AGGRESSION

With the clear premise that aggression should no longer be considered as a single entity, and that the different types of aggressive responses may vary in intensity depending on a variety of factors, it should be again emphasized that all types of aggression develop from a common basis which represents the previously and frequently cited behavioral continuum of aggression. To illustrate visually, one may conceive of a mountain ridge, where variations in the altitudes

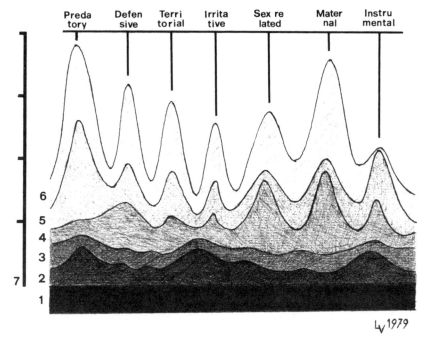

FIG. 22. Theoretical construction of an aggressive profile and its components.
1. Neuroanatomical network
2. Genetic predetermination
3. Neurochemical setting
4. Hormonal setting
5. Learning, education, culture, environment
6. Behavioral aggression
7. Arbitrary scale

of the summits and the depths of the valleys outline the characteristics of the *aggressive profile* of a subject, either animal or human (Fig. 22).

Further, in Fig. 22 the thickness of the hatched area indicates the extent to which various participating genetic, emotional, endocrine, learned, and cultural components contribute; the invariant background represents the neuroanatomical continuum underlying aggression. In broad terms, the neuroanatomical machinery of aggression can be considered to almost entirely coincide with most of the previously described structures that form the reptilian brain (Chapter 2, pp. 33 and 34) and which subserve the basic requirements for survival. Experimental evidence indicates that in pigeons, food intake, drinking, courtship, nest building, and aggression are not appreciably affected by decortication (Rogers 1923). In mammals, too, decortication does not affect nutrition, sexual activity, and aggression (Jung and Hassler, 1960); these findings have recently been confirmed in totally decorticated rats, which retain most of their locomotion, climbing, grooming, feeding, and aggressive patterns (Vanderwolf et al., 1978). Such aggressive patterns can now be actually assigned almost directly to the functions of a series of experimentally identified neuroanatomical structures.

Animal Studies

As previously noted (p. 76), the hypothalamus plays an important role in feeding behavior, and also underlies obvious and specific aggressive displays (Paxinos, 1974; Roberts and Kiess, 1964). Feeding behavior is actually displayed as predatory aggression after electrical stimulation of the lateral hypothalamus when the appropriate prey for the animal under study is available (Wasman and Flynn, 1962). Stimulation of the anterior hypothalamus causes predatory attacks which, however, are not sustained by hunger (Roberts and Kiess, 1964). However, the predatory attacks elicited by lateral hypothalamic stimulation (Adamec, 1976; Hutchinson and Renfrew, 1966) are also evoked by stimulation of some additional anatomical loci in the lower brainstem (Bandler, 1977); this may result either in eating behavior or in pure killing activity (King and Hoebel, 1968) depending on the degree of hunger, food availability, and the intensity of stimulation (Hutchinson and Renfrew, 1966). Lateral hypothalamic stimulation can lead to intraspecific fighting which depends on environmental variables, such as the presence of an object of prey and the behavior of the opponent; a stimulated subordinate never initiates fighting against a dominant conspecific (Koolhaas, 1978).

Stimulation of the dorsomedial hypothalamus produces a different type of aggression, such that a cat may ignore the presence of an available rat, and viciously attack the experimenter or any other person (Egger and Flynn, 1963). Clear reinforcement of a similar irritative response is also obtained in the rat (Stachnik et al., 1966*a*), in which stimulation of the posterior hypothalamus causes the animal to display consistent aggressive approaches, even toward a cat or a squirrel monkey (Stachnik et al., 1966*b*).

Several other anatomical loci in the midline thalamic, hypothalamic, and midbrain areas have been implicated in the regulation of predatory attacks (Adamec, 1976; Bandler and Fatouris, 1978; Berntson, 1972; Oomura et al., 1969). Additional areas that trigger predatory attacks have been found following stimulation of the posterior midline thalamus (Bandler and Flynn, 1974; MacDonnell and Flynn, 1968), lateral preoptic regions (Inselman and Flynn, 1972), midbrain ventral tegmental area (Bandler et al., 1972; Proshansky et al., 1975), dorsolateral midbrain reticular formation (Sheard and Flynn, 1967), caudal midbrain supratrochlear periaqueductal gray matter (Bandler, 1975, 1977), central pontine tegmentum (Berntson, 1973), and portions of the cerebellar fastigial nucleus (Reis et al., 1973). Irritative biting is also obtained in cats by stimulating the thalamic center median (Andy et al., 1975). Stimulation of the medial hypothalamus, ventromedial hypothalamus, and mammillary bodies stops either feeding behavior or predatory attacks (Adamec, 1974; Oomura et al., 1967*a;* Popova et al., 1976), as does stimulation of the basolateral nuclei of the amygdala (Adamec,

1975*a, b, c;* Fonberg and Delgado, 1961*a, b*). Irritative hypothalamic attacks are suppressed by stimulation of the prefrontal, periamygdaloid, or medial prepyriform cortex (Siegel et al., 1972, 1975, 1977), and of the head of the caudate nucleus (Delgado, 1974; Plumer and Siegel, 1973); delays in such attacks occur following the stimulation of the anterior cingulate gyrus (Siegel and Chabora, 1971). Stimulation of the lateral amygdala, however, greatly facilitates hypothalamic irritative aggression (Siegel et al., 1972).

At the anatomical level, the cingulate gyrus may represent a substrate for a repetitive series of bimodal functions observed for other processes. Older experiments reported that lesions of the cingulate gyrus either increase (Brutkowsky et al., 1961; Kennard, 1955) or decrease aggression in cats and monkeys (Glees et al., 1950); aside from the possible differences in the extent and specificity of surgery, the results may once more reflect differences in presurgical genetic and emotional baselines.

Lesions of the ventromedial hypothalamus enhance the so-called ventromedial hypothalamic syndrome (Hoebel, 1971; Stevenson, 1969; Tannenbaum et al., 1974) or hypothalamic savage syndrome (Glusman, 1974), in which rats manifest hyperphagia, hyperdipsia, consequent obesity, and overtly increased aggression (Colpaert, 1975). Cats which display clear defensive behaviors in a shock-box situation show, following such lesions, strong irritative and defensive aggressive responses (Glusman, 1974). Lesions confined to the lateral hypothalamus abolish territorial aggression (Adams, 1971), whereas lesions of the ventrobasal thalamus in rats instead abolish defensive aggression (Kanki and Adams, 1978).

Bilateral lesions of the amygdala tame a variety of innately hostile and vicious animals, which, immediately following surgical intervention, can be handled without gloves (Schreiner and Kling, 1956; Woods, 1956); such lesions also consistently decrease competitive aggression in reptiles such as caiman and iguana (Keating et al., 1970; Tarr, 1977). The same operation blocks intermale competitive aggression, dominance, and the muricidal behavior of laboratory rats made artificially aggressive (Bunnell, 1966; Karli, 1974; Karli and Vergnes, 1964*a, b;* Miczek et al., 1974; Vergnes and Karli, 1964, 1965). In contrast, the bilateral destruction of either the basal or lateral nuclei of the amygdala produces opposite results: previously domesticated and friendly animals will postsurgically attack without provocation (Wood, 1958). Bilateral amygdalectomy also dramatically reduces irritative aggression (Schreiner and Kling, 1953) and fear-induced aggression (Galef, 1970*b;* Rosvold et al., 1954; Schreiner and Kling, 1953, 1956; Ursin, 1965), and also suppresses predatory aggression (Adamec, 1975*b;* Summers and Kaelber, 1962; Woods, 1956). Since the basolateral and centromedial nuclei of the amygdala play an opposite role in aggressiveness, the occurrence of increased aggression after bilateral amygdalectomy reported by some investigators (Bard, 1950; Bard and Mountcastle, 1948) probably depends on a subtotal ablation; there may be an aggression-facilitating nucleus that remains *in situ* (Siegel and Flynn, 1968), but such an increased aggression may also depend on differences in the degree of the spontaneous aggressive behavior among the animals employed in the experiments (Adamec, 1975*a, b*). Thus the bimodal effects of the same kind of surgical intervention may depend on the interaction with different pre-existing emotional baselines.

A rage reaction is induced in normal laboratory rats by lesions of the bed nuclei and/or the tracts of the stria terminalis, and is inhibited by lesions of the ventrobasal (Kanki and Adams, 1978) and ventral thalamus (Turner, 1970); lesions of the periamygdaloid cortex, cortical nuclei of the amygdala, or bed nuclei of the stria terminalis abolish isolation-induced agression in rats (Miczek et al., 1974). A similar bimodal result is obtained when the olfactory bulbs are cut *(bulbotomy)* or removed *(bulbectomy)*: this induces muricidal behavior, violent intermale aggression, and rat-pup killing by previously normal laboratory rats (Bandler and Chi, 1972; Bernstein and Moyer, 1970; Cain, 1974; Di Chiara et al., 1971; Didiergeorges et al., 1966*a;* Karli et al., 1969; Malick, 1970; Myer, 1964), while eliminating spontaneous frog killing, mouse killing, and rat-pup killing in rats (Bandler and Chi, 1972), and intermale competitive aggression in spontaneously vicious mice and hamsters (Murphy, 1970*b;* Ropartz, 1968; Rowe and Edwards, 1971). Stimulation of the dorsomedian nucleus of the thalamus inhibits interspecific muricidal aggression in rats (Vergnes and Karli, 1972), whereas bilateral destruction of the dorsomedian and paraventricular nuclei of the thalamus induces aggression (Eclancher and Karli, 1968).

Other brain areas are further involved in modulation of the various types of aggressiveness. Lesions of the dorsolateral frontal lobe performed in normal monkeys do not change dominance but decrease threatening behavior and produce a concomitant increase in aggressive displays (Miller, 1976). Increased aggression of the irritative type is also observed in cats and rats after ablation of the frontal lobes, posterior cingulum, dorsal hippocampus, and the ventral midbrain tegmentum (Bandler et al., 1972; Blanchard et al., 1970; Chi et al., 1976; Heller et al., 1962; Karli, 1955, 1956; Kim et al., 1971; Romaniuk and Gołebiewski, 1977; Vergnes and Karli, 1969).

In contrast, muricidal aggression is not consistently impaired by lesions or sections of the hippocampal fimbria fornix (De Castro and Balagura, 1975; De Castro and Marrone, 1974), whereas an electrically induced paroxysmal hippocampal discharge is capable of blocking muricidal activity (Vergnes and Karli, 1968).

Experimentally induced epileptic foci in the hippocampus or amygdala of normal rats and cats enhance irritative and defense aggression (Adamec, 1975*c;* Pinel et al., 1977; Wasman and Flynn, 1966), the interaction between the two anatomical structures being such that, in normal conditions, the ventral hippocampus inhibits the behavioral effects of heightened amygdaloid excitability (Adamec, 1975*c*).

Bilateral lesions of the septal nuclei of normal laboratory rats, depending on lesion size (Kenyon and Krieckhaus, 1965) and the specific nuclei involved (Golda et al., 1977; Stark and Henderson, 1966; Wallace and Thorne, 1978), induce hyperirritability (Heller et al., 1962; Stark and Henderson, 1966; Zeman and Innes, 1963), which often is transformed into episodes of irritative aggression (Brady and Nauta, 1953; Golda et al., 1977; King, 1958; Miczek and Grossman, 1972; Thompson, 1975; Wallace and Thorne, 1978). Similar lesion-induced irritative aggression has been observed in other species (Bunnell and Smith, 1966; Bunnell et al., 1966; Slotnick and McMullen, 1972; Sodetz and Bunnell, 1970), whereas dominance and competitive aggression are decreased by the same kind of surgical intervention (Blanchard et al., 1977; Gage et al., 1978; Lau and Miczek, 1977; Sodetz et al., 1967). Depending on a clear strain-linked susceptibility for the development of muricidal aggression, septal damage may produce a limited facilitation of such anomalous behavior (Lathman and Thorne, 1978), probably as a result of an increased hyperirritability (Wallace and Thorne, 1978). With continued reference to irritative aggression, the previously described effect of dominance abolition as a consequence of stimulation of the head of the caudate nucleus in monkeys (Chapter 3, p. 55) has found an interesting parallel in the observation that electrical stimulation of this region in charging bulls immediately halts the attack (Delgado, 1974).

Insofar as maternal and sex-related types of aggression are concerned, the hypothalamus, amygdala, hippocampus, hippocampal fimbria and fornix, cingulate gyrus, and septal nuclei, all anatomical structures related to mechanisms of aggression (Chapter 3, p. 54; and Table 1, p. 59), are also variously implicated in pleasure, grooming, sexual, and parental-care behaviors (Delgado, 1967b; Elwers and Critchlow, 1961; Kawakamy et al., 1967; Klüver and Bucy, 1938, 1939; MacLean, 1954, 1957a, b, 1958; Sainsbury and Jason, 1976; Smith and Holland, 1975; Ward, 1948).

The brain structures related to different types of aggression are tentatively summarized in Table 2.

After even a superficial glance at the anatomical trigger or suppressor loci subserving aggression listed in Table 2, one realizes that most of them variously interact in regulating more than a single aggressive action, thereby suggesting a fundamental neuroanatomical continuum on which a need-based structural hierarchy with functional relevance is produced dynamically and plastically. Figure 23 outlines the components of the neuroanatomical mechanisms for aggression.

Human Studies

Data on the neuroanatomy of aggression in man are limited. They obviously come from reports dealing with surgical intervention aimed at the relief of either mental or organic brain disease no susceptible or responsive to other clinical treatment, such studies differ from animal experiments in that the presurgical baseline is pathological or at least abnormal to some extent. Nevertheless, the data available in this area, derived from ablative or electrical stimulation techniques, are abundant. Moreover, they are more informative than those reported

TABLE 2. *Anatomical correlates of the different kinds of animal aggression*

Kind of aggression	Brain structures involved as	
	Triggers	Suppressors
Predatory	Anterior hypothalamus Lateral hypothalamus Lateral preoptic nuclei Posterior medial thalamus Ventral midbrain tegmentum Ventromedial periaqueductal gray matter	Prefrontal cortex Ventromedial hypothalamus Basolateral amygdala Mammillary bodies
Competitive	Laterobasal septal nuclei Centromedial amygdala Ventrolateral posterior thalamus Stria terminalis	Dorsolateral frontal lobe Olfactory bulbs Dorsomedial septal nuclei Head of caudate
Defensive	Centromedial amygdala Fimbria fornix Stria terminalis Ventrobasal thalamus	Ventromedial hypothalamus Septal nuclei Basolateral amygdala Ventral hippocampus
Irritative	Anterior hypothalamus Ventromedial hypothalamus Dorsomedial hypothalamus Posterior hypothalamus Anterior cingulate gyrus Thalamic center median Ventrobasal thalamus Ventral hippocampus Ventral midbrain tegmentum Ventromedial periaqueductal gray matter Cerebellar fastigium	Frontal lobes Prefrontal cortex Medial prepyriform cortex Ventromedial hypothalamus Septal nuclei Head of caudate Dorsomedian nucleus of thalamus Stria terminalis Dorsal hippocampus Posterior cingulate gyrus Periamygdaloid cortex
Territorial	Lateral hypothalamus	Basolateral amygdala
Maternal-protective	Hypothalamus Ventral hippocampus	Septal nuclei Basolateral amygdala
Sex-related	Medial hypothalamus Fimbria fornix *(male)* Ventral hippocampus	Septal nuclei Fimbria fornix *(female)* Cingulate gyrus Dorsolateral amygdala

by Plotnik (1974) to support his position that neuroanatomical triggers for aggression do not occur inside the human brain.

King (1961) reported a female patient with an electrode implanted in the amygdala (probably the centromedial area), who became angry, verbally hostile, and threatening when stimulated with a 5 mA current, but became friendly again and apologized for her aggressive behavior when the stimulation was turned off. Similar results after the electrical stimulation of the amygdala have been reported by others (Heath et al., 1955; Sweet et al., 1969). Fairly good control of even seriously violent, injurious, and homicidal outbursts has been

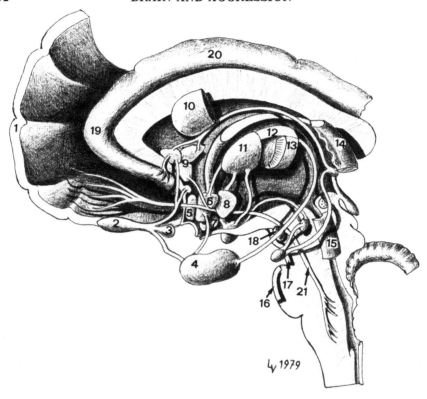

FIG. 23. Neuroanatomy of aggression.

1. Frontal cortex
2. Olfactory bulb
3. Olfactory tubercle
4. Amygdala
5. Preoptic nucleus
6. Anterior hypothalamus
7. Ventromedial hypothalamus
8. Posterior hypothalamus
9. Septal nuclei
10. Head of caudate
11. Anterior thalamus
12. Dorsomedial thalamus
13. Ventrolateral posterior thalamus
14. Dorsal hippocampus
15. Periacqueductal grey matter
16. Ventral pontine tegmentum
17. Ventral mesencephalic tegmentum
18. Red nucleus
19. Anterior cingulate gyrus
20. Medial cingulate gyrus
21. Median forebrain bundle

routinely achieved in previously intractable psychiatric patients through stimulation of electrodes implanted deep in the amygdala, hippocampus, septal region, and cerebellar nuclei and lobes (Delgado et al., 1968; Heath, 1954, 1977; Heath and Harper, 1974, 1976; Riklan et al., 1974). Clinical pre- and postsurgical evaluations (Delgado et al., 1968; Heath, 1963, 1972, 1975, 1977; Mahl et al., 1964; Nashold and Slaughter, 1969; Riklan et al., 1974) have shown that septal regions, corticomedial amygdala, and several cerebellar loci in man are associated with pleasurable emotion and aggression-inhibiting activity, whereas the medial hypothalamus, thalamic center median, cingulate gyrus, periaqueductal sites, and mesencephalic tegmentum subserve unpleasant feelings, ranging from anxiety, fear, and rage, to intense aversion and aggression.

Unilateral or bilateral lesions of the amygdala have also been described as abolishing or reducing pharmacotherapy-resistant outbursts of destructive violence by patients who had previously shown threatening behavior with overt, repetitive aggressive acts such as assaults on other persons, and/or a marked tendency to self-mutilation and suicidal attempts (Balasubramaniam and Ramamurthi, 1968, 1970; Chatrian and Chapman, 1960; Heimburger et al., 1966; Kiloh et al., 1974; Mempel, 1971; Narabayashi and Mizutani, 1970; Narabayashi and Uno, 1966; Narabayashi et al., 1963; Sawa et al., 1954; Schwab et al., 1965; Sweet et al., 1969; Turnbull, 1969; Ursin, 1960; Vaernet and Madsen, 1970). One of these patients reported that after the operation he could not become angry even if he wanted to (Sawa et al., 1954).

However, with regard to the antiaggressive effects obtained through stimulating the amygdala and the similarly antiaggressive effect of amygdala ablation, such seemingly antithetical therapeutic effects of amygdaloid ablation are most likely possible because each amygdala is a complex cluster of nuclei with several different functions (Gloor, 1967). Separate areas within the amygdala are responsible for either initiating or suppressing aggressive behavior (Kaada, 1965), as already observed in animal experiments (p. 88).

Lesions of the dorsomedial thalamus (Spiegel et al., 1951), thalamic center median (Andy, 1970), thalamic lamella medialis (Poblete et al., 1970), posteromedial hypothalamus (Kalyanaraman, 1975; Sano, 1962; Sano et al., 1966, 1970; Sramka and Nadvornik, 1975), anterior cingulum (Tow and Whitty, 1953), and temporal lobe (Pool, 1954; Scoville and Milner, 1957; Terzian, 1958) have also proved successful in reducing uncontrolled hostility and agression in man. Further, cingulectomy has been especially recommended as being consistently effective in controlling intractable and compulsive cases of permanent agitation, anger, aggressiveness, and violence (LeBeau, 1952).

Stereotactic coagulation of the ventromedial hypothalamic nucleus, of the tuberomammillar complex, and of the anterior hypothalamus abolishes either homosexual pederastic activity or hypersexual violent behavior (Dieckmann and Hassler, 1975; Roeder, 1966). As previously noted, stimulation of the septal region reduces agitated, psychotic, and violent human behavior, also almost instantly altering disorganized rage to happiness and mild euphoria (Heath, 1963). In contrast, bilateral stimulation of the anterior medial amygdala in man elicits overt assaultive behavior (Delgado et al., 1968).

Brain tumors with an irritative focus frequently induce increased irritability and attacks of rage; examples are tumors in the temporal lobe (Kreschner et al., 1936; Mulder and Daly, 1952; Vonderahe, 1944), frontal lobe (Strauss and Kreschner, 1935), or hypothalamus (Alpers, 1940; Sano, 1962).

Although reports of aggressive behavior in patients with temporal lobe epilepsy are common, it has proven difficult in clinical settings to achieve the experimental control necessary to systematically investigate temporal lobe aggression or even to provide unequivocal evidence of its existence (Kligman and Goldberg, 1975). In animal experiments, increased aggression has been reported (p. 89) in rats with experimentally induced foci in temporal lobe structures (amygdala and hippocampus) but not in those with foci in the caudate nucleus (Pinel et al., 1977). In general, however, there is ample evidence that spontaneous or abnormal firing in the human temporal lobe results in feelings of irritation, anger, or rage, giving rise to actual violence when such firing activity is sufficiently intense. Gibbs (1951, 1956) has estimated that about half the epileptic patients with an anterior temporal lobe focus have psychiatric disorders, and these patients have the propensity to be provoked into explosive and violent behavior (Gloor, 1960). Antisocial violent behavior is also present in patients with posterior temporal foci (Fenton et al., 1974); some clinical observations have further suggested that the side of localization of the temporal lobe focus may affect the features of the psychotic state (Brazier, 1970; Sigal, 1976). In addition, Blumer et al. (1976) reported on 36 patients with a history of violent aggressive behavior associated with temporal lobe epilepsy, whereas behavior disorders and outbursts of rage and aggression concomitant to epilepsy have also been reported by others (Cherlow and Serafetinides, 1977; Nuffield, 1961; Ounsted, 1969; Serafetinides, 1965).

In the broad context, bilateral temporal lobe ablation performed in monkeys (Klüver and Bucy, 1937, 1938, 1939) induces a recently reconfirmed syndrome (Horel and Misantone, 1974) involving a peculiar docility that seems to depend on an apparent loss of fear, since the animals repeatedly expose themselves to known harmful situations (Pribram and Bagshaw, 1953).

Bilateral removal of the temporal lobes, including most of the uncus and the anterior hippocampus, in a patient with frequent attacks of aggressive and violent, overtly homicidal behavior reproduced exactly the Klüver-Bucy syndrome, including serious memory deficiencies, loss of recognition of people with whom the patient had been well acquainted—even of relatives—increased sexual activity, bulimia, and a consistent reduction of fear and rage; the patient thus became completely resistant to any attempt to arouse aggressiveness and violent reactions in him (Terzian and Dalle Ore, 1955). Temporal lobe lesions or ablation to different degrees, both unilaterally and bilaterally, has been extensively employed in man to relieve uncontrollable epilepsy. The side effect most frequently reported is a general reduction of anger, hostility, overt aggression, and violence (Bailey, 1958; Falconer et al., 1955, 1968; Green et al., 1951; Pool, 1954; Scoville and Milner, 1957; Terzian, 1958).

Most of the therapeutically utilized ablative surgery has, at present, been abandoned and in part substituted by less traumatic and more selectively targeted techniques; these include deep electrode implantation because temporal lobectomy, even with complete seizure relief, does not always guarantee full clinical remission of psychiatric symptoms (James, 1960; Simmel and Counts, 1958; Walker, 1973). Based on the asymmetry of the human brain, a comparison of patients with unilateral right or left anterior temporal spiking foci has indicated that the right focus is associated with elation and optimism, while the left focus facilitates the emergence of such attributes as humorlessness, sadness, obsessiveness, anger, and aggressiveness (Bear, 1977; Bear and Fedio, 1977; Mandell, 1978).

In conclusion, Table 3 tentatively summarizes the anatomical structures of the human brain which relate to some emotional-affective and aggressive issues, as indicated by the pertinent literature. The absence of some "animal" aggressive items in the list should not imply their absence in man. This generalization depends largely on the facts that (1) available data necessarily and sharply reflect human pathology of the listed items, and (2) most of the authors refer to their results on aggressiveness, hostility, anger, and other similar issues without further indications or applications.

TABLE 3. *Anatomical correlates of some emotional-affective and aggressive behaviors of man*

Feelings of behavioral patterns	Brain structures involved as	
	Triggers	Suppressors
Pleasure	Corticomedial amygdala Septal region Cerebellar fastigium	(see Aversion triggers)
Anxiety	Cingulate gyrus	Cerebellar lobes Cerebellar fastigium
Aversion	Cingulate gyrus Hippocampus Mesencephalic tegmentum Periaqueductal gray matter	(see Pleasure triggers)
Sex-related aggression	Anterior hypothalamus Ventromedial hypothalamus Tubero-mammillar complex	
Irritative aggression	Medial hypothalamus Posteromedial hypothalamus Thalamic center median Thalamic lamella medialis Dorsomedial thalamus Anterior cingulum Anterior *(ventral)* hippocampus Centromedial amygdala	Frontal lobes Septal nuclei Cerebellar lobes Cerebellar fastigium

.

6

The Physiological Basis of Aggression

NEUROCHEMISTRY OF AGGRESSION

In the framework of studies concerned with behavior, the challenge to neurochemistry is to explain how and which biochemical principles and reactions in the brain give rise to different behavioral patterns. Generally speaking, although many difficulties still exist in directly relating brain biochemistry to specific behavioral issues, some progress is being made toward clarifying certain possible neurochemical correlates of behavior. However, the continuing trend to relate specific types of behavior to a single brain chemical principle is a great oversimplification; it does not take into account that neuronal circuits regulating specific behavior utilize several different neurotransmitters, and that any specific brain chemical principle can play a role in a variety of nervous structures controlling disparate behaviors.

This general criticism also applies to aggression, thereby questioning the validity of any rigid explanation and classification for different kinds of aggression, as dependent on any single "specific" neurotransmitter (Valzelli, 1977b). A similar consideration has also been raised by others (Reis, 1974) in relating specific neurotransmitters to specific behaviors or diseases. Nevertheless, some attempts to associate the different kinds of aggression described by Moyer to various neurochemical parameters have already been made (Eichelman and Thoa, 1973). A further obstacle to conclusions in this area is that such studies are necessarily performed with experimentally induced aggression, which is only slightly comparable to the various types of spontaneous aggression.

The most elementary criterion for attributing functional relevance to endogenous chemicals—such as serotonin (5-HT), norepinephrine (NE), dopamine (DA), acetylcholine (ACh), and many others (Chapter 2, p. 25)—in the activity of the brain in general and of the limbic system in particular is the uneven distribution of these neurotransmitters in the different structures comprising this region (Brownstein et al., 1974; Paasonen et al., 1957; Palkovits et al., 1974a,b; Saavedra et al., 1974; Valzelli and Garattini, 1968a). However, as stated previously (Chapter 3, p. 44), perhaps the greatest recent advance in brain neurochemistry is identification of biochemically specific neural pathways in the brain. These pathways consist of anatomically defined neural projections, each of which operates specifically on a single neurotransmitter diffusely innervating several structures and nuclei of the limbic system of both animals and man.

Also within the general framework of "chemical coding of behavior" (Miller, 1965), considerable but indirect evidence implicates various brain amines in the regulation of emotion and affect, including aggression (Valzelli, 1978a). However, to avoid misinterpretations of the relevant experimental data, it is important to note that there are biochemical or functional interactions among the various neurotransmitters (Blondeaux et al., 1973; Héry et al., 1973; Kostowski et al., 1974; Johnson et al., 1972; Jouvet, 1973; Lichtensteiger et al., 1967). The biosynthesis of neurotransmitters within the brain normally reflects the rate of impulse flow in the aminergic neurons (Costa and Meek, 1974); such a synthesis rate is further subject to change as a result of several factors which, within limits, are under the control of the central nervous system. Neurotransmitter synthesis is also regulated by a presynaptic inhibition (Eccles et al., 1963; Schmidt, 1971) depending on the presence of presynaptic receptors or self-receptors on the membrane of the presynaptic terminals (Fig. 12, p. 25), or on a system of interneurons which reverberate at the intensity of the postsynaptic receptor excitation on the presynaptic endings (Aghajanian and Bunney, 1974; Andén, 1974). This latter mechanism is prevalent in the modulation of serotonergic activity (Costa and Meek, 1974).

Serotonin

Changes of serotonin level and/or metabolism have been related to changes in affective behavior in general, and to aggressive behavior in particular. Spontaneous daily fluctuations from a peak to a nadir, or circadian rhythms, in serotonin content have been widely described either in the whole brain or for selected cerebral structures (Albrecht et al., 1956; Essman, 1975a; Friedman and Walker, 1968; Héry et al., 1972; Morgan et al., 1975; Quay, 1965, 1967; Quay and Meyer, 1978; Reis et al., 1969; Scheving et al., 1968; Valzelli et al., 1977); such areas have included the inferior colliculus, a region of the red nucleus, the midbrain tegmentum (Reis and Wurtman, 1968; Reis et al., 1968), periaqueductal gray matter, interpeduncular nucleus, olfactory tubercle, anterior hypothalamus, posterior hypothalamus, tuber cinereum, septal area, caudate nucleus, and substantia nigra (Reis et al., 1968, 1969). All of these structures belong to the anatomical network for aggression. Circadian variations in brain serotonin content depend on corresponding daily variations of brain tryptophan; these are independent of daily changes of food intake (Miller et al., 1979; Morgan and Yndo, 1973), but they do suggest the presence of daily variations in brain serotonin metabolism (Héry et al., 1972).

The circadian fluctuation of serotonin is related to a similar variability of brain acetylcholine (Saito et al., 1975) and a diverse and sometimes opposite trend in brain norepinephrine level (Asano, 1971; Graziani and Montanaro, 1966; Reis and Wurtman, 1968; Reis et al., 1968, 1969). Such variations correspond with daily fluctuations in basal emotional behavior and in the tendency for aggressive responses (Reis, 1971). The circadian rhythms in brain serotonin also reflect differences in response to various behavioral (Rusak and Zucker, 1975), pharmacological (Evans et al., 1973), and biochemical parameters that occur daily as well as seasonally. This last issue concerns the fact that seasonal variations, or ultradian rhythms, in whole brain serotonin level have been described; these show a maximum during the winter and lower concentrations in the summer (Valzelli and Garattini, 1968 b; Valzelli et al., 1977). Conversely, an exactly opposite trend in the seasonal changes of serotonin uptake has been recently described for the suprachiasmatic nucleus of the hypothalamus, which represents an important center for the modulation of several endocrine activities (Meyer and Quay, 1977).

The assumption that circadian and ultradian or circannual rhythms of brain serotonin may reflect homologous fluctuations of various behavioral patterns is in general agreement with studies relating brain serotonin content to the emotional behavior of different strains of rats and mice (Maas, 1962, 1963; Sudak and Maas, 1964a, b). These experiments have shown that animals displaying greater exploratory behavior and more emotional stability have less serotonin in the brainstem and hippocampus than highly emotionally labile mice (Maas, 1962, 1963; Sudak and Maas, 1964b; Wimer et al., 1973). Further, in those rat strains chosen on the basis of the same criteria, pooled tissues from the hippocampus, fimbria of the hippocampus, fornix, pyriform cortex, and the amygdala contain significantly less serotonin in emotionally stable rats than in emotionally unstable ones (Sudak and Maas, 1964b); the latter also show a decreased forebrain serotonin turnover (Rosecrans, 1970).

According to these observations, it may be generally assumed that: (1) daily variations in brain serotonin concentration may result in more stable emotional behavior during the morning hours; (2) seasonal fluctuations in brain serotonin concentration may result in more stable emotional behavior during the summer months; (3) if emotional lability facilitates the emergence of anxiety and fear, and thus of irritability, probably serotonin is more directly involved than norepinephrine in the regulation of these affective components and in their possible aggression-promoting effect; and (4) because emotional traits can be genetically predetermined (Chapter 2, p. 29), the serotonin concentration in the limbic structures may largely reflect a gene-dependent characteristic.

Daily fluctuations in emotional susceptibility, motor activity, and avoidance learning of rodents (Ader et al., 1967; Aschoff et al., 1973; Davies et al., 1973; Zucker and Stephan, 1973), ultradian rhythms in motor activity and in social behavior of monkeys (Delgado-Garcia et al., 1976; Maxim et al., 1976), and even seasonal variations in learning capability of fishes (Agranoff and Davis,

1974) have been shown. With respect to aggression, wild male mice show circan-nual cycles in the level of their spontaneous competitive aggression (Turner and Iverson, 1973). Further, a circadian rhythm has been demonstrated for experimentally induced irritative aggression with a peak intensity during late afternoon (Sofia and Salama, 1970), when brain serotonin or tryptophan levels are low (Miller et al., 1979; Valzelli et al., 1977). A tentative integration of these independently obtained experimental findings is outlined in Fig. 24.

Interestingly, a circadian fluctuation in alertness and specialized performance in man (Fröberg et al., 1975), and seasonal variations of some patterns of human affective behavior, as reflected by the rate of marriage and divorce, also exist (Kop, 1974). Moreover, a correlation between the time of year and the frequency of hospitalization for mentally disturbed subjects has been found (Eastwood and Peacocke, 1976; Faust, 1974; Faust and Sarreither, 1975). With respect to human aggression, a seasonal pattern for suicide has long been recognized (Durkheim, 1897), and this has recently been reconfirmed with peak rates in the spring and autumn for the inhabitants of several European countries, the United States, and Canada (Dublin, 1963; Eastwood and Peacocke, 1976; Lester, 1971). An additional and statistically significant monthly periodicity was demon-strated for homicide (Liber and Sherin, 1972). Further, a similar "lunar synodic cycle" has recently been suggested to govern homicides, and also to correlate statistically with suicides, fatal traffic accidents, aggravated assaults, and psychi-atric emergencies (Lieber, 1978).

Needless to say, no specific speculation about the above findings in terms of brain serotonin or other neurotransmitter change in the brain can at present be advanced. Nevertheless, in healthy human volunteers, the concentration of serum free tryptophan is significantly higher at midnight than at noon (Taglia-monte et al., 1974), and, since serum free tryptophan is assumed to reflect the rate of brain serotonin synthesis (Tagliamonte et al., 1973), this finding may suggest that serotonin synthesis in the human brain varies comparably. However, evidence that personality traits can influence and even desynchronize the circa-dian rhythms of man (Defayolle et al., 1966; Lund, 1974), further confuse such studies.

In animal experiments, it is intriguing that those biochemical results that have been obtained in this field necessarily relate to artifactually induced aggres-sive displays in previously docile animals. This suggests that there may be no direct correspondence with the neurochemical correlates of spontaneous aggres-sion. In this same sense, the validity of isolation-induced, chemically induced, shock-induced, or surgically induced aggression as a model for spontaneous species-specific aggressive behaviors has been questioned (Daruna, 1978; Miczek and Barry, 1976). The aggression induced by prolonged isolation has been further suggested to represent a behavioral abnormality probably useful in investigating the pathology of aggression (Valzelli, 1973a, 1978a; Valzelli and Bernasconi, 1976). The widespread notion that serotonin, as responsible for behavioral seda-tion (Brodie and Costa, 1962; Brodie and Shore, 1957), must consequently be

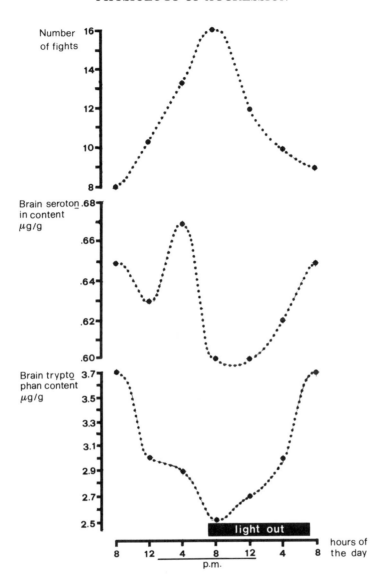

FIG. 24. Behavioral and neurochemical circadian changes. Circadian alternations of shock-induced irritative aggression, and of brain serotonin and tryptophan contents in mice.

an inhibitor of aggression is not always consistent with the bulk of the evidence (Daruna, 1978; Valzelli, 1978a). It should be made clear once more that changes of brain serotonin are not necessarily to be assumed as directly responsible for aggression (Valzelli, 1967a), but rather they may be considered as an indication that changes have occurred in the functions of serotonergic neural projections; these variously modulate either aggression or other behavioral patterns.

Further, the changes in brain serotonin must be viewed as balanced by changes in norepinephrine, dopamine, acetylcholine, and probably many other neurochemical substances, in affecting either the facilitation or the inhibition of the various kinds of aggression (Valzelli and Bernasconi, 1979).

With exercise of some caution, competitive aggression for food induces more rapid brain serotonin turnover in the area rostral to the anterior commissure, and in the hypothalamus-midbrain-medulla oblongata area (brainstem) of winner rats; no serotonin changes occurred in the amygdala, striatum, hippocampus, and the overlying cortex (Daruna and Kent, 1976). Para-chlorophenylalanine (PCPA), a potent blocker of serotonin synthesis, induces both a substantial depletion in brain serotonin and an increase in irritative aggression in previously docile laboratory rats (Koe and Weissman, 1966); the competitive performance of winner rats is unaffected, whereas brain norepinephrine or dopamine is, respectively, decreased or increased in previously defeated and naive rats that become victors (Masur et al., 1974). An initial decline in serotonin and a comparable initial rise in norepinephrine in the hypothalamus and amygdala has been observed in the brain of mice defeated after exposures to competitive confrontations; this is followed by a progressive and sustained (16 days) increase and decrease of these amines in the same anatomical areas. In the frontal cortex serotonin continuously decreases and norepinephrine increases (Eleftheriou and Church, 1968). Correspondent changes in the monoamine-oxidase (MAO) activity in these brain areas have been previously described (Boehlke and Eleftheriou, 1967; Eleftheriou and Boehlke, 1967), the overall neurochemical profile of defeated mice likely being indicative of defensive aggression.

Interestingly, ants have a high concentration of both brain serotonin and epinephrine. The concentrations of both of these amines is further and significantly increased after competitive fights, while brain norepinephrine is significantly decreased (Kostowski and Tarchalska-Krynska, 1975; Tarchalska et al., 1975). The administration of either serotonin or its precursor, 5-hydroxytryptophan, consistently increases competitive aggression among ants, while decreasing the aggression displayed by ants against the beetle (Kostowski and Tarchalska, 1972; Kostowski et al., 1962).

Shock-induced irritative aggression in rats has been shown to increase norepinephrine turnover and release (Bliss et al., 1968; Thierry et al., 1968), but has no effect on dopamine, though it enhances serotonin synthesis (Thoa et al., 1972a). Microinjection of serotonin into the lateral hypothalamus reduces shock-induced irritative aggression, probably by increasing the pain threshold (Leroux and Myers, 1975; Lints and Harvey, 1969), whereas methysergide, a serotonin antagonist, increases this kind of experimentally induced irritative aggression (Bell and Brown, 1977). As previously mentioned, PCPA depletes forebrain serotonin and induces irritative aggression in rats. This drug has been shown to increase the sensitivity to painful foot-shock, thereby increasing the resulting aggressive displays (Butcher and Dietrich, 1973; Tenen, 1967). PCPA also facilitates irritative aggression produced by ventromedial hypothalamic stimulation in cats (Katz and Thomas, 1976). On the contrary, PCPA administration decreases irritative aggression established by septal lesions in rats (Dominguez and Longo, 1969; Jones et al., 1976). This bimodal effect, also observed for other variables (Chapter 5, p. 81, 89), may reflect an interaction between drug activity and differences in either the emotional or neurochemical baseline of the animal.

Evidence that differences in emotionality can influence the effects of psychoactive drugs, and conversely, that the response to these drugs may vary considerably depending on the baseline activity state of the recipient's central nervous system ("law of initial value," Wilder, 1967), has been obtained (Valzelli, 1977a; Valzelli and Bernasconi, 1976). Further support for emotional status–drug–neurotransmitter interactions is derived from data that septal lesions produce a 14% decrease in brain serotonin (Heller et al., 1962). These considerations may explain the contrasting results obtained for the effects of PCPA on the shock-induced irritative aggression (Conner et al., 1970b) in Long-Evans, but not Wistar, Sprague-Dawley, or other rat strains.

In the case of irritative aggression, serotonin-depleting lesions of the midbrain raphe nuclei (Aghajanian et al., 1969; Rosecrans and Sheard, 1967; and Fig. 19, p. 45) also affect norepinephrine metabolism (Kostowski et al., 1974) and induce a slight but significant increase in brain norepinephrine; grouped male Wistar rats with such lesions fight among themselves, especially during the evening (Kostowski et al., 1975). However, as indicated above for PCPA, the same type of lesion fails to consistently increase the foot-shock-induced irritative aggression in Wistar male rats (Vergnes and Penot, 1976). Further, either the intracisternal or intracerebral administration of 5,6-dihydroxytryptamine (5,6-DHT), which causes a sustained depletion of brain serotonin due to the degeneration of serotonergic fibers (Baumgarten et al., 1972a, b, c), also induces hyperirritability, irritative aggression, and muricidal behavior in rats (Breese et al., 1974; Paxinos and Atrens, 1977; Vergnes et al., 1977); this latter peculiar type of aggression will be further considered in a later discussion.

Catecholamines

A considerable number of experiments have implicated brain catecholamines in aggression, often raising conflicting results. Mice fed supplements of dietary L-tyrosine, the metabolic precursor of catecholamines, display a markedly intense territorial aggression; such behavior is also increased, though to a lesser degree, by dietary supplements of L-phenylalanine, the essential amino acid precursor of tyrosine (Thurmond et al., 1977). Repeated immobilization stress in rats results in an increased activity of hypothalamic tyrosine hydroxylase, the rate-limiting enzyme in the biosynthesis of norepinephrine, thereby resulting in increased catecholamine synthesis. A significant increase in irritative aggression displayed by such animals when exposed to a painful foot-shock situation (Lamprecht et al., 1972) has been ascribed to such modified catecholamine metabolism.

The lateral hypothalamus has been shown to relate to shock-induced aggression (Panksepp, 1971), but the infusion of neither norepinephrine nor dopamine into this structure modified shock-elicited fighting in rats (Bell and Brown, 1975a, 1976a). Further, a consistent depletion of brain catecholamines, induced in rats by the administration of α-methyl-p-tyrosine (α-MPT), which inhibits the enzymatic activity of tyrosine hydroxylase (Rech et al., 1966), failed to reduce the intensity of shock-induced aggression (McLain et al., 1974). Interestingly, shock-induced aggression by itself increases brain norepinephrine synthesis in fighting rats; such increased norepinephrine synthesis was also positively correlated with the number of fighting episodes (Stolk et al., 1974). In contrast, in another model of irritative aggression in a different animal species—sham-rage produced in the cat, either by brain electrical stimulation or by acute transection of the brainstem—the magnitude of aggression is directly related to the decrease of brainstem norepinephrine (Reis and Fuxe, 1969; Reis and Gunne, 1965; Reis et al., 1967).

These experiments have been mainly concerned with brain norepinephrine changes, with only little or no direct data concerning the possible involvement of brain dopamine. In this regard, rats reared in enriched colony environments, following a selective depletion of striatal dopamine produced by a radio-frequency lesion of the substantia nigra, pars compacta, from which the nigrostriatal dopaminergic projections originate (Faull and Laverty, 1969; Ungerstedt, 1971; and Fig. 21, p. 47), became hyperactive and displayed intense competitive aggression; there was also an increased display of dominance (Eison et al., 1977), similar to that shown by rats with lesions of the caudate nucleus (Coyle and Kirkby, 1975). In this experiment, the destruction of noradrenergic pathways, leading to brain norepinephrine depletion resulted in inactivity and submissive behavior, even to a greater extent than observed in intact control animals (Eison et al., 1977). Competitive aggression and the number of winner rats are increased when the level of brain dopamine is unchanged with decreased norepinephrine or in situations in which, with a decrease in both these amines, the concentration of the former remains higher than that of the latter (Masur et al., 1973); in both cases brain serotonin content remains unchanged.

With further regard to irritative aggression, the infusion of low concentrations of dopamine into the lateral ventricle of unrestrained rats greatly increases shock-induced fighting, whereas norepinephrine infusion markedly decreases the attacks (Geyer and Segal, 1974). 6-Hydroxydopamine (6-OHDA) is a chemical which, when injected intraventricularly, chronically depletes brain catecholamines through the selective destruction of catecholaminergic terminals (Bloom et al., 1969; Uretsky and Iversen, 1970). 6-OHDA depletes brain norepinephrine to a larger extent (85% less than control values) than dopamine (53% less than control values), leaving the level of brain serotonin unchanged (Samanin et al., 1972); thus treated rats are primarily "serotonergic" and partly "dopaminergic" in their central functioning. Male rats treated with 6-OHDA become hyperactive, hyperemotional, and show a significant increase in irritative aggression induced by foot-shock (Eichelman et al., 1972a; Geyer and Segal, 1974; Sorenson and Gordon, 1975; Thoa et al., 1972a, b). The intraventricular infusion of norepinephrine reduces the frequency of fighting responses, suggesting that dopamine has a facilitatory effect and norepinephrine and inhibitory activity in the regulation of shock-elicited aggression in rats (Anand et al., 1977; Antelman and Caggiula, 1977; Geyer and Segal, 1974). This position is further supported by the fact that the administration of 6-OHDA, which provides for a selective depletion of brain norepinephrine without any effect on brain dopamine concentration, greatly increases shock-induced fighting in mice (Thoa et al., 1972c).

Related to those observations are the findings of Senault (1970) showing that apomorphine, a dopaminergic agonist, is capable of evoking irritative aggression in laboratory animals. Furthermore, even in the absence of any aversive electrical stimulation, 6-OHDA induced an overt "rage" reaction not only in male rats (Coscina et al., 1973, 1975) but also in females; in the latter case, during pregnancy irritative aggression is changed to an enhanced maternal protective aggression (Sorenson and Gordon, 1975).

In contrast, in the "rage syndrome" or in irritative aggression resulting from surgical damage to the septal nuclei in rats, L-DOPA administration, which results in an increased level of brain catecholamines, as well as the administration of dopaminergic agonists, including apomorphine, both "dramatically" abolish the syndrome (Gage and Olson, 1976; Marotta et al., 1975, 1977). Different results with apomorphine-injected septal rats were obtained by Senault (1973), probably as a consequence of a diverse extension of the septal lesion and of adjacent areas, or of the strain of animal employed.

Acetylcholine

Cholinergic stimulation of various brain limbic structures triggers either various types of aggression or muricidal behavior.

Irritative aggression in cats, including vicious attacks of the experimenter, has been reported following the intracerebral injection of acetylcholine and cholinergic drugs (Allikmets, 1974; Hull et al., 1976; Myers, 1964), and the cholinergic mediation of irritative shock-induced aggression has also been shown (Bell and Brown, 1976b; Powell et al., 1973; Rodgers and Brown, 1973). In the medial hypothalamus, mesencephalon, and amygdala, cholinergic stimulation produces the same effects as electrical stimulation: defensive aggression (Romaniuk, 1974; Romaniuk et al., 1973, 1974) or irritative and competitive displays (Gianutsos and Lal, 1976, 1977; Thomas et al., 1978). In contrast, local application of a cholinesterase inhibitor to the septal area or to the basolateral amygdala simulates a surgical lesion of the nuclei, with enhanced hyperactivity and irritative aggression (Igić et al., 1970).

No direct data are available concerning the effect of altered brain acetylcholine concentration in relation to various aggressive displays; injection of acetylcholine into the lateral ventricle of rats has been shown to alter the level of either serotonin or norepinephrine in discrete brain areas, without inducing any correlated behavioral modifications (Herman et al., 1972). In turn, the modulation of brain acetylcholine by dopaminergic and serotonergic projections has also recently been shown (Butcher et al., 1976).

Others

Few data concerning aggression and other neurochemical substances are available. For the sake of completeness, microinjections of L-glutamate were reported to mimic the effects of electrical stimulation of areas of hypothalamus which are responsible for attack or flight, whereas γ-hydroxybutyric acid (GABA) blocked these effects (Brody et al., 1969).

Lithium and repeated administration of rubidium salts both increased shock-elicited irritative aggression, which was unchanged by potassium or cesium chloride (Eichelman and Thoa, 1973). Interestingly, even acute lithium administration has been shown to lower the availability of brain norepinephrine (Schanberg et al., 1967; Schildkraut et al., 1967), to increase norepinephrine turnover (Schildkraut et al., 1969; Stern et al., 1969), and also to increase both brain tryptophan uptake (Knapp and Mandell, 1973, 1975) and serotonin synthesis (Essman, 1975b; Poitou et al., 1974). Increased glutamate was observed in the hypothalamus and amygdala together with an increased GABA content exclusively in the hypothalamus after acute lithium treatment (Gottesfeld et al., 1971); in *in vitro* experiments, lithium has also been reported to inhibit acetylcholine synthesis inside the nerve terminals (Vizi et al., 1972). Repeated administration of rubidium also greatly enhances brain norepinephrine turnover (Eichelman and Thoa, 1973).

For obvious reasons there are no comparable data concerning neurochemical correlates of human aggression. Several indirect accounts, however, indicate that in 83% of patients studied, hyperactive and aggressive behaviors are associated with a fall of blood 5-hydroxyindoles (Greenberg and Coleman, 1976); lithium also exerts an antiaggressive effect in mentally defective patients (Worrall et al., 1975). Aggression in depressed patients has also been reported following large doses of L-DOPA (Goodwin et al., 1970, 1971).

Table 4 presents a tentative summary of some of the neurochemical correlates of various types of aggression.

SENSORY CONTROL

All living organisms are equipped with a series of specific sensors which continuously communicate data to the central nervous system about the status of its internal and external environment, thus allowing the subject to respond appropriately to incoming stimuli. In a narrow sense, which incidentally corresponds to a behavioristic view, it may be thought that sensory input rather

TABLE 4. *Brain neurochemical correlates of the various types of aggression*

Type of aggression	Serotonin		Norepinephrine		Dopamine		Acetylcholine	
	T	L	T	L	T	L	T	L
Competitive	↑	↑		↑↓		↓↑		↑
Defensive		↑		↓				↑
Irritative	↑	↓↓↓	↑↑↑	↓↑↓↓↓↓	↑	↑↑↑↓↓		↑
Territorial		—		↑		↑		
Maternal protective		—		↓		↓		

T = turnover; L = level.
The number of arrows approximately corresponds to the experiments quoted in the text, while their size broadly refers to the intensity of changes reported.

exclusively governs behavior. However, as frequently cited, all behavior represents an adaptive search for the satisfaction of internal needs which govern homeostasis; sensory input plays only a partial role, since the conscious and unconscious processing of ongoing information includes prior learned experiences, education, culture, expectation, and purpose, all organized into an anticipatory response to the forthcoming behavior (Livingston, 1974). This condition is common to both animals and man, and sensory exploration merely subserves an assembly of new and revised images by which behavior can be suitably organized in the presence of unfamiliar objects and environments. An incorrect sensory input produces a distortion in information processing which, in turn, evokes inappropriate behaviors; this general situation necessarily reflects on aggression as well.

Changes in visual responsiveness as a result of intercollicular lesions influence the attack-avoidance system, a reduced responsiveness to startling visual stimuli depressing attack, defensive aggression, and escape, and facilitate avoidance in different animal species (Andrew, 1974; Andrew and DeLanerolle, 1975; Hunsperger, 1963; Schaefer, 1970). Paired rats with visual impairment display less intense irritative aggression to foot-shock than unimpaired control animals, and visual impairment combined with vibrissae removal produces an even greater decrease in aggressive displays (Flory et al., 1965). Anesthesia or removal of vibrissae is sufficient to induce a substantial decrease of irritative aggression in rats (Bugbee and Eichelman, 1972; Ghiselli and Thor, 1975; Thor and Ghiselli, 1973*a*, *b*, 1975*a;* Thor et al., 1974), and anesthesia of the mystacial vibrissae also greatly reduces either competitive or territorial aggression in these animals (Flannelly et al., 1974; Thor and Ghiselli, 1975*b*). In contrast, muricidal rats are not, or are only minimally impaired in their killing activity by vibrissae removal or anesthesia (Bugbee and Eichelman, 1972; Karli, 1961; Thor and Ghiselli, 1975*a, b*).

It is apparent that the facial tactile receptors and intact and functional vibrissae relay important sensory information to the central nervous system for spatial position, location, and movement, not only when engaged in fighting but also in other crucial situations; devibrissaed rats die within a few minutes after being placed in a swimming tank, apparently due to impaired coordination (Bugbee and Eichelman, 1972; Richter, 1957). It is interesting that the mystacial vibrissae are anatomically represented in the subcortical trigeminal centers (Belford and Killackey, 1979), and that in the newborn rat and mouse vibrissal damage affects the organization of the thalamocortical projections and disrupts the formation of the corresponding cortical neuronal architecture (Killackey et al., 1976).

Visual control effects are further considered, in that dim light increases irritative foot-shock-elicited fighting, without changing the intensity of competitive aggression (Thor and Ghiselli, 1974). Blinded hamsters show greatly increased territorial aggression, based on an enhanced duration of the attacks (Murphy, 1976*a*). The influence of territory on most animals, notably in the male Syrian golden hamster, is largely in terms of olfactory cues. This effect is almost completely suppressed by centrally or peripherally produced sensory impairment (Devor and Murphy, 1973; Murphy, 1976*b*); the termination of aggression, however, likely depends on visual stimuli. Specifically, those postures which display or conceal the black marking on hamster chest have been assumed to indicate submission or impending aggression (Grant and Mackintosh, 1963; Grant et al., 1970), so that blinded animals fight more ardently than normal because of impaired recognition of any signs of submission or threat from their opponent.

Additional evidence suggests that the olfactory system plays an important role in aggressive behavior of macrosmatic animals, and in mice as well (Lee and Brake, 1971; Mugford and Nowell, 1970; Ropartz, 1968; Rowe and Edwards, 1971); a significant inhibition of fighting behavior occurs when one member of the fighting pair is treated with an odor neutralizer (Lee and Brake, 1971; Ropartz, 1968; Valzelli, 1969). The importance of olfactory cues and the anatomical structures involved in aggression have been cited in the preceding chapters (Chapter 3, p. 48; Chapter 5, p. 78). In this general context, mice rendered anosmic by removal of the olfactory bulbs will not engage in fighting, and consistently exhibit submissive behavior even in response to a biting attack (Denenberg et al., 1973; Ropartz, 1968). Olfactory bulb removal also eliminates pain-induced irritative aggression (Fortuna and Gandelman, 1972); the treatment of intact mice with a deodorant reduces competitive aggression but does not affect irritative fighting (Fortuna, 1977; Rowe and Edwards, 1971), and also induces cannibalism when bulbectomized males are housed in pairs (Fortuna, 1977).

Concerning this last issue, it is important to note that the integrity of the olfactory system is essential for the initiation of maternal behavior in mice (Gandelman et al., 1971*b*) and that, when lactating mice are bulbectomized at different times after delivery, they cannibalize their pups with decreasing intensity following earlier bulbectomy (3 days) or later (14 days) (Gandelman et al., 1971*a*). Further, bulbectomized male and female mice have significantly decreased brain tryptophan-hydroxylase activity, leading to a decreased rate of whole brain serotonin synthesis, whereas tyrosine hydroxylase activity remains unchanged (Neckers et al., 1975). After bilateral neonatal bulbectomy, adult female golden hamsters show marked decrease in maternal protective aggression against males during lactation (Chapter 5, p. 82); such lesions also result in a 30% incidence of pup cannibalization (Leonard, 1972). Removal of the olfactory bulbs in rats suppresses spontaneous killing behavior but, as previously mentioned (Chapter 5, p. 89), induces killing among rats that do not naturally do so (Bandler and Chi, 1972).

Olfactory bulb surgery has also been described to induce either hyperemotionality (Alberts and Friedman, 1972; Thorne et al., 1973, 1974) or irritability (Cain and Paxinos, 1974; Huss and Homan, 1975); these effects, as with muricide, are greatly dependent on the strain of animal used for the experiments (Thorne et al., 1973; Valzelli, 1971*a*, 1978*d*) and on previous social and environmental variables (Bernstein and Moyer, 1970). Further, as already seen for other animals, olfactory damage in female rats can induce pup-killing and cannibalism (Thorne et al., 1974).

Pure anosmia, not involving direct damage to olfactory bulbs or other anatomical olfactory structures (Karli, 1955, 1956; Karli and Vergnes, 1963; Thorne et al., 1974), as may be produced by lesions of the nasal mucosa with zinc sulfate (Alberts and Friedman, 1972; Alberts and Galef, 1971; Cain and Paxinos, 1974) or by surgical removal of the nasal mucoperiosteum, thereby transecting the fila olfactoria (Spector and Hull, 1972), does not produce killing behavior in rats, even though they still display increased emotionality. This suggests that the anatomical olfactory system should not be considered as exclusively involved in sensory olfaction but also contributes to the limbic balance of emition and affect through connections with the preoptic nuclei, lateral hypothalamus, and the corticomedial amygdala (Adey, 1953; Allison, 1954; Barraclough and Cross, 1963; Pfaff and Pfaffman, 1969; Powell et al., 1965; Valverde, 1965; Valzelli, 1979*a*); these interconnections explain the influence of the olfactory bulbs on other previously cited factors such as mating, sexual behavior, fertility, and pregnancy (Bruce, 1959, 1960; Dominic, 1966; Franck, 1966*a*, *b*; Heimer and Larson, 1967; Murphy and Schneider, 1970).

No direct evidence is available to relate sensory control to human aggression, even though reduced sensory stimulation or complete sensory deprivation is known to induce a series of emotional, affective, and psychological changes (Suedfeld, 1975) in children and adults; these include anxiety, irritability, hyperactivity, hostility, and outbursts of aggression and violence (Chapter 3, p. 50

TABLE 5. *Interaction between sensory control and some behavioral patterns*

Types of sensory disablement	Behavioral patterns	
	Increased or *induced*	Decreased or *impaired*
Visive acuity	Irritative aggression	
Visive responsiveness	Avoidance	Attacks Defensive aggression
Blindness	Territorial aggression	Irritative aggression
Tactile vibrissae		Irritative aggression Competitive aggression Territorial aggression Movement coordination
Odor adulteration	*Conspecific cannibalism*	Competitive aggression
Olfactory mucosa	*Hyperemotionality* *Hyperirritability*	
Olfactory bulbectomy	*Hyperemotionality* *Hyperirritability* *Eterospecific killing* *Pup-killing* *Pup-cannibalism* Submission	*Mating* Sexual Fertility Pregnancy *Maternal* *Maternal protective aggression* Irritative aggression *Spontaneous killing*
Overall deprivation	*Irritability* Anxiety *Depressive ideation* *Hostility* *Aggression* *Violence*	Appetite Emotional stability Self-awareness

and 51). Interestingly, olfactory alterations are present in several behavior disorders (De Maio, 1966; Pryse-Phillips, 1975).

A tentative outline of the relationship between sensory control and various types of behaviors and aggressions is given in Table 5.

GENETIC COMPONENTS

Animal Studies

There is general agreement that genetic factors play an important role in aggressive behavior. The most obvious and basic genetically predetermined anatomical characteristic is sex, with the predominant aggressive repertoire being manifested by male vertebrates, at least in the animal kingdom.

Of the various reasons for this, one may have to do with physical size, with the average male of most species being larger and stronger than the average female (Vernon, 1969). This rule also applies to nonhuman primates, among which the heaviest males are the most aggressive of the group (Carpenter, 1934, 1940, 1958). Certainly, differences in the size of several physical structures between males and females are likely related to obvious hormonal differences, which are also involved in aggression (Beeman, 1947; Bevan et al., 1960), as will be discussed later. An interesting inverse relationship appears to exist between brain weight and aggression. Behavioral differences in open-field activity, some types of learning abilities, and motor and sensory patterns were found in mice genetically selected for high and low brain weights (Fuller and Geils, 1973; Wimer and Prater, 1966; Wimer et al., 1969); such animals also showed differences in aggression (Collins, 1970). In more recent and detailed experiments, low brain weight mice from different strains were shown to be more aggressive and to fight more intensely than their high brain weight litter-mates (Hahn et al., 1973); there was also an interaction between the brain weight genotype and the behavioral changes induced by septal lesions (Gonsiorek et al., 1974).

From a clinical standpoint, hypertension has been frequently associated with aggressive behavior in man (Oken, 1960; Schachter, 1957; Shapiro, 1960), but in animal experiments genetically hypertensive rats and mice have been shown to be less prone to fighting than genetically normotensive or hypotensive animals of the same species (Ben-Ishay and Welner, 1969; Eichelman et al., 1973; Elias et al., 1975). However, spontaneously hypertensive rats become muricidal as a result of socioenvironmental isolation (Eichelman et al., 1973; Rifkin et al., 1974).

The dependence of the level of aggression on hereditary variables has been further emphasized in data on selective breeding obtained by Lagerspetz (1961, 1964). Interestingly, chemical comparison between selectively bred nonaggressive and aggressive strains of mice revealed higher forebrain serotonin concentration in the former than in the latter (Lagerspetz et al., 1968); these were more susceptible to increased aggression after prolonged social isolation (Lagerspetz and Lagerspetz, 1971). A recent study with several strains of mice indicated that brain serotonin turnover decreases significantly only in those strains which react to isolation with a consistent degree of aggressiveness; therefore there appears to be an inverse relationship between these two parameters (Valzelli and Bernasconi, 1979). This is in agreement with previously cited data of decreased serotonin turnover in emotionally unstable strains of rats (p. 99; Rosecrans, 1970), and may also be consistent with the inverse relationship between brain serotonin content and emotional stability in mice and rats described by others (p. 99; Maas, 1962, 1963; Sudak and Maas, 1964a, b; Wimer et al., 1973). The aforementioned study (Valzelli and Bernasconi, 1979) also partially confirms the observation of Lagerspetz and his colleagues (1968) of a somewhat higher forebrain serotonin content in those strains of mice more likely to develop aggression following prolonged isolation.

Isolation-induced fighting behavior in mice, which is known to be sex and strain dependent (Bevan et al., 1951; Karczmar and Scudder, 1969; Valzelli,

1967*b,* 1978*d*), has recently been shown to be influenced by at least two homozygous loci and by a possible cytoplasmic factor of the ovum (Eleftheriou et al., 1974). In rats it has been possible to achieve the genetic selection of winner and loser animals in a competitive situation (Masur and Benedit, 1974). Different strains of rats are differentially susceptible to developing muricidal behavior as a result of either olfactory bulbectomy (Thorne et al., 1973) or prolonged isolation (Valzelli, 1978*d*).

Both the tendency for aggressive attacks and a high brain level of cyclic adenosine monophosphate (cAMP) have been described to be inherited as recessive traits which maintain a close association in segregated generations (Orenberg et al., 1975); considerable evidence has suggested that cAMP is a second messenger for several neurotransmitters for synaptic transmission in the mammalian nervous system. The interaction of genotype with population number, environmental conditions, and exogenous neonatal androgen or estrogen administration in female and male mice as determinants of aggression has also been described (Vale et al., 1971, 1972, 1973); these findings confirm the current concept of the sexual differentiation of mammalian brain functioning (Bronson and Dejardins, 1968, 1970; Harris, 1964; Southwick and Clark, 1968; Tata, 1966).

Finally, although still in question (Hay, 1975), evidence has been offered for the contribution of a Y chromosome to an aggressive phenotype in mice, supporting the hypothesis that there may be heritable variations of an additional Y chromosome, variably associated with some types of animal aggression (Selmanoff et al., 1975*a, b*). Generally, aggression undoubtedly has multiple genetic determinants, and the distribution in any population of an additional Y chromosome leading to an XYY karyotype would likely reflect only a variable degree of genetic predisposition to aggressive behavior.

Human Studies

One of the major differences between man and other species is the influence of cultural evolution on human behavior. Since cultural changes are to a wide extent dependent on socioenvironmental variables which, in turn, are affected by specific socio-ideological trends, the interaction between genotype and such external variables has provided an enormous number of human phenotypes.

It is a matter of concern to note that sexual differentiation in physical size and strength, which provides for a predominance of male animal aggression (p. 111), has not in recent years operated at the human level; this does not depend on the availability of lethal but light weapons, but on a motivated change in feminine attitudes. The scientific significance of the concept of human "genetic decay," which Lorenz (1973) and others predicted as a result of "liberal permissiveness," allows defective mutations to spread through society. Further, the fact that each one of us differs from his fellow not only in bodily structure but also in mental attitude, character, and ability (Garrod, 1929) reflects both the genetic predetermination of the consequent behavior (Anastasi, 1958; Childs, 1972; Erlenmeyer-Kimling and Jarvik, 1963) and the multiple interactions of genotype with the environment (Anastasi, 1958; Cavalli-Sforza, 1975; Linn, 1974; Plomin et al., 1977), despite any stubborn belief that all men are born equal, and that their moral defects appear exclusively due to defects in their environment.

Racial differences in psychiatric impairment among young adults of the lower socioeconomic classes (Schwab et al., 1973) confirm the existence of gene-environment interactions in psychiatric disorders (Cavalli-Sforza and Feldman, 1973; Kidd and Matthysee, 1978; Kinney and Matthysee, 1978). Moreover, there is ample evidence that genetic factors are involved in alcoholism, manic-depression, schizophrenia, and suicide (Gotesman and Shield, 1972; Tsuang, 1975, 1977; Winokur and Tsuang, 1975).

With regard to human aggression, a recent study of Australian Western Desert aborigines emphasizes striking analogies between mechanisms utilized by these people for handling aggression and those utilized by lower animals (Jones, 1971). Man has lived in Australia for at least 25,000 years (McCarthy, 1970) and, although it is not known for how long the Western Desert men have been socially adapted to the region, the experts believe that there has been little change in their behavior for thousands of years. Australian aborigines have acquired patterns of behavior that enable them to live in a peculiar adverse environment and, although they are not acquainted with such simple tools as the bow and arrow, their aggressive behavior is prominent, and fighting and punishment represent a substantial part of their discussions and activities (Basedow, 1925; Elkin, 1964; Warner, 1937). The latter, as described in the animal kingdom (Lorenz, 1967; Tinbergen, 1973; Chapter 4), include stereotyped codes of aggressive behavior such as ritualized spearing, clubbing, and magical acts, which elicit stereotyped placatory gestures (Jones, 1971). These observations indicate that a fairly stable environment, shared in common with other similar subjects, leads to the development of similar behavior; equally adaptive but different patterns are not followed because of a genetic disposition to learn in this way, especially when the undeniable cultural substrates have no means of being changed by new influences. These studies also are inconsistent with the previously noted tendency to consider man as "anthropologically" free from the sin of aggression (Chapter 4, p. 70).

Some 14 years ago, Jacobs and her colleagues (1965) carried out a chromosomal survey in a maximum security State Hospital at Castairs, Scotland, and detected 9 XYY karyotyped individuals among 315 subjects. This report supplemented some previous studies which reported figures from 1% (MacLean et al., 1962) to 2% (Forssman and Hambert, 1963) or more of such an unusual karyotype in large populations of mentally subnormal and "hard to manage" or frankly aggressive individuals with violent and criminal tendencies (Jacobs et al., 1965).

The XYY, XXY, or XXYY human karyotypes are characterized by unusual tallness, low IQ, an occasionally documented history of grand mal seizures, and a high level of plasma testosterone (Welch et al., 1967); other investigators have conversely indicated that patients with tall stature have a comparatively pronounced disposition to character disorders and to criminal acts (Nielsen and Tsuboi, 1970). This obviously does not mean that an extraordinary height will have an obligatory association with antisocial behavior, thus implicitly requiring selective commitment to maximum security facilities, as has been already made clear (Baker et al., 1970; Borgaonkar et al., 1972; Hook and Kim, 1971; Witkin et al., 1977). Nevertheless, the XYY karyotype subjects, who represent 0.1% of the general population, also comprise an increased percentage (2%) of patients admitted to security settings for acts of aggression and violence (Benezech, 1975; Bioulac et al., 1978; Griffiths, 1971; Hook, 1973; Noël and Benezech, 1977; Meyer-Bahlburg, 1974). The only acceptable conclusion is that XYY individuals merely have a considerably higher than average risk for displaying antisocial, aggressive behavior. Then, as already seen in mice (p. 112), the human XYY karyotype probably reflects no more than a variable degree of genetic predisposition to aggression, but this issue requires further clarification (Goldstein, 1974).

From a biochemical standpoint, determination of cerebrospinal fluid 5-HIAA level suggests the existence of decreased brain serotonin turnover in XYY individuals (Bioulac et al., 1978). Contrary to previous findings, recent studies indicate that both plasma testosterone concentration and the rate of testosterone synthesis are significantly lower in XYY subjects than in normal controls (Sharma et al., 1975).

This last issue introduces the issue of the role of hormones in the modulation of aggression.

HORMONAL CONTROL

There is substantial evidence that aggression is variably influenced by hormones; the experiments in this area, however, are sometimes contradictory. Although in many cases differences may be accounted for by some procedural and methodological problems, most discrepancies arise when one tries to correlate the hormonal substrates of different kinds of aggression; these are controlled by partially overlapping but largely independent physiological systems. It becomes quite possible, therefore, that some kinds of aggressive behavior may be preferentially elicited by a given steroid level while others are not.

Some issues, implicitly underlying obvious hormonal components of aggression such as some aspects of olfactive sensory control (p. 108), sex differences (p. 111), and genetic components (p. 112), have already been described, so that attention will now be focused on the possible hormonal correlates of different kinds of aggression.

Animal Studies

In general terms, the fact that the neurological system for aggression is sensitized by hormones derives from evidence that a reduced androgen level raises the threshold for aggressive displays. In male animals castration reduces fighting behavior (Beeman, 1947; Payne and Swanson, 1971a, 1972a; Seward, 1945; Sigg, 1969), as do anti-androgenic substances and estrogens (Clark and Birch, 1945; Edwards, 1970; Kislak and Beach, 1955; Suchowsky et al., 1969). Similarly, ovariectomy decreases spontaneous aggression of female hamsters, and progesterone, but not estradiol or testosterone, restores it (Payne and Swanson, 1972b). The chronic administration of thyroxine to isolated mice shortens the period of isolation necessary to induce aggression (Yen et al., 1962), whereas thyrotropin-releasing hormone has been recently shown to be an extremely potent antagonist of isolation-induced aggression in male mice (Malick, 1976).

When adult male albino rats are screened for their spontaneous frog killing or ranacidal behavior, which is assumed to represent a natural form of predatory aggression (Bandler and Moyer, 1970), neither exogenous testosterone administration nor castration is capable of substantially changing this behavior (Bernard, 1974). Such predatory activity is not changed by testosterone given to female rats, although such treatment selectively increases their brain norepinephrine level (Bernard, 1976) and increases foot-shock-induced irritative aggression (Bernard and Paolino, 1974, 1975). Predatory behavior, in contrast with other types of aggression, can be considered as largely independent of any androgen-bound initiation or facilitation.

The aforementioned irritative aggression is clearly androgen dependent, so that male rats of different strains respond with more intense fighting than females to foot-shock (Milligan et al., 1973). Further, male castration significantly decreases the pain-induced aggressive displays, which are restored by supplementation with testosterone (Bernard and Paolino, 1974, 1975; Conner et al., 1969; Milligan et al., 1973). As for the participation of other hormonal components in this type of aggression, shock-induced fighting in rats results in an elevation of plasma corticosterone and ACTH levels (Conner et al., 1970a, 1971). More recent experiments indicate that this kind of aggression is not mediated by the pituitary-adrenal axis (Erskine and Levine, 1973). Other studies have indicated that urine from male mice contains an aggression-promoting pherhormone, which is androgen dependent (Mugford and Nowell, 1970b), and that painful shock induces the secretion of such a fight-eliciting material from the preputial glands of the animals (Mugford and Nowell, 1971a).

Androgenic pherhormonal cues are likely to participate in eliciting competitive aggression, too; these aggression inhibiting or promoting cues (p. 108) also differentially influence the degree of "attackability" by different androgens in different strains of castrate male mice (Brain and Evans, 1974a, b). In male mice, housed in pairs, the dominant animal shows a lower adrenal weight, greater prostate and preputial gland weights than the subordinate (Brain, 1972); the lower weights of accessory sexual glands of subordinates in colonies of male mice have been also observed by others (Brain and Nowell, 1970; Lloyd, 1971). The marked suppression of gonadal function and androgen production by subordinates, as indicated by a decreased weight of their sexual accessory glands, has been interpreted as dependent on increased adrenal function (Desjardins and Ewing, 1971). In mouse colonies, a significant inverse correlation was found between adrenal weight and social rank, with dominant animals having smaller adrenals than low-ranking mice (Davis and Christian, 1957). A reciprocal relationship between adrenocortical activity and gonadotropic secretion is observable during the early phases of the establishment of dominance-submission interactions (Bronson, 1973). In competitive confrontations, hormonally normal male mice display reduced aggression against stranger females or castrated male opponents, but the intensity of fighting greatly increases when either females or castrated males are injected with testosterone (Evans and Brain, 1978; Lee and Brake, 1972; Mugford, 1974).

Paired male rats show more intense competitive aggression than females in the same experimental situation, and males castrated at birth also attack other males less frequently than normal animals (Barr et al., 1976a; Conner et al., 1969). The administration of testosterone at birth to neonatally castrated male rats, however, restores fighting activity to control levels in adulthood (Barr et al., 1976a). In addition, the corticosterone level in submissive and subordinate rats of a colony has been found to be twice as high as in dominant animals (Popova and Naumenko, 1972), suggesting that, in contrast to its role in irritative aggression, the pituitary-adrenal axis may be involved in the maintenance of a dominance-submission relationship (Taylor and Costanzo, 1975). In monkey colonies, the male who becomes dominant shows a progressive increase in plasma testosterone, which reaches peaks of 238% by 24 hr after successful defense of his group; males who become subordinate show an 80% decrease in their basal level of circulating testosterone (Rose et al., 1975).

In contrast with most mammals, the golden hamster is unusual in that both sexes are overtly aggressive, the female in addition being naturally dominant over the male (Dieterlen, 1959; Payne and Swanson, 1970). Castrated male hamsters regularly show less competitive aggression than intact males, to which they are submissive. They become considerably more aggressive and dominant over intact males, however, after ovarian implants; in this respect they resemble intact females (Payne and Swanson, 1971a). In agreement with this observation, spayed adult female hamsters become less aggressive and unusually submissive to intact males following progesterone administration, but not after treatment with estradiol or testosterone, where they become overtly aggressive and dominant over other males (Payne and Swanson, 1971b, 1972a, b).

As previously reported (p. 115), the urine of dominant male mice contains a pherhormonal factor which proves aversive to both subordinates and other dominant subjects, and attracts female mice in estrous; such a factor is of presumably androgenic preputial origin, also having a role in territorial marking (Jones and Nowell, 1973a, b, 1974). Thus testosterone as well regulates territorial aggression (Bell and Brown, 1975b). In this framework, immature rats intruding on the territory of an adult male are totally spared from any overt aggression (Thor and Flannelly, 1976a) which is instead displayed vigorously against any foreign adult species member (Flannelly and Thor, 1976; Luciano and Lore, 1975; Thor and Flannelly, 1976*b*). Rats castrated prepubertally, similar to sexually immature animals, do not elicit any aggression by the territory owner, whereas postpuberal castrated animals are subjected to transient low-intensity attacks. This suggests that the influence of gonadal hormones on preputial gland development and secretion represents an important factor for the olfactory cues identifying males to other species members (Flannelly and Thor, 1978). In gerbils, territorial marking is regulated by the secretion of midventral sebaceous glands, which also depends on circulating androgen (Anisko et al., 1973; Thiessen et al., 1968); castration in infancy (Turner, 1975) or in adulthood (Thiessen et al., 1968) impairs territorial scent-marking, which is restored by testosterone administration (Thiessen et al., 1968, 1973). Olfactory bulb removal also reduces territorial marking and aggression in gerbils, and the restoration of marking capacity in bulbectomized males requires substantially higher doses of testosterone than those necessary to restore marking in castrated animals (Baran and Glickman, 1970; Thiessen et al., 1970). Further, neonatal bulbectomy, alone or in combination with castration, abolishes both scent-marking and fighting behavior, and the administration of testosterone in adulthood only induces aggression without restoring scent-marking; this latter activity is fully recovered, together with the former, by androgen administration when bulbectomy or castration is performed in adult animals (Lumia et al., 1975, 1977). Finally, in the male hamster, marking activity remains unchanged after castration (Whitsett, 1975).

Maternal protective aggression, described in Chapter 5 (p. 82), is obviously hormone dependent. An interesting issue related to this type of aggressive pattern is that the hormones secreted during pregnancy, which provide for growth of the nipples, enabling the mother to feed her offspring, also regulate maternal protective aggression. In fact, suckling stimulation is likely responsible for the initiation of this kind of behavior, since the frequency of suckling together with exteroceptive stimuli from the pups are suggested to produce hypothalamic changes responsible for aggression (Svare, 1977). Suckling stimulation, experimentally induced in virgin female mice with foster pups and hormonal manipulation, enhances both the typical maternal behavior and overt protective aggression (Svare and Gandelman, 1976*b*). Conversely, the suppression of lactation, either by surgical ablation of the nipples (thelectomy) or by estrogen administration, abolishes the aggressive displays of normal and "virgin" mother mice (Svare, 1977; Svare and Gandelman, 1975, 1976*b*).

Lactating female hamsters also display a typical maternal protective aggression which has been shown to be prolactin dependent; ergocornine, an inhibitor of prolactin secretion, suppresses this type of aggression but increases attacks on pups (Wise and Pryor, 1977).

Estrogen has been reported to inhibit isolation-induced aggression in male mice, and the mouse-killing of rats (Banerjee, 1971*a;* Leaf et al., 1969; Suchowsky et al., 1971).

Sex-related aggression is, by definition, regulated by the hormonal setting of the animals. Consequently, female mice in estrous or diestrous, or even ovariectomized mice show only a low level of aggression toward stranger male opponents, but androgenization of ovariectomized females by testosterone injection induces male opponents to strongly attack them (Mugford and Nowell, 1970*a*, 1971*b*). This fact seems to depend on an anti-aggression substance present in the urine of estrous and diestrous female mice, and absent in the urine of spayed and testosterone-treated females (Mugford and Nowell, 1970*a*, 1971*c*). Androgen-treated female mice attack other androgenized females but not intact females (Svare and Gandelman, 1974). The evident olfactory cues that regulate these aggressive displays depend not on a urinary metabolite of testosterone, but rather on an ovary-dependent pherhormonal substance (Mugford and Nowell, 1971*c*) produced by a specific glandular apparatus of female mice (Mugford and Nowell, 1971*d*, 1972); this resembles the previously mentioned preputial secretion of pherhormones in males. It has been recently observed that castrated male mice, housed in groups, exhibit an overtly intense aggression against a lactating female intruding on the group, and this reaction depends on an imbalance of pherhormonal cues (Haug and Brain, 1978) in which attacks to lactating females by intact males are normally avoided (Evans and Brain, 1978; Mugford and Nowell, 1970*a*, 1971*b*); aggression by grouped females against lactating intruders is facilitated by such olfactory stimuli (Haug, 1973*a,b*).

In nonhuman primates, the level of aggression by both male and female monkeys changes with pregnancy, ovariectomy, and the administration of gonadal hormones. Ovariectomy increases the overall level of aggression, especially that from males, whereas either pregnancy or the administration of hormones to females both decreases aggression by males and increases their tolerance of aggression by females (Michael and Zumpe, 1970). Subcutaneous administration or brain implants of estradiol in female monkeys increase sexual receptivity and interfemale aggression, while decreasing aggression toward males. In contrast, progesterone administration, by reducing female receptivity, also decreases interfemale competition and their tolerance of male mounting attempts (Michael and Zumpe, 1970; Zumpe and Michael, 1970).

In male monkeys, the annual cycle of increased plasma testosterone from August to October corresponds with initiation of the mating season (Drickamer, 1972; Plant et al., 1974); in this interval there is an increase of intermale competitive aggression (Conaway and Sade, 1965; Wilson and Boelkins, 1970) and an increased sexual aggression of males toward their female partners (Michael and Zumpe, 1978).

An additional issue concerns the interaction of isolation and hormones in the determination of abnormal behaviors. Among isolated mice more males (40%) than females (10%) kill and eat mouse pups, whereas testosterone administration produces pup killing in 65% of intact adult female mice (Gandelman, 1972*b*). Androgenization of gonadectomized and isolated adult female mice also induces pup killing. Testosterone does not produce pup killing either in intact male mice or in males gonadectomized in adulthood, although prepubertal male castration consistently decreases this behavior which is increased by testosterone administration (Gandelman and Vom Saal, 1975). Isolated male rats also display a more intense pup-killing activity than females, and neonatal or prepubertal male castration consistently decreases pup killing; androgenized female rats do not increase their pup-killing tendency (Rosenberg et al., 1971). Once pup killing has been acquired through prior experience, it is not appreciably influenced by sex differences, castration, or testosterone administration (Rosenberg and Sherman, 1975).

In summary, it may be assumed that, predatory aggressiveness excluded, all behaviors of the male and female aggressive repertoires are sustained by sex-specific gonadal hormones. These respective hormones do not appear to specifically regulate one or another type of aggression, but rather seem to act on all aggression as an essential "fuel" for the maintenance of normal "aggressive machinery" in both males and females.

Feminized males or androgenized females tend to behave as their normal counterparts, and show alterations in their aggressive behavior which certainly reflect either their social role or their social interaction. Further, a normal or altered hormonal setting may also change or aggravate already acquired behavioral alterations.

The relevance of these general conclusions for human behavior will be considered further.

Human Studies

There is a considerable amount of clinical evidence to indicate that hormonal factors are involved in human aggressive behavior. Patients with a history of aggressiveness, when given synthetic male hormone, are greatly aroused to outbursts of rage and irritative aggression (Moyer, 1976). Although a higher level of androgen is associated with increased sex-related and irritative human aggression, a reduction of androgenic compounds may reduce some types of aggressiveness (Bell, 1978). Androgen administration has been reported to increase self-confidence and to induce aggression-like responses in subjects with feelings of insecurity and inferiority (Sands and Chamberlain, 1952). The same treatment of capable of eliciting increased irritability, hostility, and outbursts of rage in aggressive patients (Sands, 1954; Strauss et al., 1952).

A positive relationship between testosterone level, hostility, aggression, and dominance, both in young men and in criminal populations, has further relevance for the issue of androgen-sustained human aggressiveness (Ehrenkranz et al., 1974; Kreuz and Rose, 1972; Persky et al., 1971; Rada et al., 1976). A similar positive correlation also occurs between serum testosterone concentration and competitive aggression in hockey players (Scaramella and Brown, 1978). Further, in competitive hockey, outbursts of violent behavior, occasionally leading to injury, may occur, and a selective relationship between dominant individuals and their submissive partners has been described (Scaramella and Brown, 1978); this relationship reproduces the previously reported covariation of testosterone level with changes from dominance to submission in monkeys (Rose et al., 1975; p. 116).

The striking similarity between animal and human data in this case is disconcerting only to those who still maintain that animal studies are irrelevant for a better understanding of human aggression. It may be of further interest to recall that dominant mice have smaller adrenals than low-ranking mice (Davis and Christian, 1957), and it is of related interest to note that the mean level of urinary corticosteroid excretion of the two officers of a 12-man Special Forces "A" team was substantially higher than that of their subordinates. Further, corticosteroid excretion by the officers rose on the day of an anticipated attack, while the enlisted men showed a decreased corticosteroid excretion (Bourne et al., 1968).

Reconsidering the role of androgens, it may be noted that there is a positive correlation between plasma testosterone level and the psychological scoring of aggressive tendency in young volunteers (Meyer-Bahalburg et al., 1974). Although this finding has not been replicated, differences in results may depend on differences in methodology, on studying "potential" rather than "actual" aggressiveness, or on the diversity of environmental interaction and prior experiences of the subjects.

Castration has been reported to be effective in the control of some violent sex crimes in man (Bremer, 1959; Campbell, 1967; Hawke, 1950; Le Maire, 1956), and other studies suggest that stilbestrol controls aggressive behavior in adolescents and young adults (Dunn, 1941; Sands, 1954). Estrogens significantly reduce both aggression and sexual offenses (Chatz, 1972; Field and Williams, 1970), as do some potent anti-androgenic drugs such as cyproterone (Laschet, 1973) and medroxyprogesterone (Lloyd, 1964a).

In females, the "premenstrual syndrome," which includes increased irritability and feelings of hostility (Dalton, 1964; Greene and Dalton, 1953; Hamburg and Lunde, 1966), may result in actual irritative aggression, especially in individuals with inadequate behavioral control. Prison records for females show that approximately 60% of violent crimes were committed by women during the premenstrual week, whereas only 2% of the criminal violence took place at the end of the period (Dalton, 1961; Morton et al., 1953). The premenstrual syndrome is associated with a fall in progesterone level (Hamburg et al., 1968) and in estrogen concentration (Bardwick, 1971); such symptoms have been alleviated by the administration of progesterone (Dalton, 1964; Greene and Dalton, 1953; Lloyd, 1964b). Women taking oral contraceptives that contain progestational agents are significantly less irritable than those not taking the pill (Hamburg et al., 1968); the irritability-reducing effect of progestins, and their psychotropic effects (Herrmann and Beach, 1978), may be due to an action on the brain neural system related to aggression. Prolactin also may have a role in this context, since it has recently been observed that high levels of this hormone are associated with low estrogen level and may cause depression; high prolactin level with a low progesterone level may cause either anxiety or irritative hostility (Carroll and Steiner, 1978).

Hypoglycemia has also been associated with hostile and aggressive tendencies. In a series of anthropological reports, the Qolla Indians of Peru are portrayed as "perhaps the meanest and most unlikable people on the Earth" (Pelto, 1967), or *brutos y torpes,* that is, irrationally cruel, uncivilized, and dull (DeMurua, 1946). Even in recent years, Andean highland Qolla Indians have been cited as having an extreme modal personality configuration dominated by excess of hostility and aggressivity (Barnouw, 1963, 1971; Harris, 1971). This can assume a multiplicity of forms, ranging from violent encounters to an exceptionally high homicide rate (Bolton, 1973). In an interesting report on this population, Bolton demonstrated a correlation between hypoglycemia and Qolla aggression, and also noted that hypoglycemia is a widespread physiological condition in this Andean population (Bolton, 1973). Hypoglycemia has been further suggested as a possible basis for an explanation of ethnic differences in aggression (Edgerton, 1971).

In clinical studies, 89% of a total of 600 hypoglycemic patients were shown to be highly irritable, and 45% of them manifested overt unsocial, asocial, and antisocial behaviors (Frederichs and Goodman, 1969). Aggressive traits associated with hypoglycemic states were also seen in individuals convicted for homicidal threats and acts, destructiveness, and child abuse (Wilder, 1947). Low blood sugar levels also tend to intensify allergic reactions, which may in turn have as a component increased irritability ranging from argumentativeness to overt and abnormal aggressive behavior (Randolph, 1962). The basis for increased aggressive behavior during hypoglycemia is not yet clear, although it is known that the limbic system is particularly sensitive to it, and that a low brain glucose level can activate epileptic foci (Ervin, 1969). In this framework, it is important to note that the brain only stores glucose; because it is unable to convert glycogen to glucose, the brain is dependent on the blood glucose content.

As a final remark, some highly allergic subjects have also been reported to respond to specific foods with outbursts of violent rage within 30 to 60 min of ingestion (Moyer, 1971*a*).

In may be therefore concluded that, in man too, gonadal hormones represent an indispensable fuel for aggression, whereas, in a more general framework, the action of hypoglycemia on brain activity may parallel previously described effects of undernutrition (Chapter 5, p. 78) in promoting aggression.

7

The Pathology of Aggression: Violence

A DRAFT

The bulk of evidence documented in the preceding chapters seems to leave little doubt that both animals and humans have in common either various types of aggression or the neuroanatomical and the neurophysiological substrates that produce aggressive behavior (Groen, 1972; Kahn and Kirk, 1968; Moyer, 1971a, b, 1976; Roth, 1972; Visser, 1972; and references in Chapters 5 and 6). There do not appear to be major differences in animals and men in the causes and the stimuli that trigger aggression (Chapters 4 and 5); in man, however, they may be masked by infinite cultural superstructures and ingeniously prearranged justifications.

Violence appears to be peculiar to man, in that it represents an issue that has brought disrepute to mankind in general and particularly to the age in which we live. In dealing with such an issue, the problem is to avoid ideological contamination, while searching for increased understanding of those neurophysiological factors and conditions that lead man to violence. Examples of such contamination are available (Coleman, 1974; Ryan, 1972; Szasz, 1968; Valenstein, 1974) and just cloud the question without offering any consistent assistance in solving it.

On a more concrete level, Moyer (1971b) has proposed that the neurophysiological mechanisms involved in aggression can exist at any moment in one of the following states: (1) inactive and insensitive, so that aggression cannot be activated by those stimuli that usually provoke attack; (2) sensitized but inactive, due to the absence of appropriate stimulation; and (3) spontaneously (then uncontrollably) firing, even in the absence of any appropriate stimulation. I feel that the neurophysiological network for aggression can be fired by a fourth mechanism: incorrectly tuned sensory control, allowing either inappropriate stimuli or distorted perceptual input to occur and produce inappropriate or distorted aggressive output. Interestingly, the third and fourth mechanisms are rarely observed in the animal kingdom, but they can be experimentally induced in laboratory animals. In contrast, the neurophysiology of violence likely finds its roots in the same items which, in turn, sustain an incorrect, abnormal, or pathological mechanism regulating normal aggression. This may also provide a consistent basis for the numerous examples of violent and criminal behaviors associated with emotional and mental disabilities (Baron, 1977a; Beck, 1977; Cloninger and Guze, 1970; Drtil, 1969a, b; Fechter, 1971; Felthous and Yudovitz, 1977; Häfner and Böker, 1973; Lion et al., 1976; Madden, 1977; Piotrowski et al., 1976; Planansky and Johnston, 1977; Revitch, 1975; Schipkowensky, 1973; Silverman and Spiro, 1967, 1968; Thornton and Pray, 1975; Tupin et al., 1973).

It should be emphasized that not every mentally disturbed person has to be considered as a potentially violent criminal and, conversely, that not every violent offender is necessarily mentally disturbed. A series of studies has already shown the inconsistency of any predictive value of IQ level and other psychological measures in anticipating any actual violent conduct (Deiker, 1973, 1974a, b; Gunn, 1977; Häfner and Böker, 1973; Sternlight and Silverg, 1965). Instead, a high rate of violent offenders is found in schizophrenia and bipolar affective disorders, especially when associated with antisocial personality, alcoholism, or drug dependence (Blackburn, 1968a, b; Good, 1978; Häfner and Böker, 1973; Piotrowski et al., 1976; Planansky and Johnston, 1977; Silverman and Spiro, 1968; Tupin et al., 1973; Wolf, 1973). This clinical information seems to reinforce the aforementioned concept of the pathological transition from aggression to violence as possibly underlying the neurophysiological conditions previously outlined in points 3 and 4. In this framework, violence might be interpreted as the extreme pathology of irritative and perhaps competitive aggression more than an irrational exasperation with other kinds of aggression.

This does not exhaust the subject; for instance, the role of psychomotor epilepsy and temporal lobe focal diseases in engendering violent outbursts in some subjects has already been described (Chapter 5, p. 94), as well as the part played in the determination of violent conduct by some peculiar genotypes in their interaction with socioenvironmental characteristics (Chapter 6, p. 112; and Christiansen, 1974; Hutchings, 1974). Needless to say, none of these issues clarifies the meaning of a transition to violence; it rather suggests a part of the basis from which the seed of violence can bloom. The other part of the basis is that violence, like aggression, has extensive imitative and learned determinants, and it is incredibly more potent and contagious than any epidemic disease.

To minimize or deny the biomedical evidence, and to insinuate that such data are merely a means for diverting attention from our "social" dilemmas and for the development of mass behavior control is to ignore several significant needs; such views only delay any progression in this field toward an understanding of violence. To state that biomedical research is only a means for contributing to an understanding of a "political" solution to violence only serves political ideology which, under different labels, has already consistently exerted mass behavior control. To misleadingly teach children that they are "oppressed" from birth by their parents, by culture, by religion, by society, and that they have the "innate" right to behave beyond any limitation only prepares them for frustration and neurosis, and manipulates them as potentially explosive seeds in an increasing and ideologically controllable violence.

The consideration that electrical stimulation of the "pleasure centers" (Tables 1, 2, and 3, p. 59, 91, and 95) of the brain in animals abruptly stops any ongoing aggressive behavior led a developmental neuropsychologist recently to hypothesize that the principal cause of human violence is a limitation on pleasant bodily experiences during the formative period of life (Prescott, 1975). This assumption is much more than merely a hedonistic theory, and finds further experimental support in the presence of tactile receptors, the stimulation of which triggers pleasurable feelings (p. 56 and 57; Table 1, p. 59; Campbell, 1972). There is an implicit emphasis of the importance of a correctly tuned sensory input and control in properly balancing behavior. Furthermore, Prescott believes that deprivation of body touch, contact, and movement causes several emotional disturbances including depressive and autistic behaviors, hyperactivity, sexual aberration, drug abuse, aggression, and violence.

Early somatosensory deprivation, which mainly consists of a lack of parental and, especially, maternal care, and also of social grooming, has a similar result in nonhuman primates; monkeys deprived of physical contact with other monkeys from birth grow to be fearful, aggressive, and sexually abnormal adults (Mitchell, 1975). In man, an extremely high incidence of self-destructive and self-mutilative behavior (Kroll, 1978; and Chapter 3, p. 58) as well as a high incidence of childhood pathology, suggestive of deprivation and neurological impairment, was found in a group of 62 habitually violent patients (Bach-Rita and Veno, 1974). Self-mutilation and abnormally violent aggression are also observed in monkeys or other mammals after prolonged isolation (Jones and Barraclough, 1978; Mitchell, 1968; Suomi et al., 1971).

In this general framework, somatosensory deprivation is an important component of the previously cited socioenvironmental deprivation or prolonged isolation (Chapter 3, p. 51); this also has relevance for the production of abnormal behavior and aggression in the larger context of a subject-environment interaction. Thus animal experiments in this field may provide valuable models of behavioral alterations potentially useful for a better understanding of human violence.

ENVIRONMENTALLY INDUCED AGGRESSION

Behavioral Considerations

Prolonged isolation or socioenvironmental deprivation has already been described to induce several behavioral alterations in either young or adult mammals, including man (Chapter 3, p. 51). Prolonged isolation also increases or induces aggression in fish (Davis et al., 1974; Fernö, 1978) and in birds (Beach and

Jaynes, 1954; Hoffman et al., 1975), which means that social interaction is indispensable for shaping normal behavior in the vertebrate kingdom.

In adult male mice and rats of some strains, prolonged socioenvironmental deprivation produced by housing them singly in individual cages with food and water *ad libitum* induces violently compulsive and persistent aggression which predominates whenever the isolated subjects interact with other animals (Allee, 1942, Al-Maliki and Brain, 1977; Brain, 1975; Crawley et al., 1975; Essman, 1969; Johnson et al., 1972; Karczmar and Scudder, 1969; Lagerspetz, 1969; Scott, 1958*a;* Valzelli, 1967*a, b;* Yen et al., 1959); various killing activities also occur (Gandelman, 1972*b*), especially in rats (Gerall, 1963; Goldberg and Salama, 1969; Johnson et al., 1972; Kulkarni, 1968*a;* Myer, 1964; Spevak et al., 1973; Valzelli, 1971*a;* Welch et al., 1974).

In considering aggressiveness induced by prolonged isolation, it should be noted that the usual laboratory animals lose most of their aggressive repertoire as a result of successive generations of breeding in highly standardized and protected environments; cues for aggression such as competition for food, territory, and females are not operative. Isolation-induced aggression in male mice is still considered and used as a model of competitive aggression in which the behavioral, neurochemical, neuropharmacological, and hormonal correlates of such behavior are studied (Brain, 1975; Brain and Al-Maliki, 1978; Brain and Benton, 1977; Brain et al., 1978; Childs and Brain, 1979*a, b;* Clark and Schein, 1966; Crawley et al., 1975; Ebert, 1976; Svare and Leshner, 1973). This type of artificially induced aggressive reaction in mice is only one of the numerous behavioral, neuroanatomical, and neurophysiological alterations induced by prolonged socioenvironmental deprivation in these and other mammals (p. 51 and 130). It therefore becomes quite difficult to view the behavioral result of such a deprivation state as merely the re-establishment of normal competitive or territorial aggressions, just to consider it as the sole "normal" variable among a wide series of abnormalities (Valzelli, 1973*a*).

The "pathological" nature of isolation-induced aggression in mice was initially hypothesized (Valzelli, 1967*b,* 1969) on the basis that it occurs as soon as two or more previously isolated subjects are put in contact with each other (Banks, 1962; Cairns and Nakelsky, 1971; Cairns and Scholz, 1973; Valzelli, 1969), in the absence of any prior relevant or explicit training to fight (Eibl-Eibesfeldt, 1967); fierce, compulsive, and repetitive fighting begins immediately and continues until the participants are exhausted (Valzelli, 1967*b,* 1969). In addition, since such violent attacks take place either in the home-cage of one of the subjects or in a neutral cage, no convincing evidence for any territorial determinant emerges, while no obvious sexual, dominance, or food cues are present (Banerjee, 1971*b;* Valzelli, 1967*b,* 1969). This illustrates the apparent aimlessness and variation from normal aggression shown by this behavioral reaction, which dominates social interaction (see Chapter 2, p. 30). In this framework, the repetitiveness and automatic nature of the aggressive response are not appreciably delayed by the code of submission, which is almost never displayed by isolated and aggressive mice when fighting among themselves.

Such characteristics may obviously change to a great extent when isolated aggressive animals are engaged in fights against normal and naive opponents; still, the behavioral normality of naive opponents may influence the aggression of the former, which remains dominant over the latter. The intercalation of such cues as early undernutrition, food deprivation, sex, and prior experiences on isolation (Brain and Al-Maliki, 1978; Brain et al., 1978; Halas et al., 1977; Randt et al., 1975; Rohles and Wilson, 1973; Whelton and O'Boyle, 1977) accordingly will change the aggressive outcome.

Aggression induced by isolation has been reported to develop gradually (Valzelli, 1969) with a transition from initially increased general stimulus reactivity, through overt hyperirritability, to a stable degree of hyperexcitability supporting aggression (Balazs et al., 1962; Bourgault et al., 1963; King et al., 1955; Scott and Fredericson, 1951; Takemoto et al., 1975; Valzelli, 1969; Yen et al., 1959). This increased brain excitability, together with the decreased microsomal activity present in isolated aggressive animals (Essman et al., 1972), lead aggressive mice to become highly sensitive to central stimulants and resistant to sedative drugs (Balazs and Dairman, 1967; Consolo et al., 1965a, b; Del Pozo et al., 1978; Valzelli and Bernasconi, 1978; Valzelli, 1979b).

The aforementioned considerations suggest that aggressiveness induced by prolonged isolation represents a pathological exasperation of irritative aggression (p. 79), likely sustained and maintained by altered sensory control. This hypothesis seems to find further support in the fact that aggression of isolated animals is controlled by abolishing the reactivity and mobility of the opponent without directly manipulating the aggressive animals (Cairns and Sholz, 1973); such an explanation may also offer a basis for the pup killing that normally fed isolated animals often display (Gandelman, 1972b). Furthermore, just as for irritative aggression (Chapter 6, p. 79), removal of the mystacial vibrissae or an olfactory bulbectomy reduces or abolishes aggression by prolonged isolation in mice (Katz, 1976; Neckers et al., 1975; Rowe and Edwards, 1971).

Similar considerations apply to the effect of prolonged isolation on the various behavioral components of male rats in which the aggressive outcome, as in male mice, is largely dependent on the strain of animals employed for the experiments (Flandera, 1977; Valzelli, 1978d). As with mice, rats subjected to socioenvironmental deprivation demonstrate either behavioral or neurochemical and neurophysiological alterations which serve as a basis for the previously described "isolation syndrome" (Hatch et al., 1965; and Chapter 3, p. 51); this can also include muricidal aggression (mouse-killing behavior) or pup killing (Myer, 1964; Paul and Kupferschmidt, 1975).

Mouse-killing activity by rats, regardless of whether surgically or environmentally induced, has been considered as only a laboratory model of spontaneous predatory aggression (Kreiskott and Hofmann, 1975; O'Boyle, 1974), especially since rather naive and biased ethological views have held that wild rats in their natural environment attack and eat smaller animals including birds, amphibia, reptiles, and fishes (Rossi, 1975). This position is surprising in that, as for mice, it completely disregards the variety of pathological changes which accompany and probably in part promote and sustain muricidal aggression in those laboratory rats that never displayed it before surgical or environmental manipulation. Again, it seems quite misleading to select from a series of induced changes a single item, and label it as the pure restitution of a normal pattern of behavior necessarily present in the "wild ancestor" of an inbred strain of rats used for laboratory purposes.

It is widely known that a variable minority of laboratory rats of some strains, mainly Long-Evans and Wistar, will spontaneously kill mice as well as other animals such as frogs, turtles, and chicks. In the classic study by Bandler and Moyer (1970), it was shown that frogs and turtles elicit aggression in nearly 100% of rats, whereas chicks and mice elicit aggression, respectively, in 45% and only 15% of the animals. It is interesting that, although frogs and turtles also strongly induce rats to eat them, neither chicks nor mice elicit an immediate and consistent eating by killer rats (Bandler and Moyer, 1970); mice therefore cannot be considered as a specially palatable "prey" to laboratory rats. It is inappropriate to state that there is no reason why laboratory killer rats that have a choice between a "natural" prey such as a mouse and usual food should not direct their feeding activity at the mouse (Rossi, 1975). Muricidal rats appear to attack mice mostly for the sake of the act of killing in itself (King and Hoebel, 1968), since for those rats which kill and possibly eat mice (Karli, 1956; Paul, 1972), killing activity is not reduced even when they are not permitted to eat the victim; therefore, hunger cannot be considered as the determinant of muricidal activity (Karli, 1956; Myer, 1964). Moreover, functional and anatomical evidence has demonstrated that neither hunger nor a selective palate is the basis for rat mouse-killing activity (Eclancher and Karli, 1971). Other studies have also shown that relationships between attack and feeding are not important in the maintenance of killing, even if variables related to feeding may be important to the initiation of killing (Paul and Posner, 1973).

Muricidal activity in killer rats is obviously facilitated by food deprivation (Adamec and Himes, 1978; Katz and Thomas, 1977; Paul et al., 1971; Whalen and Fehr, 1964), but it is not obligatorily induced by hunger (Heimstra, 1965). On the contrary, long-term fasting even to the extent of leading to death will never induce nonkiller rats to kill mice (Karli, 1956); rats reared with mice from weaning do not kill other mice in their adulthood (Denenberg, 1968). These reports strongly emphasize the intervention of either a clear genetic predisposition or the effect of learning and prior experiences in the determination of mouse-killing behavior. On this latter issue, mouse-killing activity is self-reinforcing (Van Hemel, 1972), and punishment only temporarily inhibits rather than eliminates killing by rats (Myer, 1966). In addition, the continuation of mouse-killing activity after satiation (Van Hemel and Myer, 1970) suggests that the relationship between predation, hunger, and mouse killing is less obvious, since killing occurs continuously and consistently in the absence of starvation (Berg and Baenninger, 1974).

As previously observed (Chapter 5, p. 76), a dissociation between killing and feeding has been demonstrated from both the pharmacological (Krames et al., 1973) and neuroanatomical standpoint (Eclancher and Karli, 1971; Karli and Vergnes, 1964*b;* King and Hoebel, 1968). Further, although gonadal hormones do not influence true predatory aggression (Chapter 6, p. 115), mouse killing by male rats is inhibited by castration or by estrogen administration (Didiergeorges and Karli, 1967; Leaf et al., 1969).

It then seems possible to conclude that the bulk of evidence does not support the notion of a predatory nature for experimentally induced muricidal behavior in laboratory rats. Conversely, as with aggression induced in male mice by prolonged isolation, rat muricidal behavior seems to reflect a peculiar pathology of irritative aggression, to which an unbalanced hyperdefensiveness may contribute (Albert and Wong, 1978). This tentative interpretation emerges particularly from a wide series of studies carried out by Karli and his associates on muricidal behavior induced by olfactory-bulb ablation or other neurosurgical manipulations in previously nonkiller laboratory rats (Karli, 1955, 1960, 1968; Karli and Vergnes, 1963, 1965; Karli et al., 1969, 1972, 1975; Vergnes and Karli, 1963*a, b,* 1964, 1965, 1969*a*); removal of the olfactory bulbs releases irritative aggression from the inhibitory control of the interconnected regulatory activities of the ventromedial hypothalamus, amygdaloid nuclei, cingulum, hippocampus, and other limbic structures (Cain, 1974; and Chapter 5, p. 89). Pure anosmia induced without involving the olfactory bulbs or other olfactory structures (Chapter 6, p. 109) does not impair such inhibitory control, so that olfactory structures certainly have a nonolfactory limbic function as well, and should not be cast in a pure sensory role (Hull and Homan, 1975; Spector and Hull, 1972).

In a behavioral framework, according to the prior experience of genetically selected muricidal rats, the motion pattern of the victim, its size, and its natural or masked odor can accelerate or delay the killing attacks (Thorne and Thompson, 1976; Van Hemel and Colucci, 1973). Moreover, rats will kill newly captured wild mice in preference to docile domestic ones, and the experimental impairment of vision, audition, or vibrissal organs has no effect on the discrimination of wild and domestic mice by muricidal animals; adulteration of the victim with odors enhances the attacks, which may even be directed toward cotton swabs soaked in after-shave lotion (Thorne and Thompson, 1976). This may further suggest that muricidal activity is not subject to any specific type of control.

Muricidal behavior induced by prolonged socioenvironmental isolation in rats which, when housed in groups never displayed such violent aggression toward a mouse introduced into their cage, does not apparently differ from that induced either surgically or by genetic selection. Prolonged isolation, however, induces three different behavioral patterns which singly housed rats develop toward the mouse; and rats displaying these patterns, in addition to classic muricidal behavior, have been termed "indifferent" of "friendly," based on their general characteristics (Valzelli, 1970, 1971*a;* Valzelli and Bernasconi, 1971; Valzelli and Garattini, 1972). Indifferent rats are characterized by an almost complete lack of interest in a mouse introduced into their home cage; they remain motionless in a corner of the cage, while the mouse actively explores either the cage or its occupant. In contrast, friendly rats display a rich and dynamic behavior toward the mouse, with episodes somewhat resembling animals playing among themselves; tremor and uncoordinated activity occur and there are repetitive and compulsive patterns such as picking up the mouse and repeatedly carrying it to a distant area of the cage, or putting it in a rudimentary nest dug in the sawdust. Interestingly, a similar behavior, termed "mouse carrying," has been independently observed and described by other authors as a product of prolonged isolation (Lonowski et al., 1973a, b). Finally, the classic muricide pattern of isolated rats, even with food and water *ad libitum,* consists of constant and repetitive mouse killing by rapid cervical dislocation.

Prolonged isolation provides for 53%, 26%, and 21% of, respectively, "indifferent," "friendly," or overtly muricidal rats within 3 weeks. But after further isolation of 6 weeks or more, the percentage of indifferent animals remains fairly stable, whereas that of friendly animals decreases with time and the number of muricidal rats increases to reach 40% of consistent killers (Valzelli, 1971a; Valzelli and Garattini, 1972). Thus the "friendly" and muricidal behavior patterns seem to be interconnected, which allows the animal to shift from the former behavioral alteration to the latter as a further pathological progression of the process initiated by socioenvironmental deprivation.

It should be noted that 4 or 5 weeks of socioenvironmental deprivation corresponds to almost 3% of the mean life span of a laboratory mouse, and that 7 to 8 weeks represents 6% of the life of a rat. These figures may provide an approximate measure of the intensity of the "psychological" distress that these animals undergo, noting that a particular genetic predisposition is required for the development of such pathologically violent aggression in response to alterations in the normal socioenvironmental context.

Neurochemical Correlates of Isolation in Mice

With regard to the role of monoaminergic neurochemical transmitters in isolation-induced aggression, the speculative positions are usually divided into two main views: these suggest that changes in brain catecholamines or serotonin are responsible for such abnormal behavior (Eichelman and Thoa, 1973; Hodge and Butcher, 1974). However, as stated previously (Chapter 6, p. 97), caution should be exercised in relating any single "specific" neurotransmitter to specific behaviors (Reis, 1974; Valzelli, 1977b); the balance between various neurochemical substances is probably more critical (Hodge and Butcher, 1975). Even an obvious change in a given neurotransmitter should only be taken as an indication that the entire serotonergic or catecholaminergic machinery and the brain structures subserved by these neurotransmitters are inappropriately modulated. Such incorrect modulation in turn reflects upon the functioning of several brain areas and circuits, thereby disrupting their reciprocal balance and leading to a series of disturbed behaviors, rather than selectively affecting a single behavior (Valzelli, 1977b).

Earlier observations suggested that isolation-induced aggression in male mice does not significantly change the level of monoamines in the whole brain or in selected brain areas (Anton et al., 1968; Garattini et al., 1967; Giacalone et al., 1968; Essman, 1971*a;* Valzelli, 1967*a*). However, it was reported that isolated-aggressive mice, as compared to normally housed and nonaggressive control animals, required a significantly longer time to develop a 50% increase in their brain serotonin concentration after treatment with different monoamine oxidase inhibitors (Valzelli, 1967*a;* Welch and Welch, 1971). This biochemical change, not observable in isolated but nonaggressive female mice, had been interpreted as a decreased brain serotonin turnover in aggressive mice, and 13 years ago it was pointed out that this metabolic change is not necessarily responsible for aggression, but is simply a concomitant of the isolation procedure (Valzelli, 1967*a, b*).

Further experiments have shown that there is decreased serotonin turnover either in the whole brain (Essman, 1968, 1969; Garattini et al., 1967, 1969; Giacalone et al. 1968; Valzelli, 1967*a;* Valzelli and Garattini, 1968; Welch and Welch, 1966, 1971) or in the diencephalon and corpora quadrigemina of isolated-aggressive male mice (Garattini et al., 1967, 1969; Giacalone et al., 1968; Valzelli, 1967*a, b*). Serotonin turnover is not decreased in female mice or in those strains of mice which do not become aggressive after prolonged isolation (Garattini et al., 1969; Valzelli, 1977*b*, 1978*d*). In the latter instance, the magnitude of the decreased brain serotonin turnover has recently been reported to correlate inversely with the tendency of the strain to respond with aggression to prolonged isolation (Valzelli and Bernasconi, 1979).

In view of recent evidence that physiological or pharmacological alterations in brain tryptophan concentration result in changes of brain serotonin synthesis (Curzon and Fernando, 1976; Jacoby et al., 1975; Tagliamonte et al., 1971*a, b*), it is interesting that prolonged isolation is capable of either abolishing significant diurnal rhythms in brain tryptophan in normally housed mice or significantly decreasing tryptophan level in the brain of aggressive animals (Miller et al., 1979). This finding may be considered as an earlier stage of the biochemical sequence leading to decreased brain serotonin turnover in isolated-aggressive mice; it may also provide further relevance for the association between aggression and malnutrition (Chapter 5, p. 78), especially as concerns genetic and developmental variations in brain tryptophan hydroxylase activity in mice (Diez et al., 1976).

These data seem to indicate that a decrement of serotonergic function correlates fairly well with increased aggression, probably as a consequence of decreased inhibitory control of those behaviors which are normally reflected as a balance between serotonin and other biologically active brain chemicals (Hodge and Butcher, 1974). This speculative position is also supported by the finding the lesions of the midbrain raphe nuclei of male mice, induced prior to socioenvironmental deprivation, significantly facilitate aggression (Valzelli, 1977*b*). Similarly, an increase of isolation-induced aggression has been recently shown in mice treated with *p*-chlorophenylalanine (Matte and Tornow, 1978). Conversely, the potentiation of serotonergic activity following the administration of tryptophan, 5-hydroxytryptophan, monoamine oxidase inhibitors, and serotonin receptor stimulants reduces or abolishes aggression in isolated mice (Hodge and Butcher, 1974; Uyeno and Benson, 1965; Valzelli et al., 1967; Welch and Welch, 1969; Yen et al., 1959). Tryptophan and serotonin administration also suppresses aggression elicited by intrahypothalamic microinjections of acetylcholine (Allikmets, 1974).

Prolonged isolation also induces changes in brain catecholamine metabolism of aggressive mice in addition to decreased serotonin turnover; there is also a minor but significant decrease in norepinephrine turnover and a substantial increase in dopamine turnover (Valzelli, 1971a, 1973a). A reduction of norepinephrine turnover rate following prolonged isolation was also found by Welch and Welch (1968), and a reduced rate of serotonin and catecholamine synthesis was further reported (Modigh, 1973, 1974). Intense fighting, as a basis for a stressful situation, induces a rapid acceleration in turnover of the three monoamines (Modigh, 1973), a lowered affinity of norepinephrine reuptake into nerve endings of the cerebral cortex (Henley et al., 1973), and an increase in adrenal medullary enzymes (Maengwyn-Davies et al., 1973).

Intraventricular administration of 6-hydroxydopamine (Chapter 6, p. 104) to isolated male mice has been found to selectively facilitate aggression in those animals which, despite prolonged isolation, did not show aggressive activities; the increased aggression in these animals was inversely correlated with the magnitude of brain norepinephrine and dopamine depletion (Pöschlová et al., 1976). The lack of aggression-stimulating effect of 6-hydroxydopamine in fully aggressive mice has been tentatively explained as related to the limited space for the display of further aggression (Pöschlová et al., 1976). Another possibility may reside in an altered neurometabolic baseline and neuronal sensitivity to catecholamines for isolates aggressive and nonaggressive mice as a consequence of isolation (Pirch and Rech, 1968; Valzelli, 1967a; Welch and Welch, 1968); these may differentially modulate the effect of 6-hydroxydopamine. This experiment may also suggest that a prevailing serotonergic activity is required for the development of aggression, consistent with findings that the destruction of serotonergic terminals in fully aggressive mice, after intraventricular administration of 5,6-dihydroxytryptamine (Chapter 6, p. 103) or by lesions of the midbrain raphe nuclei, abolishes the aggressive display (Kostowski and Valzelli, 1974; Pöschlová et al., 1976). The same result was also obtained by treating aggressive mice with p-chlorophenylalanine, a selective depletor of brain serotonin, or a series of different and selective serotonergic receptor inhibitors (Malick and Barnett, 1976).

These findings conflict with the aforementioned antiaggressive activity of serotonergic compounds (p. 131), but indirectly agree with other experiments indicating a high correlation between brain norepinephrine depletion and a lowering of isolation-induced aggression in male mice after 6-hydroxydopamine administration (Crawley and Contrera, 1976). These data, in turn, are in disagreement with those already reported by Pöschlová and her associates. Thus the entire picture appears too confusing, so to justify the conclusion that isolation-induced aggression in mice has no substantial relationship with any clearly detectable metabolic change in brain serotonin or catecholamines (Goldberg et al., 1973; Rolinskí, 1975).

Nevertheless, several possible explanations for apparently conflicting results in this field occur depending on: (1) different techniques for testing and assessing aggression, i.e., isolated mice against other isolated mice, naive mice, or "standard opponents," either trained "winners" or "losers" (Brain and Poole, 1974; Kršiak and Janků, 1969; Valzelli, 1969), and others; (2) annual variations in the intensity of aggression (DaVanzo et al., 1965); (3) the number of biochemical determinations performed which may alter statistical significance when slight differences are involved; (4) different techniques employed to measure brain monoamine turnover (Morot-Gaudry et al., 1974); (5) drug interaction, which may interfere with results when multiple chemicals are used; (6) the bimodal effect (Chapter 5, p. 89) that the same experimental manipulation may develop according to earlier or later application with respect to the period of isolation (Valzelli, 1977b); (7) the age of the isolated animals, which may affect either the intensity of aggression or brain monoamine dynamics (Bernard and Finkelstein, 1975); and (8) the annual and daily changes in brain monoaminergic activity of mice and rats (Chapter 6, p. 99; and Valzelli and Garattini, 1968; Valzelli et al., 1977).

The possible relationship between isolation-induced aggression and the related changes of serotonin and dopamine metabolism in the brain appears promising. As previously reported (p. 131), isolated aggressive mice have a reduced serotonin turnover and a simultaneously increased dopamine synthesis; the result is an altered balance between brain serotonergic and dopaminergic regulation in favor of the latter. This may imply that a prevalence of dopaminergic activity is involved in sustaining isolation-induced aggression, similarly to normal irritative aggression (Chapter 6, p. 104, and Table 4, p. 107). This hypothesis has been advanced by Lycke and his associates (1969), who observed an increased brain dopamine and a decreased serotonin in encephalitic mice infected with herpes simplex virus and becoming concomitantly aggressive. A similar conclusion was reached by others for aggression and other behavioral changes induced by multiple drug treatment (Benkert et al., 1973a, b) and increased dopaminergic activity in the striatal structures of isolated aggressive mice (Hutchins et al., 1974, 1975). Furthermore, other drug studies involve a prevailing dopaminergic component as responsible for the induction of violent rage and aggression in mice (Everett, 1968; Hasselager et al., 1968; Rolinski, 1974; Ross and Ögren, 1976; Yen et al., 1970).

Other neurochemical changes are also present in isolated aggressive mice; these interact to a varied extent in sustaining either aggression or other behavioral changes produced by socioenvironmental deprivation (Valzelli, 1973a). In this context, N-acetyl-L-aspartic acid, an amino acid present in large amounts in the brain of mammals and birds (Okumura et al., 1960; Tallan, 1957), and probably involved in the synthesis of brain lipids (D'Adamo and Yatsu, 1966), was significantly decreased in whole brain and in discrete brain areas of isolated aggressive male mice; its concentration was unchanged in isolated and nonaggressive female mice (Marcucci et al., 1968) or in male rats that did not become muricidal after prolonged isolation (Marcucci and Giacalone, 1969). The interest of this finding is in that 5-hydroxytryptophan or monoamine oxidase inhibitor administration increases the brain concentration of N-acetyl-L-aspartic acid, which is decreased by reserpine or lysergic acid diethylamide (LSD) (McIntosh and Cooper, 1964, 1965). This suggests a possible correlation between altered brain serotonin metabolism in isolated aggressive male mice and the related decrease of brain N-acetyl-L-aspartic acid concentration (Garattini et al., 1969; Marcucci et al., 1968).

A large increase in O-phosphoryl-ethanolamine concentration in the limbic structures and in the cerebellar cortex of isolated aggressive mice has been recently described (Essman, 1974), associated with a reduction in the ratio of bound to free acetylcholine pools in the cerebral cortex. This change occurred as a consequence of increased acetylcholine esterase activity (Essman, 1971*b*, *c*) and in the absence of obvious changes in brain choline acetylase activity (Consolo and Valzelli, 1970). Furthermore, isolated mice have a significant elevation in the RNA content of glial fraction from cerebellar tissue, and an equivalent significant decrease in neuronal RNA (Essman, 1971*b*); in the mesencephalon, diencephalon, and cerebellum, DNA content in isolated aggressive mice is reduced (Essman, 1971*a*).

Other changes associated with socioenvironmental deprivation include a decreased content of brain γ-aminobutyric acid (Lal et al., 1972), probably depending on a reduced "binding" capacity of this inhibitory neurochemical substance (DeFeudis, 1972*c*); a decrease in brain protein content (DeFeudis, 1972*a*) as a consequence of a reduced protein synthesis (Bittman and Essman, 1970; Essman and Essman, 1969), and a decreased brain microsomal fraction (Essman et al., 1972); a decreased brain water content (DeFeudis, 1972*d*); and finally a markedly decreased utilization and incorporation of carbohydrates into the brain (DeFeudis, 1971, 1972*b;* DeFeudis and Black, 1972); this finding is suggestive of the previously noted relationship between hypoglycemia and aggression in man (Chapter 6, p. 121).

The available information does not permit any cause-effect relationship to be drawn with regard to any single neurotransmitter or brain substance responsible for the violent aggression induced by prolonged socioenvironmental deprivation in mice; it is also clear that a wide series of endogenous neurochemical changes induce the brain of isolated mice to operate on a modified functional level. It follows that the behavioral repertoire, including aggression, of such altered and severely impaired brain function produced by socioenvironmental deprivation, is clearly abnormal.

Thus, mice and other living organisms reared in isolation for long periods may be considered neurochemically and behaviorally different from normally

TABLE 6. *Neurochemical correlates of aggression by prolonged socio-environmental depriva-tion in male mice from some strains*

Brain neurochemical parameters	Variations		
	Unchanged	Increased	Decreased
Tryptophan content			*
Serotonin content	*		
turnover			*
Norepinephrine content	*		
turnover			*
Dopamine content	*		
turnover		*	
Acetylcholine content			*
GABA content			*
Glutamic acid content	*		
Aspartic acid content	*		
N-acetyl-L-aspartic acid content			*
O-phosphoryl-ethanolamine content		*	
DNA content			*
Neuronal RNA content			*
Glial RNA content		*	
Protein content			*
synthesis			*
Carbohydrate incorporation			*
Water content			*

housed members of the same species. The numerous neurochemical correlates of aggression induced by social isolation in male mice are listed in Table 6.

Neurochemical Correlates of Isolation in Rats

There is evidence that socially isolated rats develop either behavioral disinhibi-tion or neophobia (Morgan et al., 1975), and become overtly hyperemotional in a variety of experimental situations (Ader, 1965; Gill et al., 1966; Griffiths, 1960; Hatch et al., 1963; Moyer and Korn, 1965; Reynolds, 1963). There is an analogy between some of the effects induced by social isolation in the rat and those obtained by disinhibitory brain surgery (Morgan and Einon, 1975); the latter is known to result in hyperemotionality, hyperactivity, increased aggres-sion, and even muricidal behavior (Colpaert and Wiepkema, 1976; Karli et al., 1969; Miczek and Grossman, 1972). In addition, highly emotional rats have a decreased forebrain serotonin turnover (Rosecrans, 1970).

In male rats from the Wistar strain rendered muricidal as a consequence of prolonged socioenvironmental deprivation, a decreased forebrain serotonin turn-over has been demonstrated (-24%); additionally, norepinephrine turnover was increased ($+50\%$), and no substantial changes in dopamine metabolism occurred (Valzelli and Garattini, 1972). In male isolation-induced muricidal rats of the Long-Evans strain, an increase of forebrain norepinephrine turnover ($+26\%$;

Goldberg and Salama, 1969) was observed; this has been further described as temporarily related to the killing episodes (Salama and Goldberg, 1973) in the absence of significant changes of brain serotonin metabolism. Nevertheless, although not statistically significant, a decreased trend in forebrain serotonin turnover (-16%) is also present in muricidal Long-Evans rats (Goldberg and Salama, 1969); in an integrated and tentative biochemical evaluation, decreased brain serotonergic activity coincident with increased noradrenergic activity may represent a more consistent model than any isolated or absolute change in the metabolism of any individual brain amine.

Male adult mice reared in prolonged isolation show no variations in brain amine content (Goldberg and Salama, 1969; Nishikawa et al., 1976; Salama and Goldberg, 1973; Valzelli and Garattini, 1972), whereas other types of altered behavior previously mentioned—that displayed by "friendly" and "indifferent" rats (p. 130)—are accompanied by different combinations of altered brain monoaminergic activity (Valzelli and Garattini, 1972). A decreased serotonin turnover occurs only in those rat strains susceptible to muricide following isolation (Valzelli, 1971*a*, 1978*d*).

Consistent with the metabolic changes observed in muricidal rats, increased tyrosine hydroxylase activity has been found in the midbrain and striatal nuclei of isolated rats with a significant decrease in tryptophan hydroxylase activity in the septal area (Segal et al., 1973). This finding should be regarded as important, especially in view of the inhibitory role of the septal nuclei in the regulation of aggression (Chapter 3, p. 57; and Table 1, p. 59, Table 2, p. 91, and Table 3, p. 95) and of their innervation by the serotonergic pathways (Ungerstedt, 1971; Valzelli, 1979*a;* and Figure 25).

These experiments therefore suggest that decreased serotonergic control, favoring an enhanced brain catecholaminergic prevalence, correlates with environmentally induced muricidal behavior. It has also been found that the degeneration of brain serotonergic terminals induced by 5,6- or 5,7-dihydroxytryptamine (Baumgarten and Lachenmayer, 1972; Baumgarten et al., 1972*a, b, c*) produces or enhances hyperirritability and muricidal behavior (Breese et al., 1974*b;* Marks et al., 1978; Paxinos and Atrens, 1977; Vergnes et al., 1977). Further, Koe and Weissman (1966) observed that normal rats injected with *p*-chlorophenylalanine, a potent inhibitor of brain serotonin synthesis, showed hyperirritability and occasional overt aggression when handled. *p*-Chlorophenylalanine administration has also been reported to markedly increase muricidal behavior (McLain et al., 1974; Sheard, 1969), and to potentiate the effect of either prolonged social isolation or olfactory bulbectomy in producing mouse killing by rats (Di-Chiara et al., 1971; Miczek et al., 1975; Paxinos et al., 1977; Vergnes et al., 1974*b*).

Olfactory bulb ablation, by which most of normal laboratory rats are made muricidal, produces a lowering of serotonin content in the amygdala (Karli et al., 1969, 1972). The same result is obtained by lesions of the midbrain raphe nuclei (Vergnes et al., 1974*b*), from which the serotonergic projections to the

TENTATIVE RECONSTRUCTION OF THE
BRAIN SEROTONERGIC PROJECTIONS

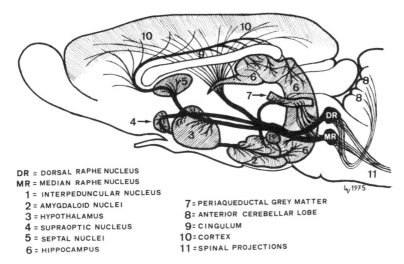

DR = DORSAL RAPHE NUCLEUS
MR = MEDIAN RAPHE NUCLEUS
1 = INTERPEDUNCULAR NUCLEUS
2 = AMYGDALOID NUCLEI
3 = HYPOTHALAMUS
4 = SUPRAOPTIC NUCLEUS
5 = SEPTAL NUCLEI
6 = HIPPOCAMPUS
7 = PERIAQUEDUCTAL GREY MATTER
8 = ANTERIOR CEREBELLAR LOBE
9 = CINGULUM
10 = CORTEX
11 = SPINAL PROJECTIONS

FIG. 25. Rat's brain serotonergic projections.

forebrain arise (Fig. 18). Furthermore, in those rats not rendered muricidal after olfactory bulbectomy, raphe nuclei lesions induce mouse killing in 73% of the treated animals, and this figure reaches 91% when p-chlorophenylalanine is additionally administered to experimental animals (Vergnes et al., 1974a).

Like p-chlorophenylalanine administration, 5,6- or 5,7-dihydroxytryptamine administration, and septal nuclei lesions, midbrain raphe nuclei lesions also increase overall reactivity to environmental stimuli in laboratory rats (Kostowski et al., 1968; Lorens et al., 1971; Srebro and Lorens, 1975). Midbrain raphe nuclei lesions may induce muricidal behavior in previously normal rats (Banerjee, 1974, Barr et al., 1976b; Vergnes and Penot, 1976; Vergnes et al., 1973, 1974a, b). Moreover, muricidal behavior produced by prolonged socioenvironmental deprivation is greatly enhanced by lesions of the midbrain raphe nuclei (Grant et al., 1973; Popova et al., 1975; Yamamoto and Ueki, 1977).

Most studies consistently and clearly indicate that muricidal behavior is enhanced when neural serotonergic inhibitory control is either environmentally or surgically and chemically impaired.

This interpretation is strongly supported by a large number of experiments showing that the potentiation of serotonergic function, through administration of tryptophan, 5-hydroxytryptophan, monoamine oxidase inhibitors, or serotonin receptor stimulants, reduces or abolished muricidal aggression (Barr et al., 1976b;

Bocknik and Kulkarni, 1974; DiChiara et al., 1971; Goldberg and Horovitz, 1978; Hodge and Butcher, 1974; Kulkarni, 1968*b;* Kulkarni et al., 1973; Marks et al., 1978; Paxinos et al., 1977; Popova et al., 1975; Rewerski et al., 1971; Sheard, 1969; Valzelli and Bernasconi, 1976; Vergnes et al., 1974*a*).

Another neurochemical factor that interacts with impaired serotonergic control in sustaining muricidal aggression in laboratory rats raised in prolonged isolation is altered cholinergic activity. It has been reported that lesions of the midbrain raphe nuclei induce a decreased forebrain acetylcholine content (Pepeu et al., 1974), and that muricidal rats have lowered cortical acetylcholine (Yoshimura et al., 1974) together with increased amygdaloid and brainstem acetylcholine content (Ebel et al., 1973; Yoshimura and Ueki, 1977). Moreover, pharmacological evidence indicates that the cholinergic drug pilocarpine either induces or enhances muricidal activity (Igić et al., 1970; Leaf et al., 1969; McCarthy, 1966; Vogel and Leaf, 1972; Wnek and Leaf, 1973), as does amygdaloid implantation of eserine (Leaf et al., 1969) or administration of muscarinic cholinomimetic or anticholinesterase drugs to the lateral hypothalamus (Bandler, 1969, 1970; Smith et al., 1970). However, other cholinergic agents—including nicotine, oxytremorine, arecoline, neostigmine, and methacholine—are ineffective over a broad dose range (Wnek and Leaf, 1973), thus not clearly establishing a definite role for the cholinergic system in muricidal aggression.

Finally the level of GABA in the olfactory lobes of muricidal rats has been found to be markedly decreased as compared with normal, nonmuricidal rats (Mandel et al., 1975).

In conclusion, as compared with various neurochemical correlates of environmentally induced aggression in male mice, the neurochemistry of muricidal aggression in socially deprived rats appears much less confused; it may be almost entirely interpreted as dependent on impairment of central inhibitory control, especially as a result of a serotonergic failure (Table 7).

TABLE 7. *Neurochemical correlates of muricidal aggression by prolonged socio-environmental deprivation in male rats from some strains*

Brain neurochemical parameters	Variations		
	Unchanged	Increased	Decreased
Tryptophan-hydroxylase activity			*
Serotonin content	*		
turnover			*
Tyrosine-hydroxilase activity		*	
Norepinephrine content	*		
turnover		*	
Dopamine content	*		
turnover	*		
Acetylcholine content			*
GABA content			*

SURGICALLY INDUCED AGGRESSION

Several data concerned with this issue have already been provided in Chapter 5. Even though some overlap with previous discussion is almost unavoidable, the purpose here is to specifically deal with those surgical manipulations capable of converting normal laboratory rats into muricidal aggressors; the neurochemical changes possibly taking place in operated animals will also be considered. Needless to say, a minority of those rats of some strains which spontaneously kill mice, independent of any experimental intervention (Karli, 1956), will only be incidentally examined.

Initially it should be observed that this interesting field of research has been almost completely studied through a wide series of important contributions of Karli and his associates. More than 16 years ago, Vergnes and Karli (1963*b*) reported that olfactory bulb ablation in previously normal and nonkiller laboratory rats was capable of inducing an overt and stable mouse-killing aggression in a percentage of both male and female animals; this effect varied according to the age at which surgery was performed (Didiergeorges et al., 1966*a*). Further studies have shown that lesions of other olfactory structures, such as the lateral olfactory tracts and partially of the praepyriform cortex, are also capable of inducing muricidal activity in normal rats (Didiergeorges and Karli, 1966; Didiergeorges et al., 1966*b*; Karli and Vergnes, 1963; Vergnes and Karli, 1969*a*). Interestingly, when olfactory bulbectomized adult rats are raised in social isolation, the percentage of consistent mouse-killing animals increases from 19 to 75 (Vergnes and Karli, 1969*a*); this suggests a positive interaction between olfactory lesions and socioenvironmental deprivation in the production of muricidal aggression (Bernstein and Moyer, 1970).

The efficacy of olfactory bulb ablation in enhancing muricidal activity has been confirmed by others (Alberts and Friedman, 1972; Bandler and Chi, 1972; Cain, 1974; Malick, 1970; Thorne et al., 1973, 1974), although the effectiveness of surgery may vary according to the strain of rats employed (Thorne et al., 1973) and the extent of the lesion (Cain, 1974; Thorne et al., 1974). In contrast, olfactory bulbectomy either abolishes some types of spontaneous aggression (Chapter 5, p. 89) or suppresses spontaneous mouse-killing activity in spontaneous killer rats (Bandler and Chi, 1972; Thompson and Thorne, 1975). This reflects upon the difference between spontaneous and induced muricidal aggression among laboratory rats. Nevertheless, recently olfactory bulbectomy has been reported not to effect a statistically significant increase in the number of muricidal rats as compared with spontaneous killers already present in a colony of laboratory animals (Thorne et al., 1978); this contrasting result probably depends on the relatively small numbers of animals compared (only 10 per group) and on technical differences from previous experiments (housing, extent of the lesion, etc.).

Olfactory bulbectomy induces a series of neurochemical changes consisting mainly of: (1) decreased and increased norepinephrine levels in the telencephalon and brainstem, respectively (Pohorecky et al., 1969), without significant modification of serotonin and dopamine levels (Eichelman et al., 1972b); (2) decreased cortical acetylcholine level (Yoshimura et al., 1974) with an increased concentration in the amygdala and brainstem (Ebel et al., 1973; Yoshimura and Ueki, 1977); and (3) increased choline acetyltransferase activity in the amygdala and pyriform cortex (Ebel et al., 1973; Mandel et al., 1975).

Electrical stimulation of the posterolateral hypothalamus facilitates mouse-killing activity in natural killers, whereas muricidal aggression can be elicited in nonkiller rats only by stimulating the anterior hypothalamus between the optic tracts and fornix (Koolhas, 1978; Vergnes and Karli, 1969b; Woodworth, 1971), or the dorsomedial and ventromedial hypothalamus (Malick, 1970; Vergnes and Karli, 1970). Interestingly, these structures normally trigger irritative aggression (Table 2, p. 91). Conversely, bilateral ventromedial hypothalamic lesions induce mouse-killing activity in a minority of laboratory rats (Eclancher and Karli, 1971; Panksepp, 1971), spontaneous muricidal aggression instead being attenuated by lateral hypothalamic lesions (Panksepp, 1971).

Another important anatomical structure involved in the regulation of muricidal behavior is the amygdala (Karli, 1955, 1974; Karli and Vergnes, 1964a, 1965a, b; Karli et al., 1975; Vergnes, 1975; Vergnes and Karli, 1964, 1965). Bilateral lesions of the amygdala abolish either spontaneous or olfactory bulbectomy-induced muricidal aggression in adult rats. Lesions of the lateral or cortico-basal amygdaloid nuclei do not influence mouse-killing activity; this is inhibited instead by lesions of the centromedial amygdala, from which ventral amygdalofugal fibers extend to the lateral hypothalamus. However, total and bilateral amygdala lesions produce a completely opposite effect depending on the age at which they are induced; this probably relates to the role that the amygdala plays in the maturation of emotional experience and social interaction. On this basis, bilateral amygdaloid lesions in rats 7 to 8 days old induce adult mouse-killing aggression in 90% of operated animals as compared with a 9% incidence among sham-operated and spontaneously evolving killer rats; when lesions are induced in 25-day-old rats, muricidal activity occurs to a somewhat lesser extent (62%; Eclancher et al., 1975).

Nonepileptogenic stimulation of the hippocampus, nonepileptogenic amygdaloid kindling, and fornix lesions do not consistently affect established muricidal behavior (DeCastro and Marrone, 1974; McIntire, 1978; Vergnes and Karli, 1968).

Septal lesions have been reported to facilitate killing in already killer rats, while such ablations do not induce muricidal aggression in preoperatively nonkiller rats (Karli, 1960; Karli et al., 1969; Malick, 1970). However, specific lesions of the posterior ventromedial septal nucleus have been shown to reliably induce muricidal aggression in previously nonkiller rats (Miczek and Grossman, 1972; Siegel and Leaf, 1969); the outcome of surgery appears dependent on either the extent of the lesion (Wallace and Thorne, 1978) or the strain of animals used in the experiments (Latham and Thorne, 1974), as previously observed with olfactory bulbectomy (Thorne et al., 1973). Furthermore, temporary lesions of the ventral anterior septum, produced by the intracranial infusion of a local anesthetic, induces irritability, hyperreactivity to handling, and muricidal aggression; the latter is especially induced by inactivation of the region posterior to the ventral septum and including the medial preoptic area, the anterior hypothalamic area, and the bed nucleus of the stria terminalis (Albert and Wong, 1978). Interestingly, septal lesions were reported to decrease forebrain serotonin and acetylcholine concentration (Lorens et al., 1970; Pepeu et al., 1971; Sorensen and Harvey, 1971).

Other anatomical structures related to muricidal aggression are represented by the ventral mesencephalic tegmentum, lesions of which produce a transient abolition of mouse-killing activity in already established muricidal rats (Chaurand et al., 1973). Destruction of the mesencephalic central gray matter induces a facilitation of muricidal aggression in killer rats and initiates muricidal activity in only a small proportion of previously nonkiller rats (Chaurand et al., 1972). Finally, in addition to the reported effect of lesions of the midbrain raphe nuclei on muricidal behavior (p. 137), lesions confined to the dorsal raphe are sufficient to produce muricide in naturally nonkiller rats (Waldbillig, 1979). Instead, bilateral lesions of the nuclei coerulei, from which most of noradrenergic fibers to the forebrain originate (Dahlström and Fuxe, 1964; Korf et al., 1973; Loizou, 1969), while increasing irritative pain-elicited and apomorphine-induced aggression, do not affect already acquired mouse-killing behavior or induce muricidal aggression in nonkiller rats (Kostowski et al., 1978).

In summary, the anatomical web involved in muricidal aggression, described in previous experiments, does not support the interpretation of this kind of violent behavior as a laboratory model for natural predation (Chaurand et al., 1972; Karli and Vergnes, 1964b; Vergnes and Karli, 1969b). Moreover, the anatomical profile of muricidal aggression (Table 8) is interestingly different from that of other kinds of aggression (Table 2, p. 91); the possibility of experimentally inducing such a violent pattern depends on either a genetic and strain-dependent predisposition or socioenvironmental components.

DRUG-INDUCED AGGRESSION

This topic has been the subject of a large number of studies which have been reviewed recently (Kršiak, 1974; Valzelli, 1978b, 1979c). It has been said that aggressive responses initiated or induced by drug administration require high and nearly toxic doses or chronic administration, and are so different from the normal kinds of aggression that they should be regarded as pathological expressions of behavior (Miczek, 1977). In addition, several drugs may have either facilitative or inhibitory effects on aggression, depending on dose level, socioenvironmental characteristics, and emotional baseline of the animals (Kršiak, 1974; Miczek, 1977; Valzelli, 1977a, 1979c; Valzelli and Bernasconi, 1971, 1976); this largely explains why the same drug may exert opposite effects

TABLE 8. *Anatomical structures involved in muricidal aggression by laboratory rats*

Muricidal aggression			
Induced by	Increased by	Decreased by	Suppressed by
—Olfactory bulb ablation	—Anterior hypothalamus stimulation	—Ventral midbrain tegmentum lesion	—Olfactory bulb ablation°
—Lateral olfactory tract lesion	—Ventromedial hypothalamus lesion		—Lateral hypothalamus lesion
—Preaepyriform cortex lesion	—Posterolateral hypothalamus stimulation		—Centromedial amygdala lesion
—Amygdala lesion	—Septal nuclei lesion		
—Ventromedial septal lesion	—Periacqueductal gray matter lesion		
—Dorsal raphe lesion	—Midbrain raphe nuclei lesion		

(°) In spontaneous muricides.

when one or more of these variables are altered. Nevertheless, due to the large extent of available data on this subject, attention will here almost exclusively be focused on those drugs which are objects of consummatory use, abuse, or habit by humans.

High to lethal doses of *amphetamine* were observed to induce wild bursts of running, fighting, and biting in normally grouped mice (Chance, 1948; Moore, 1963; Randrup and Munkvad, 1967). Such amphetamine-induced attacks were markedly more common in grouped male than female mice (Lapin and Samsonova, 1964). Both the excitatory and toxic effects of amphetamine are further increased by elevated environmental temperature (Hon and Lasagna, 1960), noise (Chance, 1946, 1947), painful stimuli (Weiss et al., 1961), the number of animals in a cage (Burn and Hobbs, 1958; Chance, 1946; Chernov et al., 1966), and reciprocal body contact (Cohen and Lal, 1964); the grouping of the animals also significantly reduces the brain drug detoxification process (Consolo et al., 1965*c*). Furthermore, isolated-aggressive mice have been shown to be much more sensitive to amphetamines and the toxicity of other central stimulants than normally grouped and nonaggressive mice (Consolo et al., 1965*a*, *b*).

Different strains of mice also show striking differences in amphetamine sensitivity (Dolfini et al., 1969*a*, *b*). Amphetamine is known to be more active in hyperthyroid animals (Askew, 1962; Dolfini and Kobayashi, 1967; Moore, 1965, 1966), and a triiodothyronine-induced hyperthyroidism greatly increases the sensitivity to amphetamine of "resistant" strains (Dolfini et al., 1970). Depending on so many variables, it is not surprising that amphetamine has been found to increase (Charpentier, 1969), decrease (Melander, 1960), or have no effect (Valzelli et al., 1967) on the aggression induced by prolonged isolation in male mice. As previously suggested, isolated-aggressive animals develop an increased central excitability; small doses of central stimulant drugs may induce a greater activity, and larger doses may apparently decrease aggression as a result of overexcitation leading to the disorganization of behavior (Valzelli, 1967*b;* Welch and Welch, 1969). In other words, there may be an optimal amphetamine dosage which, depending on the aforementioned circumstances, stimulates aggression, whereas higher doses elicit other behavioral activities which represent substitutes and mask aggression (Kršiak, 1974). Aside from dosage, amphetamine-induced aggression should also be viewed as dependent on the initial or baseline level of central excitability (Hoffmeister and Wuttke, 1969; Stille et al., 1963).

Amphetamine-induced aggression has generally been ascribed to an increased brain dopaminergic activity exerted by the drug (Fog et al., 1970; Hasselager et al., 1972; Roliński, 1974; Terada and Masur, 1973), similar to the aggressiveness induced by apomorphine, a drug which directly stimulates brain dopaminergic receptors (Andersen et al., 1975; Gotsick et al., 1975; McKenzie, 1971; Patni and Dandiya, 1974; Senault, 1970, 1971, 1972, 1973, 1974, 1976). Interestingly, apomorphine activity, which reliably induces profound, long-lasting aggression in 35 to 52% of treated rats (Senault, 1970, 1971), is enhanced by social isolation to increase aggression to 73% (Senault, 1971); it is also influenced by dose level (Senault, 1970), previous experiences (Gotsick et al., 1975), and the age of treated animals (McKenzie, 1971).

Other drugs that stimulate the central nervous system have been reported to facilitate aggressiveness and even to elicit self-directed aggression. *Magnesium pemoline* has been shown to induce self-biting of the skin of the thorax and abdomen, and automutilation of the paws and tail in mice and rats of both sexes (Genovese et al., 1969; Tedeschi et al., 1969).

Among the xanthine derivatives, different responses to *caffeine* administration can be observed in laboratory animals depending on their age and sex (Peters and Boyd, 1967*a*), a reduced food intake (Peters, 1966), the type of diet (Peters and Boyd, 1966, 1967*b*), and the time of the day at which the drug is administered (Valzelli, 1977*c*). Chronic administration of sustained doses of caffeine induce self-aggression and automutilation in previously normal rats (Boyd et al., 1965; Peters, 1967; Pfeiffer and Gass, 1962; Tonini et al., 1963), whereas acute administration consistently decreases isolation-induced aggression in mice (Valzelli and Bernasconi, 1973) and blocks muricidal aggression in rats, as do other central stimulants (Valzelli and Bernasconi, 1971, 1976). From a biochemical standpoint, caffeine has been reported to cause a sensitization of brain catecholamine receptors (Waldeck, 1973), to increase brain norepinephrine synthesis (Berkowitz et al., 1970), and to increase brain serotonin content while decreasing either brain serotonin or dopamine synthesis (Corrodi et al., 1972; Valzelli and Bernasconi, 1973). However, the decrease of brain serotonin turnover induced by caffeine is more evident in normal mice (-38%) than in aggressive mice (-23%;

Valzelli and Bernasconi, 1973). Furthermore, consistent with the increase of brain serotonin level (+16%), it has been recently shown that caffeine increases brain tryptophan content up to 73% (Valzelli et al., 1979).

Repeated administration of high doses of *theophylline* to normally grouped rats results in a sequence of behavioral changes ranging from initial stereotypy to increased aggression and the appearance of muricidal activity and automutilation (Sakata and Fuchimoto, 1973*a*, *b*). In addition to displaying muricidal aggression, theophyllinized killers vigorously attack animate and inanimate objects, and also display intense pup-killing activity without eating the "prey" (Sakata and Fuchimoto, 1973*a*).

When the strain of rats is changed, the muricidal effect induced by theophylline is no longer observed, but the facilitation of pain-elicited irritative aggression remains intact (Eichelman et al., 1978). Drug-induced modifications in brain chemistry do not substantially differ from those already described for caffeine (Berkowitz and Spector, 1971; Berkowitz et al., 1970, 1971; Sakata et al., 1975), and the overall activity of these xanthine derivatives has been largely interpreted in terms of prevalent central adrenergic activity (Eichelman et al., 1978).

Alcohol is classically considered a central nervous system depressant (Crossland, 1970), and, although it has been shown to reduce the rapidity and the consistency of some behavioral patterns, it has been also described as having stimulating properties at a low dose level and under certain circumstances (Barry and Miller, 1962; Barry et al., 1962; Chesher, 1974). In this context, moderate doses of alcohol (0.18%) have been reported to enhance, and high doses (0.35%) to decrease aggression in fish (Peeke et al., 1973; Raynes and Ryback, 1970; Raynes et al., 1968; Ryback et al., 1969); increasing doses of alcohol have also been shown to greatly increase the latency of attacks on rats by hypothalamically stimulated cats (MacDonnell and Ehmer, 1969). In rodents, alcohol consumption and its effects are related to a strain-dependent emotional baseline (Ahtee and Eriksson, 1972; Eriksson, 1971, 1972*a*, *b;* Poley, 1973; Poley and Royce, 1972); this also involves genetic differences in brain serotonin level and metabolism (Eriksson, 1972*a*, *b*) upon which alcohol in turn variously acts (Myers and Melchior, 1978).

In a similar respect, it has been recently shown that the adult offspring of female mice given alcohol during the entire gestation period are consistently more aggressive than control offspring; there is also a significant reduction in brain serotonin content among such animals (Kršiak et al., 1977). In this situation, the activity of the catecholaminergic regions of the brain is supported by an increased tyrosine hydroxylase activity in the caudate nuclei of the offspring of female rats given alcohol during pregnancy (Branchey and Friedhoff, 1973). Moreover, with respect to the multiple effects of chronic alcohol intake in the production of malabsorption and malnutrition (Seixas et al., 1975), it is of interest to note that *thiamine*-deficient rats develop severe muricidal aggression together with a disruption of avoidance learning and discrimination acquisition (Vorhees et al., 1975). Furthermore, alcoholism also usually induces a severe thiamine deficiency in man (Victor et al., 1971).

Among the hallucinogenic drugs, lysergide or *lysergic acid diethylamide* (LSD) administered intracerebrally in a low concentration (0.01 mg/kg) induces mice to attack any object with which they are confronted (Haley, 1957); it also increases the number of attacks displayed by isolated-aggressive mice (Kršiak et al., 1971) and the shock-elicited irritative aggression in mice (Kostowski, 1966) and rats (Sheard et al., 1977). Higher doses, however, do not change or even appreciably decrease isolation-induced aggression in mice (Rewerski et al., 1971; Uyeno, 1966a, 1978; Uyeno and Benson, 1965; Valzelli et al., 1967), with no effect on pain-elicited irritative aggression in mice (Sheard et al., 1977). Moreover, similar doses of lysergide block both muricidal and hyperactive behavior of "friendly" animals (p. 130) induced in rats by social deprivation (Valzelli and Bernasconi, 1971, 1976); this effect persists for several hours.

Low doses of lysergide specifically inhibit the spontaneous firing rate of presynaptic neurons of the midbrain raphe region (Aghajanian et al., 1968) and decrease the release of serotonin (Callager and Aghajanian, 1975), thereby leading to a reduction of brain serotonergic activity. Conversely, high doses of LSD inhibit the activity of the postsynaptic cells of the midbrain raphe (Haigler and Aghajanian, 1974), thus mimicking the action of serotonin on postsynaptic receptors and possibly evoking a functional serotonomimetic prevalence. These opposite neurochemical and dose-dependent effects may actually explain the bimodal dose-response activity of lysergide on aggression and other behaviors (Sheard et al., 1977).

Mescaline has been reported to elicit aggression in mice when injected intracerebrally (Haley, 1957) and to decrease an established experimentally induced aggression (Uyeno, 1966b, 1978). Recently, however, mescaline has been shown to be a potent facilitator of shock-induced irritative aggression in rats in a low dose range; higher doses induce powerful qualitative changes in rat aggression, so that treated animals engage in repeated violent attacks and vicious biting of the opponent despite submissive postures or attempts to escape (Sbordone and Carder, 1974). Mescaline-treated rats also exhibit vigorous biting attacks on a variety of targets, but show predominant fighting against an immobile dead member of the species, suggesting that the drug may release aggression from inhibitory control (Carder and Sbordone, 1975). Due to its behavioral age- and strain-independent characteristics of severe viciousness and persistent repetitiveness (Sbordone et al., 1978), the mescaline-induced exasperation of pain-elicited irritative aggression in rats has been proposed as an animal model for pathological aggression (Sbordone, 1976; Sbordone and Garcia, 1977; Sbordone, et al., 1978). As with lysergide, mescaline is classically known to antagonize serotonergic activity and to stimulate the noradrenergic regions of the brain (Crossland, 1970).

Carlini and Kramer (1965) reported that rats chronically injected with a crude extract of *marihuana* become overtly aggressive, attacking each other in their cages. Conversely, marihuana extract was shown to decrease aggression in isolated-aggressive laboratory mice (Salustiano et al., 1966; Santos et al., 1966), while increasing somewhat the tendency toward fighting in isolated male wild house mice (Matte, 1975). In an extended series of further experiments, the chronic administration of either marihuana extract or Δ^9-trans-tetrahydrocannabinol (THC) was shown to elicit aggression in previously nonaggressive laboratory rats; this activity was, however, sharply dependent on a series of variables, among which starvation of the experimental animals was prominent (Carlini, 1978; Carlini and Masur, 1969, 1970). Food deprivation alone, without hypoglycemia, acidosis, or lack of specific nutrients due to starvation, appears to be the factor which facilitates chronic marihuana aggression; such behavior is in turn strongly potentiated by relatively low temperature (14°C) and by rapid eye movement (REM) sleep deprivation (Carlini, 1977, 1978; Carlini et al., 1972, 1977).

Chronic marihuana administration has also been reported to increase shock-elicited irritative aggression in rats (Carlini and Masur, 1969), but only when the animals are unfamiliar with the experimental situation (Carder and Olson, 1972). In addition, the chronic administration of marihuana has been shown to induce muricidal aggression in previously nonkiller rats (Alves and Carlini, 1973; Miczek, 1976, 1979; Segawa et al., 1977; Ueki et al., 1972). In those strains of rats, such as Sprague-Dawley (Miczek, 1979), which are generally resistant to other muricidal-inducing experimental techniques (Thorne et al., 1973; Valzelli, 1971a, 1978d), the muricide-inducing effects of marihuana are also apparent.

Interestingly, *p*-chlorophenylalanine or DOPA administration greatly potentiates marihuana-induced aggression, and such potentiation is even greater when both agents are administered together (Palermo-Neto and Carlini, 1972). A reduction of brain serotonergic activity or a catecholaminergic prevalence, or both together, greatly enhance the aggression induced by marihuana; marihuana has itself been recently shown to induce a significant decrease in brain serotonin turnover in those rats rendered muricidal in consequence of a chronic administration with the drug (Segawa et al., 1977).

A long-lasting and vicious aggression resulting from *morphine* withdrawal was first described by Boshka and his associates (1966), who reported that morphine-dependent rats, no longer receiving the drug, start to fight for over 36 hr when placed together. Similar observations were reported by others (Gianutsos and Lal, 1978; Gianutsos et al., 1973; Lal, 1975; Lal et al., 1971; Thor et al., 1970) and were also noted in female rats (Crabtree and Moyer, 1973), in mice (Iorio et al., 1975), and in guinea pigs (Goldstein and Schulz, 1973), whereas rage reactions and difficulty in handling had been previously described in monkeys during morphine withdrawal (Seevers and Deneau, 1963). In addition, strain differences in the susceptibility to morphine-withdrawal aggression in rats have been described; such aggression has been facilitated in older animals, and seasonal variations have been observed, with higher levels of aggression during the summer and lower aggression in the winter (Borgen et al., 1970; Davis and Khalsa, 1971a). Moreover, other variables that may affect the intensity of morphine-induced fighting relate to dominant-submissive hierarchies (Gellert and Sparber, 1979).

Morphine withdrawal has been further shown to increase shock-elicited irritative aggression (Davis and Khalsa, 1971*b;* Puri and Lal, 1974); this may be dependent on the hyperalgesia which develops when morphine is withdrawn (Tilson et al., 1973). On the whole, these findings support an interpretation of increased fighting during morphine withdrawal in terms of an increase in irritative aggression that correlates initially with hyperalgesia, and then with a hyperirritable state among experimental animals (Gellert and Sparber, 1979).

Several neurochemical data agree with this suggestion. Dopaminergic receptor blockade produces decreased nociception (Männistö and Saarnivaara, 1976; Tulunay et al., 1975), as does catecholaminergic blockade within the amygdala (Rodgers et al., 1976); dopaminergic stimulation, however, results in hyperalgesia (Tulunay et al., 1975). Consistent with these results, morphine analgesia is potentiated by both dopaminergic blockade (Eidelberg and Erspamer, 1975; Paalzow and Paalzow, 1975; Tulunay and Takemori, 1974; Van der Wende and Sporlein, 1973) and a serotonergic prevalence (Samanin and Valzelli, 1971, 1972; Sigg et al., 1958); it is antagonized either by dopaminergic stimulation (Tulunay and Takemori, 1974; Tulunay et al., 1975; Van der Wende and Spoerlein, 1972, 1973) or by impaired serotonergic activity (Samanin and Bernasconi, 1972; Samanin et al., 1970, 1973). Furthermore, altered analgesic effects of morphine in rats have recently been described as dependent on intra-amygdaloid administration of dopamine (decreased analgesia) or serotonin (increased analgesia) (Rodgers, 1977); this suggests that within the amygdala aversive responding is facilitated by dopaminergic activation and inhibited by serotonergic firing (Valzelli, 1978*b*).

Considering the importance of the amygdala in the regulation of different kinds of aggression, the neurochemical changes within it must have a similarly important connection with the strong irritative aggression precipitated by morphine withdrawal. This is further suggested by other studies which have shown an overall brain dopaminergic sensitization with a serotonergic impairment among morphine-withdrawn animals (Gianutsos et al., 1974; Iwamoto et al., 1973; Shen et al., 1970); morphine withdrawal also leads to a decreased brain acetylcholine content (Bhargawa and Way, 1974; Mullin and Phillis, 1974).

In conclusion, several issues have emerged from this rapid review on drug-induced aggression, and these are summarized in Table 9. There appears to be a common ground on which decreased inhibitory control of irritative aggression is a basis for facilitation of a genetic disposition toward such behaviors.

HUMAN CORRELATES OF ANIMAL STUDIES

It seems appropriate to repeat here what has already been said about both the effect of social deprivation on human behavior (Chapter 3, p. 51; and this Chapter, p. 125) and the facilitatory effect of several genetic traits and behavior pathology on violence (Chapter 6, p. 113; and this Chapter, p. 124).

These factors variously interact among themselves to provide an especially fertile ground for violent aggression (Ellinwood, 1967; Sendi and Blomgren,

TABLE 9. *Neurochemical correlates of drug-induced aggression (°)*

Drug	Serotonergic activity	Dopaminergic activity	Noradrenergic activity	Cholinergic activity
Amphetamine		Increased	Increased	
Apomorphine		Increased		Increased
Caffeine	Decreased	Decreased	Increased	
Theophylline			Increased	
Alcohol	Decreased	Increased	Increased	
Lysergide	Decreased	Increased		
Mescaline	Decreased	Increased		
Marihuana	Decreased		Increased	
Morphine withdrawal	Decreased	Increased		

(°) Only for those doses and conditions enhancing aggression.

1975; Steward et al., 1978); on this sensitized ground uncontrolled drug use or abuse can catapult an individual with maladjusted inhibitory control into an act of violence. It has been reported that aggression is produced in man by psychostimulant drugs only after administration of doses that produce paranoid psychoses, and that aggression by drug-intoxicated persons is characteristically a defensive response to frightening delusions (Allen et al., 1975). This further emphasizes the altered arrangement of the mechanisms regulating aggression, as already suggested in point 4 on page 123.

Insofar as amphetamine is concerned, both the dose level and the pre-drug emotional baseline of the subjects are of great relevance to the final outcome, as has been observed in animal experiments. These factors may be considered as the potential basis for violent acts related to amphetamine abuse (Ellinwood, 1971, 1972; Kramer, 1969; Rawlin, 1968; Rickman et al., 1961; Smith, 1969) and for the beneficial effects of amphetamine and amphetamine-like medications in hyperactive children and adults (Allen et al., 1975; Arnold et al., 1972, 1973; Ban, 1969; Conners, 1972; Loney et al., 1978; Richmond et al., 1978).

Hyperactive and aggressive children and adolescents usually respond to stimulant medications with a clear reduction of fighting, defiance, and impulsiveness (Allen et al., 1975). Although *d*-amphetamine and *l*-amphetamine are equally effective in calming antisocial aggressive children, the former is more active than the latter in relieving the "nervousness" of overanxious and hyperkinetic subjects (Arnold et al., 1973). Moreover, based on data from animal studies in this field (Snyder, 1973; Snyder and Meyerhoff, 1973), the differences in effect of the two amphetamine isomers have suggested that childhood aggression may be a dopaminergic-dependent effect, whereas anxiety and hyperactivity could both depend on a noradrenergic mechanism (Arnold et al., 1973).

The psychostimulant effect of caffeine in man includes a number of effects, from wakefulness to alertness and mood elevation, the intensity of which varies from individual to individual (Goldstein et al., 1965*a, b*). Moreover, whereas non-coffee users react adversely to caffeine with jitteriness and nervousness, heavy users usually describe a characteristic set of dysphoric symptoms that develop with caffeine withdrawal; these include irritability, nervousness, and restlessness which change into increased alertness, decreased irritability, and feelings of satisfaction upon caffeine administration (Goldstein and Kaizer, 1969; Goldstein et al., 1969). Although overt violent behavior has never been ascribed to any coffee drinker as related to his habit, one case report described a psychotic reaction with wild, manic screaming, kicking, and biting following the ingestion of large quantities of caffeine (McManamy and Schube, 1936). More recently, the habitual high intake of caffeine, or "caffeinism" (Reimann, 1967), has been shown to produce symptoms indistinguishable from those of anxiety neurosis, such as insomnia, tremulousness, occasional muscle twitching, nervousness, hyperirritability, and anxiety (Greden, 1974; Lutz, 1978); such findings suggest that caffeinism may constitute a contaminating and probably overlooked variable in the facilitation of aggressive outbursts.

The role of alcohol in heightening human aggression has been repeatedly demonstrated, although this issue seems to be almost intentionally disregarded. Obviously, prior experiences, emotional traits, environmental circumstances, dose levels, and even the type of alcohol consumed (Taylor and Gammon, 1975) are all abligatory factors which, as in animal experiments, variously modulate the effects of alcohol on aggression.

In a study of 882 persons arrested during or immediately after the commission of felonies, 75% of such people arrested for shooting and other assaults had urine alcohol levels between 0.10% and 0.39%, whereas less than 20% of the subjects had no alcohol and only 2% of arrested persons had concentrations over 4% (Shupe, 1954). Similarly, Wolfgang and Strohm (1956), in examining the police records of homicides during a 5-year period, found alcohol to be entirely absent, or not involved at all in only 36% of the cases. Other systematic studies have further documented alcohol as being directly involved in violent assaults (Goodwin, 1973; Tinklenberg, 1973; Wolfgang, 1966), and to represent one of the major drugs involved in criminal assaults by adolescents (Tinklenberg et al., 1974; Tinklenberg and Woodrow, 1974). A close relationship between alcohol consumption and violent aggression in different experimental or family situations has also been recently reported (Byles, 1978; Gerson, 1978; Gottshalk, 1969; Kalin et al., 1965; Shuntich and Taylor, 1972).

Alcohol has been explained to affect human behavior through releasing individuals from their inhibitions, thus causing some to become disoriented and confused, to hallucinate, and even to become violent after even low doses of alcohol (Tucker, 1960). In addition, alcohol can induce paranoid-like psychoses with temporal disorganization and delusional ideation, with feelings of power, grandiosity, and persecution (Melges et al., 1974); these effects all indicate a severe impairment of either sensory or rational control.

An important issue is that some parents of severely maladjusted boys had social maladjustments in various forms including a high incidence of criminality, alcoholism, and somatic or nervous disorders; it has also been possible to trace such social maladjustment to the boys' grandparents (Jonsson, 1967). Several other available data support the view that there is a clear genetic predisposition for the development of alcohol abuse in man (Åmark, 1951; Bowman, 1978; Goodwin, 1969; Goodwin et al., 1973, 1974, 1977; Kaij, 1960; Partanen et al., 1966). In addition, Goodwin and his associates (1975) reported that many children with hyperactive syndromes have alcoholic parents; these children are also abnormally aggressive during either their childhood or adulthood, and more likely than others to become alcoholics.

There may be interesting relationships between serotonergic brain activity, hyperactive syndromes, and alcohol consumption, since on the one hand evidence suggests a reduced activity of the brain serotonergic system of hyperactive subjects (Brase and Loh, 1975), and on the other hand chronic alcohol administration to laboratory animals either decreases brain serotonin turnover (Palaić et al., 1971; Tyce et al., 1970) or increases the conversion of serotonin to 5-hydroxy-indoleacetic acid in different brain areas (Tytell and Myers, 1973). In this general context, the previously cited study by Kršiak and his associates (1977; p. 144) concerning aggression and reduced brain serotonin content of the offspring of female mice given alcohol during gestation is particularly relevant.

Interestingly, in Kalmann's or De Morsier's syndrome (congenital hypogonadal hypogonadism with anosmia), wherein ethanol preference is indicated, aggressive behavior and violence have been reported. In this disease congenital deficits in the olfactory tubercle and degenerative lesions in the ventromedial hypothalamic area have been documented on postmortem examination. Although there are no available data for serotonin in this condition, the neuropathological changes appear to imply that it is altered.

The link between human use and abuse of hallucinogenic drugs is typically obscured by a series of misleading philosophies, pseudo-religions, and ideologies which claim that the use of hallucinogens is innocuous and essential to enhance friendliness, enlarge mental powers, deepen insight, and find God. LSD is known to interact with schizoid personality features or tendencies in a complex way, rendering virtually impossible any prediction of the final outcome (Ungerleider et al., 1968a). Even though some LSD users have claimed no ill effects, but rather benefits from their lysergide intake, at least 60% have experienced adverse reactions, consisting mainly of confusion, hallucinations followed by anxiety accelerating to the point of panic, and depression often with suicidal thoughts (Ditman et al., 1962; Frosch et al., 1965; Ungerleider et al., 1968b); others have had such disastrous reactions as to induce overt psychotic episodes and even suicide (Ditman et al., 1968; Ungerleider et al., 1966). Furthermore, the effects of lysergide in modifying the habitual perception of a self-image, environment, and belief and value system are long-lasting (Ditman et al., 1962; McGlothlin et al., 1967), and include persisting and recurrent intrusions of imagery (flashbacks) long after the immediate effects of the hallucinogen have worn off. Most often this consists of frightening images and hallucinations over which the individual has no control (Horowitz, 1969). In addition, the LSD experience includes feelings of depersonalization, impaired sensory discrimination, sudden mood swings, frequent paranoid feelings, distortion of the sense of space and time, and free-floating anxiety and fear (Barter and Reite, 1969). Thus, contrary to various claims, lysergide's effects remain unpredictable and are more profound and potentially more devastating than those of alcohol (Barter and Reite, 1969).

Despite the adverse effects of LSD on the personality of its users, reports of violent aggression or homicide associated with and resulting from the use of lysergide are surprisingly rare. This may well be due either to the profound disorganization caused by the drug, which makes purposeful assaults and homicides difficult, or to the previously mentioned long-lasting effect of LSD, so that a crime committed several days after ingestion of the drug may or may not be related to lysergide intake. In this context, only a few violent and murderous aggressions have been associated with the use of this substance (Barter and Reite, 1969; Knudsen, 1964), and it may be also suggested that, based on animal experiments, the drug may induce opposite effects according to the dose level and the pre-existing emotional baseline. After studying the acute psychosis induced by several psychotomimetic drugs including lysergide (Bowers, 1972a), and their neurochemical correlates in man (Bowers, 1972b), Bowers has proposed that, similar to animal findings, the use of LSD or LSD-like drugs may result in an irreversible or slowly reversible and persistent stimulation of the serotonergic receptors in the brain (Bowers, 1972b).

The much debated association between marihuana use and crime or violent aggression has been reported as virtually nonexistent (Goode, 1974; Knudten and Meade, 1974; Tinklenberg and Murphy, 1972), and recent research indicates that cannabis decreases both human hostility and aggression (Salzman et al., 1976; Taylor et al., 1976).

Nevertheless, there is evidence that marihuana induces a temporal disintegration of sequential thought, hallucinations, delusional ideation, psychotic episodes, depersonalization, flashback phenomena, and profound panic reactions (Bromberg, 1934; Chopra and Smith, 1974; Jørgensen, 1968; Keeler, 1968; Keeler et al., 1968; Keup, 1970; Melges et al., 1970a, b, 1974; Talbot and Teague, 1969; Tennant and Groesbeck, 1972; Weil, 1970). According to Keup (1970), the mere existence of cannabis-provoked psychotic behavior makes this drug dangerous. And it has also been suggested that marihuana may actually lead to violent aggression when its use lowers the inhibitory control mechanisms that normally govern aggression (Allentuck and Bowman, 1942; Bloomquist, 1968; Chopra et al., 1942; Freedman and Rockmore, 1946a, b). Furthermore, there are certain individuals, such as those suffering from temporal lobe dysrhythmia, and certain situations of set and setting in which marihuana use may result in violence (Abel, 1977).

The individual's status before drug use is important to the behavioral outcome following marihuana consumption. Moreover, the major discrepancy between marihuana-induced aggression in laboratory animals (p. 146) and controlled human experiments in volunteers most likely depends on the lack, in the latter instance, of such important components as starvation and stress situations required for the enhancement of violence by marihuana use. It is not generally appreciated that these two components are widespread among most marihuana users, thus contributing strongly to the arousal of such violent aggression that is claimed to be nonexistent in controlled studies.

Marihuana has been observed to be associated with some criminal assaults by adolescents (Tinklenberg et al., 1974), often in combination with other drugs such as alcohol which, together with secobarbital, a barbiturate derivative, strongly contributes to the enhancement of violent aggression (Tinklenberg and Woodrow, 1974; Tinklenberg et al., 1974). Interestingly, in proportion to overall drug use, and compared with other drugs, secobarbital is overrepresented in violent crimes, also being overwhelmingly selected by pharmacologically experienced delinquents as the most reliable chemical for releasing aggression (Tinklenberg et al., 1974). Low doses of some barbiturate derivatives are known to enhance aggression in laboratory animals as well (Valzelli, 1979b).

In contrast, not much information is available on human violence as associated with morphine or other narcotic use or abstinence. Nevertheless, a general rule governing addiction from any drug is that abstinence and the related drug need can easily provoke a subject to antisocial and even violent assaults in order to obtain sufficient money to assure the drug supply. This specific issue, not reproducible in animal experiments and far beyond any socio-philosophical, clinical, or pharmacological speculation, is the adverse influence that any addictive drug has in enhancing human violence.

8

Achieving Control of Human Violence

CONSIDERING THE PROBLEM

In human behavior, the word "violence" instead of the word "aggression" has been selected for a variety of reasons given in the preceding chapters. To broadly summarize, the different expressions of normal aggression (Chapter 5, p. 75) are not particularly harmful since, by definition, they never cause injury and remain within the limits of a reasonably proper behavioral control which has either a neurophysiological or an educational basis. On the contrary, violence—the degeneration of aggression—is released either as a consequence of pathological disruption of brain control mechanisms or by an intentional oriented education that enhances ideological intolerance.

Both of these issues require more than casual comment, especially because their incidence in social interaction has now become alarmingly high. Our urgent need to prevent and control violence has been pointed out by many, but, surprisingly, almost nothing has been undertaken in this direction. This might reflect a feeling of more or less overtly resigned pessimism about the entire question, as Freud wrote to Einstein in September 1932 that "there is no likelihood of our being able to suppress humanity's aggressive tendencies" (Freud, 1933). This may be accepted as fact, yet the problem is not the suppression of aggressiveness—which remains bound to the biological roots of survival—but, once again, that of actively controlling violence. This is still possible, although there is little to waste with sophistries.

According to Baron (1977a) a survey of the psychological literature on aggression shows that most contributions deal with various social, situational, environmental, and individual antecedents of violence, and only a minority examine the means for preventing and controlling it (Kimble et al., 1977; Pisano and Taylor, 1971). In addition, many of the studies concerned with prevention and control deal with such means that prove useful to control aggression under given and specific circumstances or among special populations, without attempting to establish basic principles that might prove effective in generally lowering the incidence of violence (Bandura, 1973). Many interrelated factors contribute to this disconcerting state of affairs, and it might be useful to try to disentangle them.

Within the limits of such an attempt, there is an obstinacy in maintaining that human aggression and violence have no communality with corresponding patterns of animal behavior, and then willingly ignoring an important source of basic information, and reducing it to the general comprehension of the problem. In this context, aside from a wide body of evidence already reported, it should be emphasized that the neurophysiological machinery for aggression is entirely subcortical (Chapter 2, p. 38) or, in MacLean's terminology, reptilian and paleomammalian. Consequently, there is no consistent need for manipulation of the neocortical brain and alteration of its functions in order for the behavioral product of such mechanisms to be properly activated or suitably inhibited. This should clearly result even from elementary notions of neurophysiology of behavior, and implicitly means that man does not substantially differ from other animals in the mechanisms and the production of emotions and affects, including aggression. Thus the human neocortical brain is just an outstandingly complex superstructure which, among other capacities, allows its owner to have: (1) consciousness of his behavior and of its possible consequences; (2) the power to reinforce or block any of his ongoing behavior; (3) the freedom to choose the most proper pattern from a wide series of possible behaviors; (4) the capacity to invent new behaviors; and (5) the ability to "morally" justify or condemn certain behaviors. On these premises, the role of the normal human mind in violence could be that of releasing it by consciously reinforcing and multiplying aggressive acts beyond the limits of any naturally existing control mechanism.

A recent and provocative book by Midgley (1978), concerning essentially psychological and behavioral data, points out that man is just an animal among others, not "unique" or better, only different. In Midgley's opinion, the skills which have traditionally been cited as proof of man's superiority over other animals—such as language, rationality, and intelligence—merely make man a naturally culture-producing animal that can no longer be considered as a blank paper passively molded by his culture. This latter point also bears consideration in the context of violence.

Another factor which has delayed the adoption of reasonable measures against violence is the widespread assumption of many psychological and sociological researchers, that aggression and violence can be controlled in a "negative" way by simply removing those factors believed to elicit them. This suggestion, however, although attractive in its simplicity, is neither consistently nor generally achievable, since it is contrary to any reasonable evidence to believe that competition, frustration, malnutrition, crowding, and numerous other related factors can be suppressed or removed to any satisfactory extent. Rather, these factors are expected to greatly increase in the near future. The disappearance of major epidemic diseases, the decrease of infant mortality, and the extension of life span for both congenitally defective and normal men contribute to overcrowding and to a proportionate decrease in the resources for man's survival.

Moreover, despite the obvious inapplicability of the theory of abolishing violence by removing the cues to it, its attractiveness is ideologically utilized to make strong claims for the indiscriminate abolition of achievements, merits, hierarchies, and of any kind of other possible individual differences. This issue is of great relevance to human violence, since it is promulgated to engender feelings of isolation, frustration, and retaliation, so as to overthrow any control and to provoke actual violence against the so-called violent society.

In Baron's opinion, a third factor that has seriously impaired the adoption of proper techniques for preventing and controlling violence is that, until recently, many psychologists believed that they already knew the best means for attaining such goals. In this framework, punishment and catharsis have been assumed to be unquestionably the most effective deterrents to human violence (Berkowitz, 1962). Nevertheless, recent research in this field has shown that neither punishment nor catharsis, alone or in combination, can effectively control overt aggression (Baron, 1977a). However, it has also been shown experimentally that aggression toward a target person is considerably decreased when a reliable punishment or retaliation exists (Donnerstein and Donnerstein, 1975, 1976; Donnerstein et al., 1972; Wilson and Rogers, 1975). This finding seems to support either the aforementioned observation by Berkowitz, or the common belief that effective punishment exerts an inhibitory effect against violence. In historical perspective, it has been reported that the lowest levels of civil strife in France occurred during the years between 1850 to 1860 and 1940 to 1944, when, respectively, the rise to power of Louis Napoleon in 1851 was accompanied by strongly repressive measures, and German occupation suppressed civil rights (Tilly, 1969).

According to Baron (1977a), controlled human experiments show that punishment reduces aggression only under conditions in which the aggressor is not very angry; the magnitude of expected punishment is great; the probability of actually being punished is high; and the profit from violent acts is little. In addition, it should be recalled that winning a competition either reinforces or increases aggressive responding, especially in children (Perry and Perry, 1976), aggression then being a self-reinforcing behavioral response in man, as has been observed in animal experiments. Regarding the effect of catharsis on violence, the suggestion is that providing angry subjects with the opportunity to release their violence in some safe way either will cause them to feel better or will weaken their aggressive tendency. The origin of this concept can be traced to the writing of Aristotle, who held that the exposure to dramatic plays produces a "purging" of emotions. However, experimental evidence indicates that only attacks against the direct source of one's anger or annoyance are effective in reducing aggression (Doob and Wood, 1972; Feshback, 1970), and this fact obviously sharply reduces the applicability of cathartic procedures for controlling violence.

Although in a strictly logical context it should not be difficult to examine without bias the possible remedies to violence, a fourth factor delaying decisions is that, whenever the concept of behavioral control is raised and limited to a certain item, problems arise concerning the extent to which aggression should be controlled, by whom, and by what methods. Since the physiological methods for reducing aggression and controlling violence are powerful, and they are consequently believed to bring about a considerable potential for abuse, the ancient dilemma *Quis custodiet custodes?* (Who will control the controllers?) appears destined to emerge.

At first glance, this seems to be a reasonable matter of concern, but some doubt about the unselfishness of this position arises when it is considered that no equal concern is expressed in behalf of the victims of violence. Further, other behavioral limitations regulating social interaction are generally accepted as necessary, and it should also be considered that socialization and civilization both rest to a great extent on man's ability to control excessive aggression and violence (Chapter 4, p. 61). In this context, rather clear ideological implications arise when aggression or, even worse, violence is considered as a mere pretext or label to legitimatize coercive actions and to develop control of mass behavior by the "establishment." The radical partisans of this position construe human violence as an expression of free will, and, consequently, they consider any corrective measure as having degrading effects on human dignity. This view is both inappropriate and deceitful, not because free will is to be denied, but because the same degree of free will and human dignity has to be accorded to the victim of violence, and the quality of human life must be properly valued.

In conclusion, it is not surprising that the topic of violence control, fragmented into so many different and contrasting viewpoints, remains much more an arena for debate than a ground for undertaking concrete measures. Persisting on this theme, the entire question appears likely soon to reach a dead end; this may also represent a fruitful result to some, while a different approach may generate more successful perspectives.

The suggestion here is to consider separately those methods and techniques potentially useful in the control of individual violence and group violence. The subdivision of the problem of violence in two main categories is based on the assumption that individual violence derives mainly from genetic, pathological, addictive, and socioenvironmental components, whereas group violence is mainly ideologically founded, organized, instigated, imitatively learned, and financially supported. Implicitly this means that the former is a matter of medical care and concern, and the latter a concern for consistent educational or legislative measures. Obviously, as for all human affairs, in some cases a certain degree of overlap between the two categories may easily be foreseen, but this should not detract from the reliability of the proposed viewpoint in rapidly undertaking proper remedies for the control of human violence.

CONTROL OF INDIVIDUAL VIOLENCE

Psychiatrists are often accused of transforming people with problems into patients with symptoms (Rollin, 1973). This view is amusing, but its consistency is definitely not greater than to affirm that physicians transform people with a cough into patients with respiratory diseases.

Insofar as individual violence is concerned, it should be recalled that, as previously noted (Chapter 7, p. 124), not all criminals are automatically regarded as sick people, and that, conversely, no categories of mental disorders are mutually exclusive of the commission of crimes, to which mental deficiency, schizophrenia, affective disorders, organic dementia, and psychopathic personality all greatly contribute (Crowell et al., 1973; Dengerink and Bertilson, 1975; McKerracher et al., 1966; Rollin, 1973; Schless et al., 1974; Silverman et al., 1969; Steadman et al., 1978; Weinstock, 1976; Weissman et al., 1973; Westermeyer and Kroll, 1978). Thus, within the limits of detectable mental disorders and of their facilitatory genetic components (Chapter 6, p. 113, and Chapter 7, p. 124), which can also be worsened by drug addition (Chapter 7, p. 124), either preventive or curative medical care is obligatory. In complete agreement with Lion and Monroe (1975), while humility and social consciousness must sustain either the evaluation or the diagnosis of any violent individuals, it would in fact be clinically irresponsible to disregard those subjects overtly troubled by urges to violence, who often look for appropriate treatment.

Changes in social milieu, environmental interaction, and imitative cues, and even dietary modifications may greatly reduce the need for additional and more specific therapeutic approaches, especially in younger people who, due to their cerebral plasticity, may remarkably alter their quality of life.

Sociofamilial Milieu

The French word *milieu,* which coincides entirely with the meaning of "environment" already widely used, is sometimes preferred by clinicians to accentuate the classically implied concept of social relations and experiences (Fisher, 1970; Friedmann et al., 1975; Klerman, 1963). Thus the importance of this variable in shaping either animal or human behavior needs no further illustration, having already been discussed in its negative or positive contexts (Chapter 2, p. 24; Chapter 3, p. 49–53; Chapter 4, p. 69; Chapter 6, p. 112, and Chapter 7, p. 125).

Focusing on the immediate family environment of most violent individuals, despite the generally accepted belief that family is a social group committed to nonviolence among its members, it can be observed that, under several circumstances, violence within the family is quite widespread (Bard and Zacker, 1971; Goode, 1971; Kopernik, 1964; Lystad, 1975; Steinmetz and Straus, 1973; Straus, 1973). Such an anomalous milieu clearly provides the child with both profound affective frustration and a poor behavioral model which, in favorable circumstances, promotes his later aggressive and violent behavior; this gives rise to a sort of transmissible closed circle (Bennie and Sclare, 1969; Curtis, 1963; Duncan and Duncan, 1971; Gallenkamp and Rychlak, 1968; Sargent, 1971a, b; Silver et al., 1969). In other words, evidence indicates that parents who punish inappropriately and violently produce children who are aggressive and who will, in turn, punish their future children as violently and inappropriately.

Still (1902) was probably the first to report an association between antisocial behavior in children and psychiatric disorders in their relatives; more recently, it has been shown that children admitted to a psychiatric ward for aggressive antisocial behavior had parents who were often psychotic, neurotic, or psychopathic (Morris et al., 1956; Zalba, 1966). Children's antisocial behavior and juvenile delinquency have been further found associated with sociopathic criminality and alcoholism in parents (Gluek and Gluek, 1950; Robins, 1966; Stewart and Leone, 1978; West and Farrington, 1973), and in grandparents (Jonsson, 1967).

Sociological and contextual variables, such as low socioeconomic status, low educational achievement, unwanted pregnancies, broken family boundaries, and large family size, also contribute to the development of violent families (Becker et al., 1962; Gelles, 1973; Healy and Bronner, 1936; Hewitt and Jenkins, 1946; Rutter, 1966, 1971). These conditions lead to the parents' poor and inconsistent supervision of their children, who have to live in an either physically or emotionally impoverished and delusional atmosphere in which they do not find the opportunities for learning prosocial behaviors. On this background, a critical role of the lack of a successful development of cognitive skills emerges and a reliance on feeling to interpret the world develops (King, 1975). In such youths, violence is frequently related to serious difficulties in mastering reading, language skills, social symbols, and general comprehension, so that, incapable of coping with normal social interaction, they easily become alienated, addicted, reactive, violent, and homicidal (King, 1975). Consequently, in such young people aggression and violence, although wrong, become compensatory adjustive behaviors, bearing a complex relationship to their personality and environment (Dembo, 1973; Miller, 1958; Miller et al., 1961; Wolfgang and Ferracuti, 1967).

From even a superficial analysis of studies on this issue, four main factors clearly emerge as potent producers of violence: (1) violence inside the family, (2) lack of parental care, (3) a genetically transmitted predisposition, and (4) low educational and cultural achievements. When these factors coincide in the same individual, they produce a situation of socioenvironmental deprivation which is close to that capable of inducing violent aggression and killing activity in laboratory animals. Possibly, as in the latter instance, violence stems from a type of functional decortication (Chapter 5, p. 86), depending on a disintegration of the higher cortical mechanisms which, especially in man, normally control or redirect aggresssion. This interpretation also agrees well with the evidence that, under certain circumstances, each one of the main components of the human triune brain may operate independently of the others (MacLean, 1976; and Chapter 2, p. 33).

Since successful violence is self-reinforcing, the explosive potential of the association of the aforementioned factors is greatly increased when it develops in conditions of overcrowding; this allows for a spread of violence, thus providing a perfect basis for political exploitation and an ideal reservoir for ideological violence. Then, in the context of this general picture, the following issues emerge as possibly effective for correcting and recasting violence: (1) to intensify medical care to either defective parents or defective children; (2) to prevent children from spending their formative period with defective families and environments; (3) to promote their overall learning, culture, and prosocial education; (4) to provide them with positive models, achievable tasks, and desirable cues; (5) to make the "payoff" of prosocial behaviors consistently more attractive than that of violence; and (6) to avert habitation in overcrowded districts. Obviously, the points (1) and (2) must be carefully evaluated, humanely balanced, and wisely implemented, whereas the others remain a matter of choice, expenditure, and willingness.

Regardless of whatever other and better measures can be suggested, the quality of future human life will to a great extent depend on a consistent reversal of the increasing trend to violence.

Nutritional Components

Years ago it was reported that elevated or reduced dietary intake of cereals may accordingly worsen or ameliorate schizophrenic behavior (Dohan, 1966). Subsequent studies have further shown that, in some cases, cereal-free and milk-free diets were capable of either inducing rapid improvement in schizophrenic symptoms or considerably potentiating the therapeutic efficacy of neuroleptic administration (Dohan and Grasberger, 1973; Dohan et al., 1969; Singh and Kay, 1976). These observations, which have led to the hypothesis that one of the etiologic components of schizophrenia may be associated with that of celiac disease (Dohan, 1969a, b, 1970), are also indicative of the possible implications of malabsorption; thus malnutritional factors may enhance or sustain schizophrenic symptoms and violence, possibly as a consequence of an imbalance in the supply of proteins and amino acids to the brain (Valzelli and Sarteschi, 1977).

Prenatal nutritional deficiency, undernutrition during early life, and an increased intake of L-tyrosine with food have been reported to increase aggression in laboratory animals (Chapter 5, p. 78; Chapter 7, p. 103). Variations of dietary tryptophan intake are also known to accordingly modulate both serotonin synthesis and levels in the brain (Culey et al., 1963; Dickerson and Pao, 1975; Miller et al., 1977a, b; Neckers et al., 1977); decreased serotonergic activity of the brain has already been discussed as an important component of violent aggression (Chapter 7).

Even in the absence of or in addition to an altered availability of these and other essential amino acids, other dietary nutrients may interact by either increasing or decreasing their absorption and/or utilization by the organism. Interestingly, periods of famine and general protein and carbohydrate malnutrition have historically been associated with a profound increase of criminality and violence, and it has been recently reported that countries above the median in corn consumption have significantly higher homicide rates than those countries which consume corn below the median (Mawson and Jacobs, 1978). In relation to this observation, corn consumption by rats is associated with a marked reduction of brain serotonin; since large numbers of people subsist on corn as their major protein source, they may have an altered cerebral serotonergic activity (Fernstrom and Wurtman, 1971). Furthermore, according to Mawson and Jacobs (1978), several studies suggest that populations consuming corn-based diets may have elevated homicide rates depending on the consequent reduction of brain tryptophan and/or serotonin. In this context, the previously reported relationship between chronic hypoglycemia and homicidal violence (Chapter 6, p. 121) may also assume further significance.

Defective tryptophan metabolism, which persists for months after alcohol withdrawal, has been found in apparently healthy but chronic social drinkers (Olson et al., 1960); it should also be recalled that chronic alcohol consumption, aside from its effect on brain serotonin, induces malabsorption and malnutrition, which aggravate both the neurochemical imbalance and the propensity to violence. Moreover, as for the drugs of addiction, the chronic and heavy consumption of coffee, tea, or of other complementary foods and possibly the chronic intake of some food preservatives or traces of other chemical additives, may be relevant either to an impaired neurochemical balance or to a series of continuously spreading neurological and psychiatric diseases leading to abnormal behavior and violence.

In this broader context, it may be recalled that evidence of defective tryptophan metabolism has been reported in children with mongolism (Gershoff et al., 1958; Jerome et al., 1960; O'Brien and Groshek, 1962; O'Brien et al., 1960), whereas in the mental retardation of the Lesch-Nyhan syndrome (Lesch and Nyhan, 1964), 5-hydroxytryptophan administration specifically and effectively relieves the intense self-mutilation behavior characteristic of this sex-linked recessively inherited disease (Mizuno and Yugari, 1975). In other inherited diseases, such as Klinefelter's syndrome (47, XXY), there are aggressive physical displays (Funderburg and Ferjo, 1978). XYY karyotype aggressive subjects (Chapter 6, p. 113) have been suggested as having decreased brain serotonin turnover (Bioulac et al., 1978). Moreover, hyperactive aggressive patients also show decreased blood levels of 5-hydroxyindoles (Greenberg and Coleman, 1976). Lithium administration has been shown either to exert a clear antiaggressive effect in mentally defective patients (Worrall et al., 1975) or to increase both tryptophan uptake by the brain (Knapp and Mandell, 1973, 1975) and brain serotonin synthesis (Essman, 1975b; Poitou et al., 1974).

The bulk of evidence seems to indicate that in man, too, a decreased brain serotonergic activity, either genetically predetermined or chemically and dietarily induced, correlates to a great extent with violence. Not only the absolute level of famine and malnutrition, but also the quality of specific nutritional deficiencies may have the major responsibility for breeding violence; a dietary supplementation of foods rich in tryptophan may have curative and/or prophylactic effects against violent behaviors. In addition, avoidance of drugs, chemicals, food preservatives, and complementary foods potentially capable of inducing defective brain tryptophan metabolism could prove efficient in controlling human violence.

Psychopharmacological Notes

Although there is currently no drug which is specifically antiaggressive, there is ample evidence to indicate that available psychotropic drugs may help to control pathological aggression. The fact that no "selective" antiaggressive or antiviolent drugs have been discovered indicates that pathological aggression is the symptom of a more complex and extended behavioral and neurological pathology. In other words, and according to Itil and Mukhopadhyay (1978), subjects troubled solely by violent conduct are rare. Nevertheless, an ideal drug control of human violent and/or destructive behavior must theoretically maintain full respect for the human mind and personality. On the other hand, in most instances uncontrollable aggression greatly impairs either the social setting of the subject or his professional performance and achievement; aggression and violence have been shown to profoundly impair precision learning and later achievement in children (Gardner, 1971). The previously reported beneficial effect of psychostimulants in managing hyperactivity and aggression in children and adolescents (Chapter 7, p. 148) has been particularly emphasized by Corson (1975).

The benzodiazepine derivatives represent another group of drugs relevant to the therapeutic treatment of human aggression (Essman, 1978; Lion, 1979; Valzelli, 1979b); these apparently have more specific effects than other drugs on selective areas of the limbic system involved in the regulation of aggression. However, tricyclic antidepressants, phenothiazines, butyrophenones, antiepileptics, and antiandrogen derivatives in specific cases of assaultive sexual violence all proved useful in managing human pathological aggression; the difference in therapeutic efficacy was dependent on the differences in brain dysfunctions leading to violent outbursts.

Detailed data on the effect of various psychoactive drugs on aggression are readily available in the current literature. Furthermore, a series of reviews (Cook and Kelleher, 1963; Kršiak, 1974; Valzelli, 1967b, 1978c) include special reference to human aggression and violence (Itil and Mukhopadhyay, 1978; Resnik, 1971) and need not be repeated here. On the whole, the possibility of effectively controlling destructive behavior is within our reach as never before; in addition, the treatment of violent and potentially destructive behavior with chemicals remains the first choice, from a therapeutic point of view.

It is of interest to consider that, up to now, a largely planned trend has deterred social acceptance of any controlled and therapeutically oriented psychopharmacological intervention. This has been accomplished either in the name of the defense of the rights of the mentally disturbed—which evidently exclude the right to be cured—or in the name of a fictional mass behavior control by the "establishment." Paradoxically, however, much less effort has been undertaken to inform societies of the widespread use and abuse of addictive drugs, which represent the greatest danger to the human psyche. For example, there was recently a call for licensing one million U.S. farmers to grow $2,500 million worth of marihuana, on the basis that there are more than 40 million marihuana users in the United States and Canada, who in 1977 spent this amount in South American and Mexico (Galbraith, 1978).

The matter of social concern for mental health seems to require no further comment.

Psychosurgical Approach

The present difficult controversy over psychosurgery as the last resort treatment for violent behavior, according to Lion and Monroe (1975) reflects the political overtones ascribed to this argument by those with a societal point of view. It has been recently reported (Donnelly, 1978) that "activist" psychiatrists and others have repeatedly made highly publicized accusations that thousands of individuals are being submitted to brain operations needlessly or for exclusively experimental purposes, or even for purely political reasons.

The objective evaluation of the efficacy of a therapeutic procedure in relieving individual violence is biased, to a major degree, on nonscientific factors. A small group of influential psychiatrists, neurologists, and neurosurgeons have been accused of promoting psychosurgery exclusively for political purposes— including its use to treat rioters, protestors, and leaders of urban disorders (Breggin, 1976). This initially "one-man crusade" against psychosurgery (Culliton, 1976) fits well into the more general assumption that biomedical research serves a powerful latent social function, the affirmation of the status quo (Nassi and Abramowitz, 1976).

It has, however, been shown that psychosurgical procedures are strikingly beneficial to patients who would otherwise be lost to themselves, their families, and society (Cattell, 1969), and who, as a consequence of their operation, can again enjoy their lives (Kalinowsky, 1969). Positive evaluations have also been made by others (Brill, 1969; Freyhan, 1969). Moreover, the U.S. National Commission for the Protection of Human Subjects of Biomedical and Behavioral Research (1977) has issued a favorable report on the potential therapeutic efficacy of psychosurgery. In addition, a recent report shows that in the United States from 1971 to 1973 no more than 575 surgical procedures have been performed for the management of pharmaco-resistant psychiatric diseases; this number sharply conflicts with the claim that thousands of lobotomies are performed annually for nefarious purposes (Donnelly, 1978).

The development of public opinion about brain operations and electrical stimulation must be duly considered. Public awareness and concern have been increased in recent years, either through novels dealing explicitly with psychosurgery (Critchton, 1972; Kesey, 1962; Vonnegut, 1959) or through the writings of neuroscientists and clinicians (Delgado, 1969*d;* Mark and Ervin, 1972; Skinner, 1971). Considerable information is thus broadly available, and a legitimate concern about such powerful techniques for behavioral control comprehensibly emerges. In this context, it is imperative to: (1) elaborate a code of reference criteria that govern the choices for the application of psychosurgical therapy to patients; (2) properly evaluate the clinical efficacy of different types of operations currently available (Ransohoff, 1969; Templer, 1974); (3) adhere rigidly to ethical issues which govern the medical profession (Halleck, 1974; Mark and Neville, 1973; Older, 1974; Shevitz, 1976), and which should even disregard any legal and political directions not fully respecting and warranting human rights; and (4) establish proper and reliable control mechanisms (Dow et al., 1974; Older, 1974). Within these limits, and when other therapeutic approaches have proved useless, psychosurgery may be considered, especially for the alleviation of certain kinds of violent behavior caused by organic brain diseases.

CONTROL OF GROUP VIOLENCE

As previously observed, group violence finds its major reserve in defective families, inappropriate social environments, and subcultures which breed individual violence. This basic situation is especially augmented by parental and social reinforcers (Feshback, 1970; Johannesson, 1974; Ulehla and Adams, 1973), and among the latter the influence of mass media appears to be quite potent.

Although expected differences exist in individual susceptibility to the influence of aggressive film models, it has been shown that children respond imitatively to them with a significant increase in their aggressive baseline (Kniveton and Stephenson, 1973). The importance of the models of behavior to which the developing child is exposed becomes once more clearly evident in the acquisition of aggressive habits which, especially in boys, project into late adolescence (Eron et al., 1974*a, b*). In addition, it is important to consider that the effect of film-mediated aggressive models on behavior has proved to be most influential in mentally retarded boys (Talkington and Altman, 1973); and that subliminally presented aggressive stimuli promote an increase of pathological thinking and aggression in schizophrenic patients, also arousing aggression, pathological thinking, and libidinal imagery in nonpsychiatric subjects (Silverman, 1966). Highly erotic films have been shown to increase aggression against women, especially when male subjects are given the opportunity to aggress immediately after seeing the film (Donnerstein and Hallam, 1978).

Due to the wide dissemination of violence presented on television, there is a considerable emotional impact on viewer personality, especially in children (Heller and Polsky, 1971). The adverse effects of television's massive daily diet of sex, crime, and violence on the behavior of youth has been the matter of a recent and concerned report by Somers (1976). Although apparently disregarded, several studies have in fact already shown not only that the learning of televised aggressive techniques occurs, but also that continued exposure leads to acceptance of violence and aggression as a mode of action for oneself and others (Dominick and Greenberg, 1972; Kenny, 1972; Liebert, 1974; Neale and Liebert, 1973). In addition, the acquisition of aggressive habits by children who will later become violent adolescents has been shown to be predictable in a highly consistent manner by the amount of television violence that the child viewed, so as to indicate a cause-effect relationship between the content of violence in television programs and overt aggressive behavior (Lefkowitz et al., 1972; Steuer et al., 1971).

The potent effect of mass media in general and of television in particular in influencing behavior results from the extreme plasticity of the child's imitative learning (Chapter 2, p. 33 and 34; Chapter 5, p. 78). Childhood behavior has not yet been affected by those critical capacities that allow a child to retain good models and discard poor ones. Furthermore, imitative learning develops from a precise neurophysiological arrangement which conveys 40% of the entire bulk of information to the brain through the visual system, with an additional 1% through auditory pathways (Valzelli, 1979a). It is not surprising that, on such a basis, audiovisual input can have considerable relevance in shaping behavior, and television has such a great responsibility.

The widespread phenomenon of terrorism, which consists of highly organized group violence, finds great support either in an already predisposed basis for violence or in the awesome persuasive power of television that, more or less overtly, aligns itself with the anarchic activity of terrorists (Evert, 1979). In addition, it has also been recognized by experts in the field that almost inevitably television is one-sided in treating revolutionary conflicts, and this bias is inescapably on the side of violence against the establishment (Evert, 1979). Thus prearranged ideological distortions can camouflage overt criminal acts and represent them as brave enterprises against fictitious oppression, or even broadcast assassination and violence as the exclusive means of overthrowing society.

The techniques adopted are reminiscent of Goebbel's statement that repeated and emphasized falsifications of the truth soon become truths; and a dangerous level of refinement has already been achieved by some newspaper, movies, and television programs. Persuasively presented, one-sided truths and insinuations are highly mind polluting to most people; mind pollution is obviously more easily achievable among ignorance and social underdevelopment. Models of violent protests easily influence the behavioral choices of youth; young people seldom realize that they are only the raw material serving the advantage of ambiguous powers which profit in converting discordance into protest, protest into friction, friction into violence, and violence into terrorism.

It is important to consider that ideologies are incompatible with culture, which by its very nature is traditionally conservative, in that through centuries it transmits the growing bulk of acquired human knowledge; culture provides men with a world-wide basis for possible convergences. Consequently, as has been said for violence (Wolfgang and Ferracuti, 1967), ideology too is, by definition, subcultural, and it requires ignorance and mystification to become pervasive. Since cultural cohesion increases the high-intensity interpersonal bonding which is believed to be the prerequisite for increasing social synergy and for decreasing either aggression or mental illness (Aring, 1973; Gorney, 1971), ideology strongly conflicts with the achievement of such a goal.

The problem is not television *per se,* but what it teaches. Contrary to the adverse modeling of violent programs, it has in fact been shown that altruistic and cooperative modeling produces more cooperative, altruistic children, and those exposed to adults and peers who exhibit self-control become more capable of controlling themselves (Liebert and Poulos, 1972; Liebert et al., 1974). Empathy with parent and adult behavior, which is related to the emergence of moral development in the child (Hoffman, 1970; Hogan, 1973; Kohlberg, 1969; Piaget, 1932; Staub, 1972), can greatly counteract aggression and violence (Feshbach, 1974*a;* Hogan, 1973). Consequently, more than casual comment should be given to recall the importance of religion in counteracting violence; religion further acts more as a resource of life rather than a simple compensation (Hadaway, 1978).

On the whole, it may be said that mass media in general, and television in particular, have the opportunity to potently operate along these lines to disseminate both culture and positive models of behavior, in such a way as to materially contribute to the disruption of the adverse spiral to violence which endangers mankind.

References

Abel, E. L. (1977): The relationship between cannabis and violence. A review. *Psychol. Bull.,* 84:193–211.

Adamec, R. (1974): Neural correlates of long term changes in predatory behavior in the cat. Ph.D. dissertation, McGill University, Montreal.

Adamec, R. (1975*a*): The behavioral bases of prolonged suppression of predatory attack in cats. *Aggressive Behav.,* 1:297–314.

Adamec, R. (1975*b*): The neural basis of prolonged suppression of predatory attack. 1. Naturally occurring physiological differences in the limbic systems of killer and non-killer cats. *Aggressive Behav.,* 1:315–330.

Adamec, R. (1975*c*): Behavioral and epileptic determinants of predatory attack behavior in the cat. *Can. J. Neurol. Sci.,* 2:457–466.

Adamec, R. E. (1976): Hypothalamic and extrahypothalamic substrates of predatory attack. Suppression and the influence of hunger. *Brain Res.,* 106:57–69.

Adamec, R. E., and Himes, M. (1978): The interaction of hunger, feeding and experience in alteration of topography of the rat's predatory response to mice. *Behav. Biol.,* 22:230–243.

Adams, D. B. (1968): Cells related to fighting behavior recorded from midbrain central gray neuropil of cat. *Science,* 159:894–896.

Adams, D. B. (1971): Defence and territorial behaviour dissociated by hypothalamic lesions in the rat. *Nature,* 232:573–574.

Adams, H. B., Robertson, M. H., and Cooper, G. D. (1966): Sensory deprivation and personality change. *J. Nerv. Ment. Dis.,* 143:256–265.

Ader, R. (1965): Effects of early experience and differential housing on behavior susceptibility to gastric erosions in the rat. *J. Comp. Physiol. Psychol.,* 60:233–238.

Ader, R. (1975): Competitive and noncompetitive rearing and shock-elicited aggression in the rat. *Anim. Learning Behav.,* 3:337–339.

Ader, R., Friedman, S. B., and Grota, L. J. (1967): "Emotionality" and adrenal cortical function: Effects of strain, test and the 24 hour corticosterone rhythm. *Anim. Behav.,* 15:37–44.

Adey, W. R. (1953): An experimental study of the central olfactory connections in a Marsupial (Trichosurus vulpecula). *Brain,* 76:311–330.

Adey, W. R. (1959): Recent studies of the rhinencephalon in relation to temporal lobe epilepsy and behavior disorders. *Int. Rev. Neurobiol.,* 1:1–46.

Aghajanian, G. K., Bloom, F. E., and Sheard, M. H. (1969): Electron microscopy of degeneration within the serotonin pathway of rat brain. *Brain Res.,* 13:266–273.

Aghajanian, G. K., and Bunney, B. S. (1974): Central dopaminergic neurons: Neurophysiological identification and responses to drugs. *Biochem. Pharmacol.,* Suppl. 2:523–528.

Aghajanian, G. K., Foote, W. E., and Sheard, M. H. (1968): Lysergic acid diethylamide: Sensitive neuronal units in the midbrain raphe. *Science,* 161:706–708.

Agranoff, B. W. (1975): Neurotransmitters and synaptic transmission. *Fed. Proc.,* 34:1911–1914.

Agranoff, B. W., and Davis, R. E. (1974): More on seasonal variations in goldfish learning. *Science,* 186:65.

Ahtee, L., and Eriksson, K. (1972): 5-Hydroxytryptamine and 5-hydroxyindol-acetic acid content in brain of rat strains selected for their alcohol intake. *Physiol. Behav.,* 8:123–126.

Akert, K. (1961): Diencephalon. In: *Electrical Stimulation of the Brain,* edited by D. E. Sheer, pp. 288–310. University of Texas Press, Austin, Texas.

Akhtar, S., and Hastings, B. W. (1978): Life threatening self-mutilation of the nose. *J. Clin. Psychiatry,* 39:676–677.

Albert, D. J., and Wong, R. C. K. (1978): Hyperreactivity, muricide, and intraspecific aggression in the rat produced by infusion of local anesthetic into the lateral septum or surrounding areas. *J. Comp. Physiol. Psychol.,* 92:1062–1073.

Alberts, J. R., and Friedman, M. I. (1972): Olfactory bulb removal but not anosmia increases emotionality and mouse killing. *Nature,* 238:454–455.

Alberts, J. R., and Galef, B. G., Jr. (1971): Acute anosmia in the rat: A behavioral test of a peripherally-induced olfactory deficit. *Physiol. Behav.,* 6:619–621.

Alberts, J. R., and Galef, B. G., Jr. (1973): Olfactory cues and movement: Stimuli mediating intraspecific aggression in the wild Norway rat. *J. Comp. Physiol. Psychol.,* 85:233–242.

Albrecht, P., Visscher, M. B., Bittner, J. J., and Halberg, F. (1956): Daily changes in 5-hydroxytryptamine concentration in mouse brain. *Proc. Soc. Exp. Biol. Med.,* 92:703–706.

Alexander, B. K., Coambs, R. B., and Hadaway, P. F. (1978): The effect of housing and gender on morphine self-administration in rats. *Psychopharmacology,* 58:175–179.

Allee, W. C. (1942): Group organization among vertebrates. *Science,* 95:289–293.

Allen, R. P., Safer, D., and Covi, L. (1975): Effects of psychostimulants on aggression. *J. Nerv. Ment. Dis.,* 160:138–145.

Allen, W. F. (1932): Formatis reticularis and reticulo spinal tracts: Their visceral functions and possible relationships to tonicity and clonic contractions. *J. Washington Acad. Sci.,* 22:490–495.

Allentuck, S., and Bowman, K. M. (1942): The psychiatric aspects of marihuana intoxication. *Am. J. Psychiatry,* 99:248–251.

Allikmets, L. H. (1974): Cholinergic mechanisms in aggressive behavior. *Med. Biol.,* 52:19–30.

Allison, A. C. (1954): The secondary olfactory areas in the human brain. *J. Anat.,* 88:481–488.

Allison, G. E. (1967): Psychiatric implications of religious conversion. *Can. Psychiatr. Assoc. J.,* 12:55–61.

Al-Maliki, S., and Brain, P. F. (1977): Influences of social rank and isolation on two forms of murine aggression. *ICRS Med. Sci.,* 5:290.

Alpers, B. J. (1940): Personality and emotional disorders associated with hypothalamic lesions. *Assoc. Res. Nerv. Ment. Dis.,* 20:725–748.

Altman, J. (1965): *Organic Foundations of Animal Behavior.* Holt, Rinehart & Winston, London.

Alves, C. N., and Carlini, E. A. (1973): Effects of acute and chronic administration of *Cannabis sativa* extract on the mouse-killing behavior of rats. *Life Sci.,* 13:75–85.

Åmark, C. (1951): A study in alcoholism: Clinical, social-psychiatric and genetic investigations. *Acta Psychiatr. Neurol. Scand.,* Suppl. 70:1–271.

Anand, B. K. (1957): Structure and functions of the limbic system ("visceral brain"): A review. *Indian J. Physiol. Pharmacol.,* 1:149–184.

Anand, B. K., and Brobeck, J. R. (1951): Hypothalamic control of food intake in rats and cats. *Yale J. Biol. Med.,* 24:123–140.

Anand, B. K., and Dua, S. (1955): Stimulation of limbic system of brain in waking animals. *Science,* 122:1139.

Anand, B. K., and Dua, S. (1956): Electrical stimulation of the limbic system of brain (visceral brain) in the waking animals. *Indian J. Med. Res.,* 44:107–119.

Anand, B. K., Malhotra, C. L., Singh, B., and Dua, S. (1959): Cerebellar projections to limbic system. *J. Neurophysiol.,* 22:451–457.

Anand, M., Gupta, G. P., and Bhargava, K. P. (1977): Modification of electroshock fighting by drugs known to interact with dopaminergic and noradrenergic neurons in normal and brain lesioned rats. *J. Pharm. Pharmacol.,* 29:437–439.

Anastasi, A. (1958): Heredity, environment, and the question "how"? *Psychol. Rev.,* 65:197–208.

Andén, N.-E. (1974): Catecholamine receptor mechanisms in vertebrates. *Biochem. Pharmacol.,* Suppl. 2:539–543.

Andersen, H., Braestrup, C., and Randrup, A. (1975): Apomorphine-induced stereotyped biting in the tortoise in relation to dopaminergic mechanisms. *Brain Behav. Evol.,* 11:365–373.

Anderson, P. K. (1961): Density, social structure, and non-social environment in house-mouse populations and the implication for regulation of numbers. *Trans. N.Y. Acad. Sci.,* 23:447–451.

Andrew, R. J. (1974): Changes in visual responsiveness following intercollicular lesions and their effects on avoidance and attack. *Brain Behav. Evol.,* 10:400–424.

Andrew, R. J., and De Lanerolle, N. (1975): The effects of muting lesions on emotional behaviour and behaviour normally associated with calling. *Brain Behav. Evol.,* 10:377–399.

Andy, O. J. (1970): Thalamotomy in hyperactive and aggressive behavior. *Confin. Neurol.,* 32:322–325.

Andy, O. J., Giurintano, L., Giurintano, S., and McDonald, T. (1975): Thalamic modulation of aggression. *Pavlovian J.,* 10:85–101.

Anisko, J. J., Christenson, T., and Buehler, M. G. (1973): Effects of androgen on fighting behavior in male and female mongolian gerbils *(Meriones unguiculatus). Horm. Behav.,* 4:199–208.

Antelman, S. M., and Caggiula, A. R. (1977): Norepinephrine-dopamine interactions and behavior. *Science,* 195:646–653.

Anton, A. H., Schwartz, R. P., and Kramer, S. (1968): Catecholamines and behavior in isolated and grouped mice. *J. Psychiatr. Res.,* 6:211–220.

Archer, J. (1970): Effects of population density on behaviour in rodents. In: *Social Behaviour in Birds and Mammals,* edited by J. H. Crook. Academic Press, London.

Archer, J. (1974): Sex differences in the emotional behavior of three strains in laboratory rat. *Anim. Learning Behav.,* 2:43–48.

Aring, C. D. (1973): Aggression and social synergy. *Am. J. Psychiatry,* 130:297–298.

Arnold, L. E., Kirilcuk, V., Corson, S., and Corson, E. O'L. (1973): Levoamphetamine and dextroamphetamine: Differential effect on aggression and hyperkinesis in children and dogs. *Am. J. Psychiatry,* 130:165–170.

Arnold, L. E., Strobl, D., and Weisenberg, A. (1972): Hyperkinetic adult. Study of the "paradoxical" amphetamine response. *J.A.M.A.,* 222:693–694.

Asano, Y. (1971): The maturation of the circadian rhythm of brain norepinephrine and serotonin in the rat. *Life Sci.,* 10:883–894.

Aschoff, J., Figala, J., and Pöppel, E. (1973): Circadian rhythms of locomotor activity in the golden hamster *(Mesocricetus auratus)* measured with two different techniques. *J. Comp. Physiol. Psychol.,* 85:20–28.

Askew, B. M. (1962): Hyperpyrexia as a contributory factor in the toxicity of amphetamine to aggregated mice. *Br. J. Pharmacol. Chemother.,* 19:245–257.

Azrin, N. H., Hake, D. F., and Hutchinson, R. R. (1965): Elicitation of aggression by a physical blow. *J. Exp. Anal. Behav.,* 8:55–57.

Azrin, N. H., and Hutchinson, R. R. (1967): Conditioning of the aggressive behavior of pigeons by a fixed-interval schedule of reinforcement. *J. Exp. Anal. Bahav.,* 10:395–402.

Azrin, N. H., Hutchinson, R. R., and Hake, D. F. (1966): Extinction-induced aggression. *J. Exp. Anal. Behav.,* 9:191–204.

Azrin, N. H., Hutchinson, R. R., and Sallery, R. D. (1964): Pain-aggression toward inanimate objects. *J. Exp. Anal. Behav.,* 7:223–228.

Babb, T. L., Mitchell, A. G., and Crandall, P. H. (1974): Cerebellar influences on the hippocampus. In: *The Cerebellum, Epilepsy, and Behavior,* edited by I. S. Cooper, M. Riklan, and R. S. Snider, pp. 37–56. Plenum Press, New York.

Bach-y-Rita, G. (1974): Habitual violence and self-mutilation. *Am. J. Psychiatry,* 131:1018–1019.

Bach-y-Rita, G., and Veno, A. (1974): Habitual violence: A profile of 62 men. *Am. J. Psychiatry,* 131:1015–1017.

Baenninger, R. (1967): Contrasting effects of fear and pain on mouse killing by rats. *J. Comp. Physiol. Psychol.,* 63:298–303.

Baenninger, R. (1970): Suppression of interspecies aggression in the rat by several aversive training procedures. *J. Comp. Physiol. Psychol.,* 70:382–388.

Bailey, P. (1958): Discussion. In: *Temporal Lobe Epilepsy,* edited by M. Baldwin and P. Bailey, Charles C Thomas, Springfield, Ill.

Bain, A. (1885): *The Senses and the Intellect, 3rd Ed.* Appleton, New York.

Baker, D., Telfer, M. A., Richardson, C. E., and Clark, G. R. (1970): Chromosome errors in men with antisocial behavior. Comparison of selected men with Klinefelter's syndrome and XYY chromosome pattern. *J.A.M.A.,* 214:869–878.

Baker, J. R. (1954): What is the Golgi controversy? *J. R. Micros. Soc. Sci.,* 74:217–221.

Balasubramaniam, V., and Ramamurthi, B. (1968): Stereotaxic amygdalotomy. *Proc. Aust. Assoc. Neurol.,* 5:277–278.

Balasubramaniam, V., and Ramamurthi, B. (1970): Stereotaxic amygdalotomy in behavior disorders. *Confin. Neurol.,* 32:367–373.

Balazs, T., and Dairman, W. (1967): Comparison of microsomal drug-metabolizing enzyme systems in grouped and individually caged rats. *Toxicol. Appl. Pharmacol.,* 10:409.

Balazs, T., Murphy, J. B., and Grice, H. C. (1962): The influence of environmental changes on the cardiotoxicity of isoprenoline in rats. *J. Pharm. Pharmacol.,* 14:750–755.

Ball, G. G., Micco, D. J., and Berntson, G. G. (1974): Cerebellar stimulation in the rat: Complex stimulation-bound oral behaviors and self-stimulation. *Physiol. Behav.,* 13:123–127.

Ban, T. A. (1969): The use of amphetamines in adult psychiatry. *Semin. Psychiatry,* 1:129–143.

Bandler, R. J., Jr. (1969): Facilitation of aggressive behaviour in rat by direct cholinergic stimulation of the hypothalamus. *Nature,* 224:1035–1036.

Bandler, R. J., Jr. (1970): Cholinergic synapses in the lateral hypothalamus for the control of predatory aggression in the rat. *Brain Res.,* 20:409–424.

Bandler, R. J. (1975): Predatory aggression: Midbrain-pontine junction rather than hypothalamus as the critical structure. *Aggressive Behav.,* 1:261–266.

Bandler, R. J. (1977): Predatory behavior in the cat elicited by lower brain stem and hypothalamic stimulation: A comparison. *Brain Behav. Evol.,* 14:440–460.

Bandler, R. J., Jr., and Chi, C. C. (1972): Effects of olfactory bulb removal on aggression: A reevaluation. *Physiol. Behav.,* 8:207–211.

Bandler, R. J., Jr., Chi, C. C., and Flynn, J. P. (1972): Biting attack elicited by stimulation of the ventral midbrain tegmentum of cats. *Science,* 177:364–366.

Bandler, R., and Fatouris, D. (1978): Centrally elicited attack behavior in cats: Post-stimulus excitability and midbrain-hypothalamic inter-relationships. *Brain Res.,* 153:427–433.

Bandler, R. J., Jr., and Flynn, J. P. (1974): Neural pathways from thalamus associated with regulation of aggressive behavior. *Science,* 183:96–99.

Bandler, R., Jr., and Moyer, K. E. (1970): Animals spontaneously attacked by rats. *Commun. Behav. Biol.,* 5:177–182.

Bandura, A. (1973): *Aggression: A Social Learning Analysis.* Prentice-Hall, Englewood Cliffs, N.J.

Bandura, A. (1977): *Social Learning Theory.* Prentice-Hall, Englewood Cliffs, N.J.

Bandura, A., and Walters, R. H. (1963): *Social Learning and Personality Development.* Holt, Rinehart and Winston, London.

Banerjee, U. (1971a): Influence of some hormones and drugs on isolation-induced aggression in male mice. *Commun. Behav. Biol.,* 6 pt. A:163–170.

Banerjee, U. (1971b): An inquiry into the genesis of aggression in mice induced by isolation. *Behaviour,* 50:86–92.

Banerjee, U. (1974): Modification of the isolation-induced abnormal behavior in male Wistar rats by destructive manipulation of the central monoaminergic systems. *Behav. Biol.,* 11:573–579.

Banin, A., and Navrot, J. (1975): Origin of life: Clues from relations between chemical compositions of living organisms and natural environments. *Science,* 189:550–551.

Banks, E. M. (1962): A time and motion study of prefighting behavior in mice. *J. Genet. Psychol.,* 101:165–183.

Baran, D., and Glickman, S. E. (1970): "Territorial marking" in the Mongolian gerbil: A study of sensory control and function. *J. Comp. Physiol. Psychol.,* 71:237–245.

Barclay, A. M., and Haber, R. N. (1965): The relation of aggressive to sexual motivation. *J. Pers.,* 33:462–475.

Bard, M., and Zacker, J. (1971): The prevention of family violence: Dilemmas of community intervention. *J. Marriage Fam.,* 33:677–682.

Bard, P. (1950): Central nervous mechanisms for the expression of anger in animals. In: *Feelings and Emotions,* edited by M. L. Reymert. The Moseheart Symposium. McGraw-Hill, New York.

Bard, P., and Mountcastle, V. B. (1948): Some forebrain mechanisms involved in expression of rage with special reference to suppression of angry behavior. *Res. Publ. Assoc. Res. Nerv. Ment. Dis.,* 27:362–404.

Bardwick, J. M. (1971): *Psychology of Women.* Harper and Row, New York.

Barnes, H. E. (1935): *The History of Western Civilization.* Hartcourt, Brace, New York.

Barnett, S. A. (1972): The ontogeny of behavior and the concept of instinct. In: *Brain and Human Behavior,* edited by A. G. Karczmar and J. C. Eccles, pp. 375–392. Springer-Verlag, Berlin.

Barnett, S. A. (1975): *The Rat: A Study in Behavior, Rev. Ed.,* p. 52. University of Chicago Press, Chicago.

Barnouw, V. (1963): *Culture and Personality,* Homewood, Ill.

Barnouw, V. (1971): *An Introduction to Anthropology: Ethnology. Vol. 2.* Homewood, Ill.

Baron, R. A. (1971): Aggression as a function of audience presence and prior anger arousal. *J. Exp. Soc. Psychol.,* 7:515–523.

Baron, R. A. (1974a): Aggression as a function of victim's pain cues, level of prior anger arousal, and exposure to an aggressive model. *J. Pers. Soc. Psychol.,* 29:117–124.

Baron, R. A. (1974b): The aggression-inhibiting influence of heightened sexual arousal. *J. Pers. Soc. Psychol.,* 30:318–322.

Baron, R. A. (1977a): *Human Aggression.* Plenum Press, New York.

Baron, R. A. (1977b): Heightened sexual arousal and physical aggression: An extension to females. Unpublished manuscript, Purdue University.

Baron, R. A., and Bell, P. A. (1973): Effects of heightened sexual arousal on physical aggression. *Proceedings of the American Psychological Association,* 81st. Annual Convention, pp. 171–172.

Baron, R. A., and Bell, P. A. (1976): Aggression and heat: The influence of ambient temperature, negative effect, and a cooling drink on physical aggression. *J. Pers. Soc. Psychol.,* 33:245–255.

Barr, G. A., Gibbons, J. L., and Moyer, K. E. (1975): The relationship between mouse-killing and intraspecific fighting in the albino rat. *Behav. Biol.,* 14:201–208.

Barr, G. A., Gibbons, J. L., and Bridger, W. H. (1976a): Neuropharmacological regulation of mouse killing by rats. *Behav. Biol.,* 17:143–159.

Barr, G. A., Gibbons, J. L., and Moyer, K. E. (1976b): Male-female differences and the influence of neonatal and adult testosterone on intraspecies aggression in rats. *J. Comp. Physiol. Psychol.,* 90:1169–1183.

Barraclough, C. A., and Cross, B. A. (1963): Unit activity in the hypothalamus of the cyclic female rat: Effect of genital stimuli and progesterone. *J. Endocrinol.,* 26:339–359.

Barrett, R. E., and Balch, T. St. (1971): Uptake of catecholamines into serotonergic nerve cells as demonstrated by fluorescence histochemistry. *Experientia,* 27:663–664.

Barrett, T. W. (1968): The relation of mind to brain. *Confin. Psychiatr.,* 11:133–153.

Barry, H., III, and Miller, N. E. (1962): Effects of drugs on approach-avoidance conflict tested repeatedly by means of a "telescope alley." *J. Comp. Physiol. Psychol.,* 55:201–210.

Barry, H., III, Wagner, A. R., and Miller, N. E. (1962): Effects of alcohol and amobarbital on performance inhibited by experimental extinction. *J. Comp. Physiol. Psychol.,* 55:464–468.

Barry, V. C., and Klawans, H. L. (1976): On the role of dopamine in the pathophysiology of anorexia nervosa. *J. Neural Transm.,* 38:107–122.

Barter, J. T., and Reite, M. (1969): Crime and LSD: The insanity plea. *Am. J. Psychiatry,* 126:531–537.

Basedow, H. (1925): *The Australian Aborigines,* Preace, Adelaide, Australia.

Baumgarten, H. G., Björklund, A., Holstein, A. F., and Nobin, A. (1972a): Chemical degeneration of indoleamine axons in the rat brain by 5,6-dihydroxytryptamine: An ultrastructural study. *Z. Zellforsch. Mikrosk. Anat.,* 129:256–271.

Baumgarten, H. G., Björklund, A., Lachenmayer, L., Nobin, A., and Stenevi, U. (1972b): Long-lasting selective depletion of brain serotonin by 5,6-dihydroxytryptamine. *Acta Physiol. Scand.,* 84:Suppl. 373:1–15.

Baumgarten, H. G., Evetts, K. D., Holman, R. B., Iversen, L. L., Vogt, M., and Wilson, G. (1972c): Effects of 5,6-dihydroxytryptamine on monoaminergic neurones in the central nervous system of the rat. *J. Neurochem.,* 19:1587–1597.

Baumgarten, H. G., and Lachenmayer, L. (1972): 5,7-Dihydroxytryptamine: Improvement in chemical lesioning of indoleamine neurons in the mammalian brain. *Z. Zellforsch. Mikrosk. Anat.,* 135:399–414.

Beach, F. A. (1937): The neural basis of innate behavior. I. Effects of cortical lesions upon the maternal behavior pattern in the rat. *J. Comp. Psychol.,* 24:393–436.

Beach, F. A. (1943): Effects of injury to the cerebral cortex upon the display of masculine and feminine mating behavior by female rats. *J. Comp. Psychol.,* 36:169–198.

Beach, F. A. (1967): Cerebral and hormonal control of reflexive mechanisms involved in copulatory behavior. *Physiol. Rev.,* 47:289–316.

Beach, F., and Jaynes, J. (1954): Effects of early experience upon the behavior of animals. *Psychol. Bull.,* 51:239–263.

Bear, D. (1977): The significance of behavioral change in temporal lobe epilepsy. *McLean Hosp. J.,* June, p. 9–21.

Bear, D. M., and Fedio, P. (1977): Quantitative analysis of interictal behavior in temporal lobe epilepsy. *Arch. Neurol.,* 34:454–467.

Bechtereva, N. P., and Bondartchuk, A. N. (1968): An optimization of stages of the surgical treatment of hyperkinesis. *Vopr. Neurokhir.,* 3:39–44.

Bechtereva, N. P., Bondartchuk, A. N., Smirnov, V. M., and Meliutcheva, L. A. (1972): Therapeutic electrostimulations of the brain deep structures. *Vopr. Neurokhir.,* 1:7–12.

Bechtereva, N. P., Bondartchuk, A. N., Smirnov, V. M., Meliutcheva, L. A., and Shandurina, A. N. (1975): Method of electrostimulation of the deep brain structures in treatment of some chronic diseases. *Confin. Neurol.*, 37:136–140.

Beck, D. (1977): Das "Koryphäen-Killer-Syndrom". *Dtsch. Med. Wochenschr.*, 102:303–307.

Becker, W. C., Peterson, D. R., Luria, Z., Schoemaker, D. J., and Hellmer, L. A. (1962): Relations of factors derived from parent interview rating to behavior problems of five-year-olds. *Child. Dev.*, 33:509–535.

Beddington, J. R., Free, C. A., and Lawton, J. H. (1975): Dynamic complexity in predator-prey models framed in difference equations. *Nature*, 255:58–60.

Beeman, E. A. (1947): The effect of male hormone on aggressive behavior in mice. *Physiol. Zool.*, 20:373–405.

Belford, G. R., and Killackey, H. P. (1979): Vibrissae representation in subcortical trigeminal centers of the neonatal rat. *J. Comp. Neurol.*, 183:305–322.

Bell, R. (1978): Hormone influences on human aggression. *Ir. J. Med. Sci.*, 147: Suppl. 1:5–9.

Bell, R., and Brown, K. (1975a): Failure of penthlamine and sotalol, injected into the lateral hypothalamus, to affect shock-induced aggression in rats: Evidence against a mediating role for norepinephrine. *IRCS Med. Sci.*, 3:623.

Bell, R., and Brown, K. (1975b): Testosterone and YG 19-256: Effects on home-cage aggression and locomotor activity in mice. *IRCS Med. Sci.*, 3:202.

Bell, R., and Brown, K. (1976a): Failure of dopamine and spiroperidol, injected into the lateral hypothalamus, to affect shock-induced aggression in rats: Evidence against a mediating role for dopamine. *IRCS Med. Sci.*, 4:527.

Bell, R., and Brown, K. (1976b): Lateral hypothalamus and shock-induced aggression—cholinergic mediation. *IRCS Med. Sci.*, 4:83–84.

Bell, R., and Brown, K. (1977): Lateral hypothalamus and shock-induced aggression. Serotonergic inhibition. *IRCS Med. Sci.*, 5:221.

Bender, L., and Yarnell, H. (1941): An observation nursery. A study of 250 children on the psychiatric division of Bellevue Hospital. *Am. J. Psychiatry*, 97:1158–1174.

Bénézech, M. (1976): *Aberration du Chromosome Y en Pathologie Médico-Legale*. Masson, Paris.

Ben-Ishay, D., and Welner, A. (1969): Sensitivity to experimental hypertension and aggressive reactions in rats. *Proc. Soc. Exp. Biol. Med.*, 132:1170–1173.

Benjamin, R. M., and Burton, H. (1968): Projection of taste nerve afferents to anterior opercular-insular cortex in squirrel monkey *(Saimiri sciureus)*. *Brain Res.*, 7:221–231.

Benkert, O., Gluba, H., and Matussek, N. (1973a): Dopamine, noradrenaline and 5-hydroxytryptamine in relation to motor activity, fighting and mounting behaviour. 1. L-Dopa and d,1-threo-dihydroxyphenylserine in conbination with R0-4-4602, pargyline and reserpine. *Neuropharmacology*, 12:177–186.

Benkert, O., Renz, A., and Matussek, N. (1973b): Dopamine, noradrenaline, and 5-hydroxytryptamine in relation to motor activity, fighting and mounting behaviour. 2. L-Dopa and d,1-threo-dihydroxyphenylserine in combination with R0-4-4602 and parachlorophenylalanine. *Neuropharmacology*, 12:187–193.

Bennett, E. L., Diamond, M. C., Krech, D., and Rosenzweig, M. R. (1964): Chemical and anatomical plasticity of brain. *Science*, 146:610–619.

Bennett, E. L., and Rosenzweig, M. R. (1968): Brain chemistry and anatomy: Implications for theories of learning and memory. In: *Mind as a Tissue*, edited by C. Rupp, pp. 63–86. Harper and Row, New York.

Bennett, E. L., Rosenzweig, M. R., and Diamond, M. C. (1969): Rat brain: Effects of environmental enrichment on wet and dry weights. *Science*, 163:825–826.

Bennett, E. L., Rosenzweig, M. R., and Diamond, M. C. (1970): Time courses of effects of differential experiences on brain measures and behavior or rats. In: *Molecular Approaches to Learning and Memory*, edited by W. L. Byrne, pp. 55–89. Academic Press, New York.

Bennett, E. L., Rosenzweig, M. R., and Wu, S. Y. C. (1973): Excitant and depressant drugs modulate effects of environment on brain weight and cholinesterases. *Psychopharmacologia*, 33:309–328.

Bennie, E., and Sclare, A. (1969): The battered child syndrome. *Am. J. Psychiatry*, 125:975–979.

Benson, A. A. (1966): On the orientation of lipsids in chloroplast and cell membranes. *J. Am. Oil Chem. Soc.*, 43:265–270.

Berg, D., and Baenninger, R. (1974): Predation: Separation of aggressive and hunger motivation by conditioned aversion. *J. Comp. Physiol. Psychol.*, 86:601–606.

Berkowitz, B. A., and Spector, S. (1971): The effect of caffeine and theophylline on the disposition of brain serotonin in the rat. *Eur. J. Pharmacol.*, 16:322–325.

Berkowitz, B. A., Spector, S., and Pool, W. (1971): The interaction of caffeine, theophylline and theobromine with monoamine oxidase inhibitors. *Eur. J. Pharmacol.*, 16:315–321.

Berkowitz, B. A., Tarver, J. H., and Spector, S. (1970): Release of norepinephrine in the central nervous system by theophylline and caffeine. *Eur. J. Pharmacol.*, 10:64–71.

Berkowitz, L. (1962): *Aggression: A Social Psychological Analysis.* McGraw-Hill, Maidenhead.

Berkowitz, L. (1965): The concept of aggressive drive: Some additional considerations. *Adv. Exp. Sociol. Psychol.*, 2:301–329.

Berkowitz, L. (1969): The frustration-aggression hypothesis revisited. In: *Roots of Aggression: A Re-Examination of the Frustration-Aggression Hypothesis,* edited by L. Berkowitz, Atherton Press, New York.

Berkowitz, L. (1974): External determinants of impulsive aggression. In: *Determinants and Origins of Aggressive Behavior,* edited by J. de Wit and W. W. Hartup, pp. 147–165. Mouton, The Hague.

Berkowitz, L., and LePage, A. (1967): Weapons as aggression-eliciting stimuli. *J. Pers. Soc. Psychol.*, 7:202–207.

Berman, A. J., Berman, D., and Prescott, J. W. (1974): The effect of cerebellar lesions on emotional behavior in the rhesus monkey. In: *The Cerebellum, Epilepsy and Behavior,* edited by I. S. Cooper, M. Riklan, and R. S. Snider, pp. 277–284. Plenum Press, New York.

Bernard, B. K. (1974): Frog killing (ranacide) in the male rat: Lack of effect of hormonal manipulations. *Physiol. Behav.*, 12:405–408.

Bernard, B. (1976): Testosterone manipulations: Effects on ranacide aggression and brain monoamines in the adult female rat. *Pharmacol. Biochem. Behav.*, 4:59–65.

Bernard, B. K., and Finkelstein, E. R. (1975): Alterations in mouse aggressive behavior and brain monoamine dynamics as a function of age. *Physiol. Behav.*, 15:731–736.

Bernard, B. K., and Paolino, R. M. (1974): Time-dependent changes in brain biogenic amine dynamics following castration in male rats. *J. Neurochem.*, 22:951–956.

Bernard, B. K., and Paolino, R. M. (1975): Temporal effects of castration on emotionality and shock-induced aggression in adult male rats. *Physiol. Behav.*, 14:201–206.

Berne, E. (1964): *Games People Play.* Grove Press, New York.

Bernstein, H., and Moyer, K. E. (1970): Aggressive behavior in the rat: Effects of isolation, and olfactory bulb lesions. *Brain Res.*, 20:75–84.

Bernstein, I. S. (1964): Role of the dominant male rhesus monkey in response to external challenges to the group. *J. Comp. Physiol. Psychol.*, 57:404–406.

Bernstein, I. S., and Gordon, T. P. (1974): The function of aggression in primate societies. *Am. Sci.*, 62:304–311.

Bernstein, L. A. (1952): A note on Christie's "Experimental naïveté and experiential naïveté," *Psychol. Bull.*, 49:38–40.

Berntson, G. G. (1972): Blockade and release of hypothalamically and naturally elicited aggressive behaviors in cats following midbrain lesions. *J. Comp. Physiol. Psychol.*, 81:541–554.

Berntson, G. G. (1973): Attack, grooming and threat elicited by stimulation of the pontine tegmentum in cats. *Physiol. Behav.*, 11:81–87.

Berntson, G. G., and Micco, D. J. (1976): Organization of brainstem behavioral systems. *Brain Res. Bull.*, 1:471–483.

Berntson, G. G., Potolicchio, S. J., Jr., and Miller, N. E. (1973): Evidence for higher functions of the cerebellum: Eating and grooming elicited by cerebellar stimulation in cats. *Proc. Natl. Acad. Sci. USA,* 70:2497–2499.

Bevan, W., Jr., Bloom, W. L., and Lewis, G. T. (1951): Levels of aggressiveness in normal and amino acid deficient albino rats. *Physiol. Zool.*, 24:231–237.

Bevan, W., Daves, W. F., and Levy, G. W. (1960): The relation of castration, androgen therapy and pre-test fighting experience to competitive aggression in male C57 BL/10 mice. *Anim. Behav.*, 8:6–12.

Bhargava, H. N., and Way, E. L. (1974): Brain acetylcholine (ACh) and choline (Ch) changes during acute and chronic morphinization and during abrupt and naloxone precipitated withdrawal in morphine tolerant-dependent mice and rats. *Proc. West. Pharmacol. Soc.*, 17:173–177.

Bindra, D. (1959): *Motivation: A Systematic Reinterpretation.* Ronald Press, New York.

Bioulac, B., Benezech, M., Renaud, B., and Noël, B. (1978): Biogenic amines in 47, XYY syndrome. *Neuropsychobiology*, 4:366–370.

Bishop, G. H. (1956): Natural history of the nerve impulse. *Physiol. Rev.*, 36:376–399.

Bishop, G. H., and Clare, M. H. (1955): Organization and distribution of fibres in the optic tract of the cat. *J. Comp. Neurol.*, 103:269–304.

Bishop, M. P., Elder, S. T., and Heath, R. G. (1963): Intracranial self-stimulation in man. *Science*, 140:394–396.

Bittman, R., and Essman, W. B. (1970): 5-Hydroxytryptamine-nucleic acid interactions. Implications for physical and in vivo studies as a model for neural function. Abstract Papers, 3rd Annual Winter Conference on Brain Research, Snowmass-at-Aspen, Colo.

Blackburn, R. (1968a): Emotionality, extraversion and aggression in paranoid and nonparanoid schizophrenic offenders. *Br. J. Psychiatry*, 115:1301–1302.

Blackburn, R. (1968b): Personality in relation to extreme aggression in psychiatric offenders. *Br. J. Psychiatry*, 114:821–828.

Blacker, K. H., and Wong, N. (1963): Four cases of autocastration. *Arch. Gen. Psychiatry*, 8:169–176.

Blanc, A. C. (1961): Some evidence for the ideologies of Early man. In: *Social Life of Early Man*, edited by S. L. Washburn. Viking Fund Publications in Anthropology, No. 31.

Blanchard, D. C., Blanchard, R. J., Takahashi, L. K., and Takahashi, T. (1977): Septal lesions and aggressive behavior. *Behav. Biol.*, 21:157–161.

Blanchard, R. J., and Blanchard, D. C. (1968): Limbic lesions and reflexive fighting. *J. Comp. Physiol. Psychol.*, 66:603–605.

Blanchard, R. J., and Blanchard, D. C. (1969a): Crouching as an index of fear. *J. Comp. Physiol. Psychol.*, 67:370–375.

Blanchard, R. J., and Blanchard, D. C. (1969b): Passive and active reactions to fear-eliciting stimuli. *J. Comp. Physiol. Psychol.*, 68:129–135.

Blanchard, R. J., and Blanchard, D. C. (1970a): Dual mechanisms in passive avoidance. I. *Psychonomic Sci.*, 19:1–2.

Blanchard, R. J., and Blanchard, D. C. (1970b): Dual mechanisms in passive avoidance. II. *Psychonomic Sci.*, 19:3–4.

Blanchard, R. J., and Blanchard, D. C. (1971): Defensive reactions in the albino rat. *Learning Motivation*, 2:351–362.

Blanchard, R. J., and Blanchard, D. C. (1977): Aggressive behavior in the rat. *Behav. Biol.*, 21:197–224.

Blanchard, R. J., Blanchard, D. C., and Fial, R. A. (1970): Hippocampal lesions in rats and their effect on activity, avoidance, and aggression. *J. Comp. Physiol. Psychol.*, 71:92–102.

Blanchard, R. J., Blanchard, D. C., and Takahashi, L. K. (1978): Pain and aggression in the rat. *Behav. Biol.*, 23:291–305.

Blanchard, R. J., Fukunaga, K., Blanchard, D. C., and Kelley, M. J. (1975): Conspecific aggression in the laboratory rat. *J. Comp. Physiol. Psychol.*, 89:1204–1209.

Blanchard, R. J., Kelley, M. J., and Blanchard, D. C. (1974): Defensive reactions and exploratory behavior in rats. *J. Comp. Physiol. Psychol.*, 87:1129–1133.

Blanchard, R. J., Mast, M., and Blanchard, D. C. (1975): Stimulus control of defensive reactions in the albino rats. *J. Comp. Physiol. Psychol.*, 88:81–88.

Blick, N., Gros, C. M., and Ropartz, P. (1971): Effet de la densité du groupement sur l'agressivité et les fonctions reproductrices des souris femelles. *C. R. Acad. Sci. [D] (Paris)*, 272:293–296.

Blinkov, S. M., and Glezer, I. I. (1968): *The Human Brain in Figures and Tables. A Quantitative Handbook*. Plenum Press, New York.

Bliss, E. L., Ailion, J., and Zwanziger, J. (1968): Metabolism of norepinephrine, serotonin and dopamine in rat brain with stress. *J. Pharmacol. Exp. Ther.*, 164:122–134.

Blondaux, C., Juge, A., Sordet, F., Chouvet, G., Jouvet, M., and Pujol, J.-F. (1973): Modification du métabolisme de la sérotonine (5HT) cérébrale induite chez le rat par administration de 6-hydroxydopamine. *Brain Res.*, 50:101–114.

Bloom, F. E., Algeri, S., Groppetti, A., Revuelta, A., and Costa, E. (1969): Lesions of central norepinephrine terminals with 6-OH dopamine: Biochemistry and fine structure. *Science*, 166:1284–1286.

Bloomquist, E. R. (1968): *Marihuana*. Glencoe Press, Beverly Hills, Ca.

Blumer, D. (1969): Transsexualism, sexual dysfunction and temporal lobe disorder. In: *Transsexualism and Sex Reassignment,* edited by R. Green and J. Money, Johns Hopkins Press, Baltimore.

Blumer, D. (1970): Hypersexual episodes in temporal lobe epilepsy. *Am. J. Psychiatry,* 126:1099–1106.

Blumer, D. P., Williams, H. W., and Mark, V. H. (1974): The study and treatment, on a neurological ward, of abnormally aggressive patients with focal brain disease. *Confin. Neurol.,* 36:125–176.

Bocknik, S. E., and Kulkarni, A. S. (1974): Effect of a decarboxylase inhibitor (R0-4-4602) on 5-HTP induced muricide blockade in rats. *Neuropharmacology,* 13:279–281.

Boehlke, K. W., and Eleftheriou, B. E. (1967): Levels of monoamine oxidase in the brain of C57 BL/6J mice after exposure to defeat. *Nature,* 213:739–740.

Bohman, M. (1978): Some genetic aspects of alcoholism and criminality. A population of adoptees. *Arch. Gen. Psychiatry,* 35:269–276.

Bok, S. T. (1959): *Histonomy of the Cerebral Cortex.* Elsevier, Amsterdam.

Bolles, R. C. (1970): Species-specific defense reactions and avoidance learning. *Psychol. Rev.,* 77:32–48.

Bolton, R. (1973): Aggression and hypoglycemia among the Qolla: A study in psychobiological anthropology. *Ethnology,* 12:227–257.

Bolton, R., and Vadheim, C. (1973): The ecology of East African homicide. *Behav. Sci. Notes,* 8:319–342.

Boman, K. (1964): Psychological testing before and after a period of isolation. *Acta Psychiatr. Scand.,* 40 Suppl. 180:463–467.

Borden, R. J. (1975): Witnessed aggression: Influence of an observer's sex and values on aggressive responding. *J. Pers. Soc. Psychol.,* 31:567–573.

Borden, R. J., and Taylor, S. P. (1973): The social instigation and control of physical aggression. *J. Appl. Soc. Psychol.,* 3:354–361.

Borgaonkar, D. S., Unger, W. M., Moore, S. M., and Crofton, T. A. (1972): 47, XYY Syndrome, height and institutionalization of juvenile delinquents. *Br. J. Psychiatry,* 120:549–550.

Borgen, L. A., Khalsa, J. H., King, W. T., and Davis, W. M. (1970): Strain differences in morphine-withdrawal-induced aggression in rats. *Psychonomic Sci.,* 21:35–36.

Boshka, S. C., Weisman, H. M., and Thor, D. H. (1966): A technique for inducing aggression in rats utilizing morphine withdrawal. *Psychol. Rec.,* 16:541–543.

Bourgault, P. C., Karczmar, A. G., and Scudder, C. L. (1963): Contrasting behavioral, pharmacological, neurophysiological, and biochemical profiles of C57 BL/6 and SC-I strains of mice. *Life Sci.,* 8:533–553.

Bourne, P. G., Rose, R. M., and Mason, J. W. (1968): 17-OHCS levels in combat. Special forces "A" team under threat of attack. *Arch. Gen. Psychiatry,* 19:135–140.

Bowers, M. B. (1972a): Acute psychosis induced by psychotomimetic drug abuse. I. Clinical findings. *Arch. Gen. Psychiatry,* 27:437–440.

Bowers, M. B. (1972b): Acute psychosis induced by psychotomimetic drug abuse. II. Neurochemical findings. *Arch. Gen. Psychiatry,* 27:440–442.

Bowlby, J. (1952): Maternal care and mental health. *WHO Monogr. Ser. 2,* World Health Organization, Geneva.

Bowlby, J. (1953): Some pathological processes set in train by early mother-child separation. *J. Ment. Sci.,* 99:265–272.

Boyd, E. M., Dolman, M., Knight, L. M., and Sheppard, E. P. (1965): The chronic oral toxicity of caffeine. *Can. J. Physiol. Pharmacol.,* 43:995–1007.

Brady, J. V. (1955): Emotional behavior and the nervous system. *Trans. N.Y. Acad. Sci.,* 18:601–612.

Brady, J. V., and Nauta, W. J. H. (1953): Subcortical mechanisms in emotional behavior: Affective changes following septal forebrain lesions in the albino rat. *J. Comp. Physiol. Psychol.,* 46:339–346.

Brain, P. (1975): What does individual housing mean to a mouse? *Life Sci.,* 16:187–200.

Brain, P. F. (1972): Endocrine and behavioral differences between dominant and subordinate male house mice housed in pairs. *Psychonomic Sci.,* 28:260–262.

Brain, P. F., and Al-Maliki, S. (1978): A comparison of "intermale fighting" in "standard opponent" tests and attack directed towards locusts by "TO" strain mice: Effects of simple experimental manipulations. *Anim. Behav.,* 26:723–737.

Brain, P. F., and Benton, D. (1977): What does individual housing mean to a research worker? *IRCS Med. Sci.,* 5:459–463.

Brain, P. F., Benton, D., and Bolton, J. C. (1978): Comparison of agonistic behavior in individually-housed male mice with those cohabiting with females. *Aggressive Behav.,* 4:201–206.

Brain, P. F., and Evans, C. M. (1974a): Effect of androgens on the attackability of gonadectomized mice by TO trained fighter individuals: Confirmatory experiments. *IRCS Med. Sci.* 2:173.

Brain, P. F., and Evans, C. M. (1974b): Influences of two naturally occurring androgens on the attack directed by "trained fighter" TO strain mice towards castrated mice of three different strains. *IRCS Med. Sci.,* 2:1672.

Brain, P. F., and Nowell, N. W. (1970): The effects of differential grouping on endocrine function of mature male albino mice. *Physiol. Behav.,* 5:907–910.

Brain, P., and Poole, A. (1974): Some studies on the use of "standard opponents" in intermale aggression testing in TT albino mice. *Behaviour,* 50:100–110.

Branchey, L., and Friedhoff, A. J. (1973): The influence of ethanol administered to pregnant rats on tyrosine hydroxylase activity of their offspring. *Psychopharmacologia,* 32:151–156.

Brase, D. A., and Loh, H. H. (1975): Possible role of 5-hydroxytryptamine in minimal brain dysfunction. *Life Sci.,* 16:1005–1015.

Brazier, M. A. B. (1970): Regional activities within the human hippocampus and hippocampal gyrus. *Exp. Neurol.,* 26:354–368.

Breese, G. R., Cooper, B. R., Grant, L. D., and Smith, R. D. (1974a): Biochemical and behavioural alterations following 5,6-dihydroxytryptamine administration into brain. *Neuropharmacology,* 13:177–187.

Breese, G. R., Cooper, B. R., and Hollister, A. S. (1974b): Relationship of biogenic amine to behavior. *J. Psychiatr. Res.,* 11:125–133.

Breggin, P. R. (1976): Psychosurgery for political purposes. *Am. J. Psychiatry,* 133:864.

Bremer, J. (1959): *Asexualization.* Macmillan, New York.

Brill, H. (1969): Psychosurgery today. *Dis. Nerv. Syst.,* GWAN Suppl., 30:54–55.

Brizzee, K. R., Vogt, J., and Kharetchko, X. (1964): Postnatal changes in glia/neuron index with a comparison of methods of cell enumeration in the white rat. *Prog. Brain Res.,* 4:136–146.

Broadhurst, P. L. (1957): Determinants of emotionality in the rat. 1. Situational factors. *Br. J. Psychol.,* 48:1–12.

Broadhurst, P. L. (1960): Application of biometrical genetics to the inheritance of behaviour. In: *Experiments in Personality, Vol. 1, Psychogenetics and Psychopharmacology,* edited by H. J. Eysenck, pp. 3–102. Routledge & Kegan Paul, London.

Broca, P. (1878): Anatomie comparée des circomvolutions cérébrales. Le grand lobe limbique et la scissure limbique dans la serie des mammifères. *Rev. Anthropol.,* 1:385–498.

Brock, T. C., and Buss, A. H. (1962): Dissonance, aggression, and evaluation of pain. *J. Abnorm. Soc. Psychol.,* 65:197–202.

Brodie, B. B., and Costa, E. (1962): Some current views on brain monoamines. In: *Monoamines et Système Nerveux Central,* edited by J. de Ajuriaguerra, pp. 13–49. George & Cie S. A., Geneva.

Brodie, B. B., and Shore, P. A. (1957): A concept for a role of serotonin and norepinephrine as chemical mediators in the brain. *Ann. N.Y. Acad. Sci.,* 66:631–642.

Brody, E. B., and Rosvold, H. E. (1952): Influence of prefrontal lobotomy on social interaction in a monkey group. *Psychosom. Med.,* 14:406–415.

Brody, J. F., DeFeudis, P. A., and DeFeudis, F. V. (1969): Effects of micro-injections of L-glutamate into the hypothalamus on attack and flight behaviour in cats. *Nature,* 224:1330.

Bromberg, W. (1934): Marihuana intoxication: A clinical study of *Cannabis sativa* intoxication. *Am. J. Psychiatry,* 91:303–330.

Bronson, F. H. (1973): Establishment of social rank among grouped male mice: Relative effects on circulating FSH, LH, and corticosterone. *Physiol. Behav.,* 10:947–951.

Bronson, F. H., and Desjardins, C. (1968): Aggression in adult mice: Modification by neonatal injections of gonadal hormones. *Science,* 161:705–706.

Bronson, F. H., and Desjardins, C. (1970): Neonatal androgen administration and adult aggressiveness in female mice. *Gen. Comp. Endocrinol.,* 15:320–325.

Bronson, F. H., and Desjardins, C. (1971): Steroid hormones and aggressive behavior in mammals. In: *The Physiology of Aggression and Defeat,* edited by B. Eleftheriou and J. Scott, pp. 43–64. Plenum Press, London.

Brown, G. W., and Cohen, B. D. (1959): Avoidance and approach learning motivated by stimulation of identical hypothalamic loci. *Am. J. Physiol.*, 197:153–157.

Brown, H. (1976): *Brain and Behavior: A Textbook of Physiological Psychology.* Oxford University Press, New York.

Brown, J. S., and Farber, I. E. (1951): Emotions conceptualized and intervening variables—with suggestions toward a theory of frustration. *Psychol. Bull.*, 48:465–495.

Brownstein, M., Saavedra, J. M., and Palkovits, M. (1974): Norepinephrine and dopamine in the limbic system of the rat. *Brain Res.*, 79:431–436.

Bruce, H. M. (1959): An exteroceptive block to pregnancy in the mouse. *Nature*, 184:105.

Bruce, H. M. (1960): A block to pregnancy in the mouse caused by proximity of strange males. *J. Reprod. Fertil.*, 1:96–103.

Brutkowski, S., Fonberg, E., and Mempel, E. (1961): Angry behavior in dogs following bilateral lesions in the genual portion of the rostral cingulate gyrus. *Acta Biol. Exp.*, 21:199–205.

Bryson, Y., Sakati, N., Nyhan, W. L., and Fish, C. H. (1971): Self-mutilative behavior in the Cornelia de Lange syndrome. *Am. J. Ment. Defic.*, 76:319–324.

Bucher, K., Myers, R. E., and Southwick, C. (1970): Anterior temporal cortex and maternal behavior in monkey. *Neurology (Minneap.)*, 20:415.

Bueker, E. D., Schenkein, I., and Bane, J. L. (1960): The problem of distribution of a nerve growth factor specific for spinal and sympathetic ganglia. *Cancer Res.*, 20:1220–1228.

Bugbee, N. M., and Eichelman, B. S. (1972): Sensory alterations and aggressive behavior in the rat. *Physiol. Behav.*, 8:981–985.

Bunge, M. (1977): Emergence and the mind. *Neuroscience*, 2:501–509.

Bunnell, B. N. (1966): Amygdaloid lesions and social dominance in the hooded rat. *Psychonom. Sci.*, 6:93–94.

Bunnell, B. N., Bemporad, J. R., and Flesher, C. K. (1966): Septal forebrain lesions and social dominance behavior in the hooded rat. *Psychonom. Sci.*, 6:207–208.

Bunnell, B. N., and Smith, M. H. (1966): Septal lesions and aggressiveness in the cotton rat. *Psychonom. Sci.*, 6:443–444.

Burke, C. (1975): *Agression in Man.* Lyle Stuart, Inc., Secaucus, New York.

Burn, J. H., and Hobbs, R. (1958): A test for tranquilizing drugs. *Arch. Int. Pharmacodyn. Ther.*, 113:290–295.

Burns, B. D. (1956): The electrophysiological approach to the problem of learning. *Can. J. Biochem. Physiol.*, 34:380–389.

Burns, R. D. (1968): The role of agonistic behavior in regulation of density in Uinta ground squirrels, *Citellus armatus.* Masters thesis, Utah State University, Logan, Utah.

Burnstock, G. (1972): Purinergic nerves. *Pharmacol. Rev.*, 24:509–581.

Burnstock, G. (1972): Do some nerve cells release more than one transmitter? *Neuroscience*, 1:239–248.

Burnstock, G., and Costa, M. (1975): *Adrenergic Neurons: Their Organisation, Function, and Development in the Peripheral Nervous System.* Chapman & Hall, London.

Bursten, B., and Delgado, J. M. R. (1958): Positive reinforcement induced by intracranial stimulation in the monkey. *J. Comp. Physiol. Psychol.*, 51:6–10.

Burt, C. (1962): The concept of consciousness. *Br. J. Psychol.*, 53:229–242.

Buser, P. (1976): Higher functions of the nervous system. *Annu. Rev. Physiol.* 38:217–245.

Buss, A. H. (1961): *The Psychology of Aggression.* John Wiley, New York.

Buss, A. H. (1963): Physical aggression in relation to different frustrations. *J. Abnorm. Soc. Psychol.*, 67:1–7.

Buss, A. H. (1966): The effect of harm on subsequent aggression. *J. Exp. Res. Pers.*, 1:249–255.

Buss, A. H. (1971): Aggression pays. In: *The Control of Aggression and Violence. Cognitive and Physiological Factors,* edited by J. L. Singer. Academic Press, New York.

Butcher, L. L., and Dietrich, A. P. (1973): Effects of shock-elicited aggression in mice of preferentially protecting brain monoamines against the depleting action of reserpine. *Naunyn Schmiedebergs Arch. Pharmacol.*, 277:61–70.

Butcher, S. H., Butcher, L. L., and Cho, A. K. (1976): Modulation of neostriatal acetylcholine in the rat by dopamine and 5-hydroxytryptamine afferents. *Life Sci.*, 18:733–744.

Butterfield, E. C., and Zigler, E. (1965): The influence of differing institutional social climates on the effectiveness of social reinforcement in the mentally retarded. *Am. J. Ment. Defic.*, 70:48–56.

Byles, J. A. (1978): Violence, alcohol problems and other problems in disintegrating families. *J. Stud. Alcohol.,* 39:551–553.

Cafiero, G. (1979): Prontuario dei conflitti: Ed ecco le possibili guerre. *IlSettimanale,* 6:20–22.

Cain, D. P. (1974): Olfactory bulbectomy: Neural structures involved in irritability and aggression in the male rat. *J. Comp. Physiol. Psychol.,* 86:213–220.

Cain, D. P., and Paxinos, G. (1974): Olfactory bulbectomy and mucosal damage: Effects on copulation, irritability, and interspecific aggression in male rats. *J. Comp. Physiol. Psychol.,* 86:202–212.

Cairns, R. B., and Nakelski, J. S. (1971): On fighting mice: Ontogenetic and experimental determinants. *J. Comp. Physiol. Psychol.,* 74:354–364.

Cairns, R. B., and Scholz, S. D. (1973): Fighting in mice: Dyadic escalation and what is learned. *J. Comp. Physiol. Psychol.,* 85:540–550.

Cairns, R. B., and Werboff, J. (1967): Behavior development in the dog: An interspecific analysis. *Science,* 158:1070–1072.

Calhoun, J. B. (1973): Death squared: The explosive growth and demise of a mouse population. *Proc. R. Soc. Med.,* 66: pt. 2, 80–86.

Campbell, B. A., and Raskin, L. A. (1978): Ontogeny of behavioral arousal: The role of environmental stimuli. *J. Comp. Physiol. Psychol.,* 92:176–184.

Campbell, H. E. (1967): The violent sex offender: A consideration of emasculation in treatment. *Rocky Mt. Med. J.,* 64:40–43.

Campbell, H. J. (1972): Peripheral self-stimulation as a reward in fish, reptile and mammal. *Physiol. Behav.,* 8:637–640.

Carder, B., and Olson, J. (1972): Marihuana and shock-induced aggression in rats. *Physiol. Behav.,* 8:599–602.

Carder, B., and Sbordone, R. (1975): Mescaline treated rats attack immobile targets. *Pharmacol. Biochem. Behav.,* 3:923–925.

Carlini, E. A. (1977): Further studies of the aggressive behavior induced by Δ^9-tetrahydrocannabinol in REM sleep-deprived rats. *Psychopharmacology,* 53:135–145.

Carlini, E. A. (1978): Effects of cannabinoid compounds on aggressive behavior. *Mod. Probl. Pharmacopsychiat.,* 13:82–102.

Carlini, E. A., and Green, J. P. (1963): Acetylcholine activity in the sciatic nerve. *Biochem. Pharmacol.,* 12:1367–1376.

Carlini, E. A., and Kramer, C. (1965): Effects of *Cannabis sativa* (marihuana) on maze performance of the rats. *Psychopharmacologia,* 7:175–181.

Carlini, E. A., Hamaoui, A., and Märtz, R. M. W. (1972): Factors influencing the aggressiveness elicited by marihuana in food-deprived rats. *Br. J. Pharmacol.,* 44:794–804.

Carlini, E. A., Lindsey, C. J., and Tufik, S. (1977): Cannabis, catecholamines, rapid eye movement sleep and aggressive behavior. *Br. J. Pharmacol.,* 61:371–379.

Carlini, E. A., and Masur, J. (1969): Development of aggressive behavior in rats by chronic administration of *Cannabis sativa* (marihuana). *Life Sci.,* 8, pt. 1:607–620.

Carlini, E. A., and Masur, J. (1970): Development of fighting behavior in starved rats by chronic administration of $(-)\Delta^9$-trans-tetrahydrocannabinol and cannabis extracts. Lack of action of other psychotropic drugs. *Commun. Behav. Biol. [A],* 5:57–61.

Carpenter, C. R. (1934): A field study of the behavior and social relation of howling monkeys *(Alanatta polliata). Comp. Psychol. Mono. Series no. 48,* 10:1–168.

Carpenter, C. R. (1940): A field study of Siam of the behavior and social relations of the gibbon *(Hylobates lar.) Comp. Psychol. Monogr. Series no. 84,* 16:1–212.

Carpenter, C. R. (1958): Territoriality: A review of concepts and problems. In: *Behavior and Evolution,* edited by A. Roe and G. G. Simpson, pp. 224–250. Yale University Press, New Haven.

Carpenter, C. R. (1963): Societies of monkey and apes. In: *Primate Social Behavior,* edited by C. H. Southwick. Van Nostrand, Princeton, N.J.

Carroll, B. J., and Steiner, M. (1978): The psychobiology of premenstrual dysphorie: The role of prolactin. *Psychoneuroendocrinology,* 3:171–180.

Carroll, R. L. (1969): Problems of the origin of reptiles. *Biol. Rev.,* 44:393–432.

Carroll, R. L., and Baird, D. (1972): Carboniferous stem reptiles in the family Romeriidae. *Bull. Museum Comp. Zool. (Harvard),* 143:321–363.

Carthy, J. D., and Ebling, F. J. (1964): Prologue and epilogue. In: *The Natural History of Aggression,* edited by J. D. Carthy and F. J. Ebling, pp. 1–5. Academic Press, New York.

Casler, L. (1963): Maternal deprivation: A critical review of the literature. *Monogr. Soc. Res. Child Dev.,* 26:3–31.

Cattell, J. P. (1969): Results with psychosurgery. *Dis. Nerv. Syst., G WAN Suppl.,* 30:57.

Cavalli-Sforza, L. L. (1975): Quantitative genetic perspectives: Implications for human development. In: *Developmental Human Behavior Genetics,* edited by K. W. Schaie, V. E. Anderson, G. E. McClearn, and J. Money. Lexington Books, Lexington, Mass.

Cavalli-Sforza, L. L., and Feldman, M. W. (1973): Cultural versus biological inheritance: Phenotypic transmission from parents to children (a theory of the effect of parental phenotypes on children's phenotypes). *Am. J. Hum. Genet.,* 25:618–637.

Chance, M. R. A. (1946): Aggregation as a factor influencing the toxicity of sympathomimetic amines in mice. *J. Pharmacol. Exp. Ther.,* 87:214–219.

Chance, M. R. A. (1947): Factors influencing the toxicity of sympathomimetic amines to solitary mice. *J. Pharmacol. Exp. Ther.,* 89:289–296.

Chance, M. R. A. (1948): *Behavior,* 1:64–69.

Charles-Dominique, P. (1974): Aggression and territoriality in nocturnal prosimias. In: *Primate Aggression, Territoriality, and Xenophobia,* edited by R. Holloway, pp. 31–48. Academic Press, New York.

Charpentier, J. (1969): Analysis and measurement of aggressive behaviour in mice. In: *Aggressive Behaviour,* edited by S. Garattini and E. B. Sigg, pp. 86–100. Excerpta Medica, Amsterdam.

Chatrian, G. E., and Chapman, W. P. (1960): Electrographic study of the amygdaloid region with implated electrodes in patients with temporal lobe epilepsy. In: *Electrical Studies on the Unanesthetized Brain,* edited by E. R. Ramey and D. S. O'Doherty, pp. 351–373. Hoeber Publishers, New York.

Chatz, T. L. (1972): Management of male adolescent sex offenders. *Int. J. Offender Ther.,* 2:109.

Chaurand, J. P., Schmitt, P., and Karli, P. (1973): Effets de lésions du tegmentum ventral du mésencéphale sur le comportement d'agression rat-souris. *Physiol. Behav.,* 10:507–515.

Chaurand, J. P., Vergnes, M., and Karli, P. (1972): Substance grise centrale du mésencéphale et comportement d'agression interspécifique du rat. *Physiol. Behav.,* 9:475–481.

Cheal, M. (1975): Social olfaction: A review of the ontogeny of olfactory influences on vertebrate behavior. *Behav. Biol.,* 15:1–25.

Cherlow, D. G., and Serafetinides, F. A. (1977): The measurement of emotional concepts in patients with temporal lobe epilepsy. *Dis. Nerv. Syst.,* 38:613–616.

Chernov, H. I., Furness, P., Partyka, D., and Plummer, A. J. (1966): Age, confinement and aggregation as factors in amphetamine group toxicity in mice. *J. Pharmacol. Exp. Ther.,* 154:346–349.

Chesher, G. B. (1974): Facilitation of avoidance acquisition in the rat by ethanol and its abolition by α-methyl-p-tyrosine. *Psychopharmacologia,* 39:87–95.

Cheyne, W. M., and Jahoda, G. (1971): Emotional sensitivity and intelligence in children from orphanages and normal homes. *J. Child Psychol. Psychiatry,* 12:77–90.

Chi, C. C., Bandler, R. J., and Flynn, J. P. (1976): Neuroanatomic projections related to biting attack elicited from ventral midbrain in cats. *Brain Behav. Evol.,* 13:91–110.

Childs, B. (1972): Genetic analysis of human behavior. *Annu. Rev. Med.,* 23:373–406.

Childs, G., and Brain, P. F. (1979a): A videotape analysis of biting targets employed in intraspecific fighting encounters in laboratory mice from four treatment categories. *IRCS Med. Sci.,* 7:44.

Childs, G., and Brain, P. F. (1979b): A videotape analysis of behavioural strategies of attacked anosmic mice in encounters with three different types of conspecific. *IRCS Med. Sci.,* 7:80.

Chitty, D. (1964): Animal numbers and behaviour. In: *Fish and Wildlife: A Memorial to W. J. K. Harkness,* edited by J. R. Dymond. Longmans, Canada.

Chomsky, N. (1966): Quoted in Cartesian Linguistic. Harper and Row, New York.

Chopra, G. S., and Smith, J. W. (1974): Psychotic reactions following cannabis use in East Indians. *Arch. Gen. Psychiatry,* 30:24–27.

Chopra, R. N., Chopra, G. S., and Chopra, I. C. (1942): *Cannabis sativa* in relation to mental diseases and crime in India. *Ind. J. Med. Res.,* 30:155–171.

Chrichton, M. (1972): *Terminal Man.* Bantam Books, New York.

Christensen, C. W. (1963): Religious conversion. *Arch. Gen. Psychiatry,* 9:207–216.

Christiansen, K. O. (1974): The genesis of aggressive criminality: Implications of a study of crime in a Danish twin study. In: *Determinants and Origins of Aggressive Behavior,* edited by J. de Wit and W. W. Hartup, pp. 233–253. Mouton, The Hague.

Clancy, H., and McBride, G. (1975): The isolation syndrome in childhood. *Dev. Med. Child Neurol.,* 17:198–219.

Clark, G., and Birch, H. G. (1945): Hormonal modification of social behavior. The effect of sex-hormone administration on the social status of male-castrate chimpanzee. *Psychosom. Med.,* 7:321–329.

Clark, L. H., and Schein, M. W. (1966): Activities associated with conflict behaviour in mice. *Anim. Behav.,* 14:44–49.

Clarke, A. D. B., and Clarke, A. M. (1954): Cognitive changes in the feebleminded. *Br. J. Psychol.,* 45:173–179.

Clavier, R. M., and Corcoran, M. E. (1976): Attenuation of self-stimulation from substantia nigra but not dorsal tegmental noradrenergic bundle by lesions of sulcal prefrontal cortex. *Brain Res.,* 113:59–69.

Clavier, R. M., and Fibiger, H. C. (1977): On the role of ascending cateocholaminergic projections in intracranial self-stimulation of the substantia nigra. *Brain Res.,* 131:271–286.

Clavier, R. M., Fibiger, H. C., and Phillips, A. G. (1976): Evidence that self-stimulation of the region of the locus coeruleus in rats does not depend upon noradrenergic projections to telencephalon. *Brain Res.,* 113:71–81.

Cloninger, C. R., and Guze, S. B. (1970): Female criminals: Their personal familial, and social backgrounds. The relation of these to the diagnoses of sociopathy and hysteria. *Arch. Gen. Psychiatry,* 23:554–558.

Cockrum, E. L., and McCauley, W. J. (1965): *Zoology.* W. B. Saunders, Philadelphia.

Cohen, M., and Lal, H. (1964): Effect of sensory stimuli on amphetamine toxicity in aggregated mice. *Nature,* 201:1037.

Cohen, S. (1971): Peers as modeling and normative influences in the development of aggression. *Psychol. Rep.,* 28:995–998.

Cole, S. O. (1973): Hypothalamic feeding mechanisms and amphetamine anorexia. *Psychol. Bull.,* 79:13–20.

Coleman, L. S. (1974): Perspectives on the medical research of violence. *Am. J. Orthopsychiatry,* 44:675–687.

Collins, R. A. (1970): Experimental modification of brain weight and behavior in mice: An enrichment study. *Dev. Psychobiol.,* 3:145–155.

Colpaert, F. C. (1975): The ventromedial hypothalamus and the control of avoidance behavior and aggression: Fear hypothesis versus response-suppression theory of limbic system function. *Behav. Biol.,* 15:27–44.

Colpaert, F. C., and Wiepkema, P. R. (1976): Effects of ventromedial hypothalamic lesions on spontaneous intraspecies aggression in male rats. *Behav. Biol.,* 16:117–125.

Conaway, C. H., and Sade, D. S. (1965): The seasonal spermotogenic cycle in free ranging rhesus monkeys. *Folia Primatol.,* 3:1–12.

Conde, F. P., and DeFeudis, F. V. (1977): Polypeptides of cerebral subcellular fractions of differentially-housed mice. *Experientia,* 33:928–930.

Conner, R. L., Levine, S., and Vernikos-Danellis, J. (1970a): Shock-induced fighting and pituitary-adrenal activity. Proceedings of the 78th Annual Convention of American Psychological Association, pp. 201–202.

Conner, R. L., Levine, S., Wertheim, G. A., and Cummer, J. F. (1969): Hormonal determinants of aggressive behavior. *Ann. N.Y. Acad. Sci.,* 159:760–776.

Conner, R. L., Stolk, J. M., Barchas, J. D., Dement, W. C., and Levine, S. (1970b): The effect of parachlorophenylalanine (PCPA) on shock-induced fighting behavior in rats. *Physiol. Behav.,* 5:1221–1224.

Conner, R. L., Vernikos-Danellis, J., and Levine, S. (1971): Stress, fighting and neuroendocrine function. *Nature,* 234:564–566.

Conners, C. K. (1972): Psychological effects of stimulant drugs in children with minimal brain dysfunction. *Pediatrics,* 49:702–708.

Consolo, S., Garattini, S., and Valzelli, L. (1965a): Amphetamine toxicity in aggressive mice. *J. Pharm. Pharmacol.,* 17:53–54.

Consolo, S., Garattini, S., and Valzelli, L. (1965b): Sensitivity of aggressive mice to centrally acting drugs. *J. Pharm. Pharmacol.,* 17:594–595.

Consolo, S., Garattini, S., Ghielmetti, R., and Valzelli, L. (1965c): Concentrations of amphetamine in the brain in normal and aggressive mice. *J. Pharm. Pharmacol.,* 17:666.

Consolo, S., and Valzelli, L. (1970): Brain choline acetylase and monoamine oxidase activity in normal and aggressive mice. *Eur. J. Pharmacol.,* 13:129–130.

Cook, L., and Kelleher, R. T. (1963): Effects of drugs on behavior. *Annu. Rev. Pharmacol.,* 3:205–222.

Corrodi, H., Fuxe, K., and Jonsson, G. (1972): Effects of caffeine on central monoamine neurons. *J. Pharm. Pharmacol.,* 24:155–158.

Corson, S. A. (1975): The violent society and its relation to psychopathology in children and adolescents. In: *Society, Stress and Disease, Vol. 2, Childhood and Adolescence,* edited by L. Levi, pp. 132–140. Oxford University Press, London.

Corum, C. R., and Thurmond, J. B. (1977): Effects of acute exposure to stress on subsequent aggression and locomotion performance. *Psychosom. Med.,* 39:436–443.

Coscina, D. V., Goodman, J., Godse, D. D., and Stancer, H. C. (1975): Taming effect of handling on 6-hydroxydopamine induced rage. *Pharmacol. Biochem. Behav.,* 3:525–528.

Coscina, D. V., Seggie, J., Godse, D. D., and Stancer, H. C. (1973): Induction of rage in rats by central injection of 6-hydroxydopamine. *Pharmacol. Biochem. Behav.,* 1:1–6.

Costa, E., and Meek, J. L. (1974): Regulation of biosynthesis of catecholamines and serotonin in the CNS. *Annu. Rev. Pharmacol.,* 14:491–511.

Costin, A., Bergmann, F., and Chaimovitz, M. (1967): Influence of labyrinthine stimulation on hippocampal activity. *Prog. Brain. Res.,* 27:183–188.

Costin, A., Hafemann, D., Elazar, Z., and Adey, W. R. (1970): Posture and role of vestibular and proprioceptive influences on neocortical, limbic, subcortical and cerebellar EEG activity. *Brain Res.,* 17:259–275.

Cox, B. (1976): Mysteries of early dinosaur evolution. *Nature,* 264:314.

Coyle, I. R., and Kirkby, R. J. (1975): Dominance behavior in the rat following lesions of the caudate nuclei. *Physiol. Behav.,* 14:75–80.

Coyle, J. T., and Enna, S. J. (1976): Neurochemical aspects of the ontogenesis of gabaergic neurons in the rat brain. *Brain Res.,* 111:119–133.

Crabtree, J. M., and Moyer, K. E. (1973): Sex differences in fighting and defense induced in rats by shock and d-amphetamine during morphine abstinence. *Physiol. Behav.,* 11:337–343.

Crabtree, J. M., and Moyer, K. E. (1977): *Bibliography of Aggressive Behavior: A Reader's Guide to the Research Literature.* Alan R. Liss, New York.

Crawford, M. P. (1942): Dominance and social behavior, for chimpanzees, in a non-competitive situation. *J. Comp. Psychol.,* 33:267–277.

Crawley, J. N., and Contrera, J. F. (1976): Intraventricular 6-hydroxydopamine lowers isolation-induced fighting behavior in male mice. *Pharmacol. Biochem. Behav.,* 4:381–384.

Crawley, J. N., Schleidt, W. M., and Contrera, J. F. (1975): Does social environment decrease propensity to fight in male mice? *Behav. Biol.,* 15:73–83.

Creer, T. L. (1973): Hunger and thirst shock-induced aggression. *Behav. Biol.,* 8:433–437.

Crick, F. H. C., Brenner, S., Klug, A., and Pieczenik, G. (1976): A speculation on the origin of protein synthesis. *Orig. Life,* 7:389–397.

Crick, F. H. C., and Orgel, L. E. (1973): *Icarus,* 19:341.

Crissey, O. R. (1937): Mental development as related to institutional residence and educational achievement. *Univ. Iowa Stud. Child Welfare,* 13:1–81.

Critchley, M. (1960): The evolution of man's capacity for language. In: *Evolution After Darwin, Vol. 2, The Evolution of Man, Culture, and Society,* edited by S. Tax, pp. 289–308. University of Chicago Press, Chicago.

Cross, H. A., and Harlow, H. F. (1965): Prolonged and progressive effects of partial isolation on the behavior of macaque monkeys. *J. Exp. Res. Pers.,* 1:39–49.

Crossland, J. (1970): *Lewis's Pharmacology, 4th Ed.* Williams & Wilkins, Baltimore.

Crow, T. J. (1972): A map of the rat mesencephalon for electrical self-stimulation. *Brain Res.,* 36:265–273.

Crow, T. J., Spear, P. J., and Arbuthnott, G. W. (1972): Intracranial self-stimulation with electrodes in the region of the locus coeruleus. *Brain Res.,* 36:275–287.

Crowcroft, P., and Rowe, F. P. (1963): Social organization and territorial behaviour of wild house mice. *Proc. Zool. Soc. Lond.,* 140:517–531.

Crowell, R. M., Tew, J. M., and Mark, V. J. (1973): Aggressive dementia associated with normal pressure hydrocephalus. *Neurology (Minneap.),* 23:461–464.

Cuello, A. C., and Kanazawa, I. (1978): The distribution of substance P immunoreactive fibers in the rat central nervous system. *J. Comp. Neurol.*, 178:129–156.

Culley, W. J., Saunders, R. N., Mertz, E. T., and Jolly, D. H. (1963): Effect of a tryptophan deficient diet on brain serotonin and plasma tryptophan level. *Proc. Soc. Exp. Biol. Med.*, 113:645–648.

Cullinton, B. J. (1976): Psychosurgery: National commission issues surprisingly favorable report. *Science*, 194:299–301.

Cumberbatch, G., and Howitt, D. (1974): Identification with aggressive television characters and children's moral judgments. In: *Determinants and Origins of Aggressive Behavior*, edited by J. de Wit and W. W. Hartup, pp. 517–523. Mouton, The Hague.

Curti, M. W. (1935): Native fear responses of white rats in the presence of cats. *Psychol. Monogr.*, 46:78–98.

Curti, M. W. (1942): A further report on fear responses of white rats in the presence of cats. *J. Comp. Physiol. Psychol.*, 34:51–53.

Curtis, G. C. (1963): Violence breeds violence—Perhaps? *Am. J. Psychiatry*, 120:386–387.

Curzon, G., and Fernando, J. C. R. (1976): Effect of aminophylline on tryptophan and other aromatic amino acids in plasma, brain and other tissues and on brain 5-hydroxytryptamine metabolism. *Br. J. Pharmacol.*, 58:533–545.

DaVanzo, J. P., Daugherty, M., Ruckart, R., and Kang, L. (1966): Pharmacological and biochemical studies in isolation-induced fighting mice. *Psychopharmacologia*, 9:210–219.

Davenport, R. K., Jr., Menzel, E. W., Jr., and Rogers, C. M. (1961): Maternal care during infancy: It's effect on weight gain and mortality in the chimpanzee. *Am. J. Orthopsychiatry*, 31:803–809.

Davenport, R. K., Jr., Menzel, E. W., Jr., and Rogers, C. M. (1966): Effects of severe isolation on "normal" juvenile chimpanzees. Healthy, weight gain, and stereotyped behaviors. *Arch. Gen. Psychiatry*, 14:134–138.

Davidson, J. M., and Davidson, R. J. (eds.) (1978): *The Psychobiology of Consciousness*. Plenum Press, New York.

Davidson, N. (1976): *Neurotransmitter Amino Acids*. Academic Press, London.

Davies, J. A., Navaratnam, V., and Redfern, P. H. (1973): A 24-hour rhythm in passive-avoidance behaviour in rats. *Psychopharmacologia*, 32:211–214.

Davis, D. E., and Christian, J. J. (1957): Relation of adrenal weight to social rank of mice. *Proc. Soc. Exp. Biol. Med.*, 94:728–731.

Davis, R. E., Harris, C., and Shelby J. (1974): Sex differences in aggressivity and the effects of social isolation in the anabantoid fish, *Macropodus opercularis. Behav. Biol.*, 11:497–509.

Davis, W. M., and Khalsa, J. H. (1971a): Some determinants of aggressive behavior induced by morphine withdrawal. *Psychonom. Sci.*, 24:13–15.

Davis, W. M., and Khalsa, J. H. (1971b): Increased shock induced aggression during morphine withdrawal. *Life Sci.*, 10:1321–1327.

Deaux, K. (1971): Honking at the intersection: A replication and extension. *J. Soc. Psychol.*, 84:159–160.

De Castro, J. M., and Balagura, S. (1975): Fornicotomy: Effect on the primary and secondary punishment of mouse killing by LiC1 poisoning. *Behav. Biol.*, 13:483–489.

De Castro, J. M., and Marrone, B. L. (1974): Effect of fornix lesions on shock-induced aggression, muricide and motor behavior in the albino rat. *Physiol. Behav.*, 13:737–743.

Defayolle, M., Courtot, D., and Bonan, J. (1966): Influence de la personnalité sur le rythme circadien de l'excrétion urinaire de catécholamines et de l'acide 5-hydroxy indol acétique. *C. R. Soc. Biol.*, 160:2351–2355.

D'Adamo, A. F., Jr., and Yatsu, F. M. (1966): Acetate metabolism in the nervous system. N-acetyl-1-aspartic acid and the biosynthesis of brain lipids. *J. Neurochem.*, 13:961–965.

Dahl, E., Falck, B., Von Mecklenburg, C., and Myhrberg, H. (1963): An adrenergic nervous system in sea anemones. *Q. J. Microsc. Sci.*, 104:531–534.

Dahl, E., Falck, B., Von Mecklenburg, C., Myhrberg, H, and Rosengren, E. (1966): Neuronal localization of dopamine and 5-hydroxytryptamine in some mollusca. *Z. Zellforsch. Mikrosk. Anat.*, 71:489–498.

Dahlström, A. (1971): Axoplasmic transport (with particular respect to adrenergic neurons). *Philos. Trans. R. Soc. Lond. [Biol.]*, 261:325–358.

Dahlström, A., and Fuxe, K. (1964): Evidence for the existence of monoamine-containing neurons in the central nervous system. I. Demonstration of monoamines in the cell bodies of brain stem neurons. *Acta Physiol. Scand.,* 62: Suppl. 232, 1–55.

Dalton, K. (1961): Menstruation and crime. *Br. Med. J.,* 3:1752–1753.

Dalton K. (1964): *The Premenstrual Syndrome.* Charles C Thomas, Springfield, Ill.

Danielli, J. F., and Davson, H. (1935): The permeability of thin films. *J. Cell. Comp. Physiol.,* 5:495–508.

Daruna, J. H. (1978): Patterns of brain monoamine activity and aggressive behavior. *Neurosci. Biobehav. Rev.,* 2:101–113.

Daruna, J. H., and Kent, E. W. (1976): Comparison of regional serotonin levels and turnover in the brain of naturally high and low aggressive rats. *Brain Res.,* 101:489–501.

Darwin, C. (1859): *On the Origin of Species by Means of Natural Selection or the Preservation of Favoured Races in the Struggle for Life.* John Murray, London.

Darwin, C. (1872a): *The Descent of Man and Selection in Relation to Sex.* Appleton, New York.

Darwin, C. (1872b): *The Expression of the Emotions in Man and Animals.* John Murray, London.

Das, J. P. (1971): Visual search stimulus density, and subnormal intelligence. *Am. J. Ment. Defic.,* 76:357–361.

DeFeudis, F. V. (1971): Effects of environmental changes on the incorporation of carbon atoms of D-glucose into mouse brain and other tissues. *Life Sci.,* 10:pt.II:1187–1194.

DeFeudis, F. V. (1972a): A note on the subcellular distribution of brain protein in differentially-housed mice. *Experientia,* 28:1427–1428.

DeFeudis, F. V. (1972b): Effects of isolation and aggregation on the incorporation of carbon atoms of D-mannose and D-glucose into mouse brain. *Biol. Psychiatry,* 4:239–242.

DeFeudis, F. V. (1972c): Binding of ^{3}H-acetylcholine and ^{14}C-α-aminobutyric acid to subcellular fractions of the brains of differentially-housed mice. *Neuropharmacology,* 11:879–888.

DeFeudis, F. V. (1972d): Effects of isolation and aggregation on the water content of mouse brain. *Biol. Psychiatry,* 5:207–210.

DeFeudis, F. V., and Black, W. C. (1972): Effects of environment on the incorporation of radioactive carbon atoms of glucose into various structures of the brains of mice; Evidence for diurnal rhythmicity. *Exp. Neurol.,* 36:41–45.

DeFeudis, F. V., Defeudis, P. A., and Somoza, E. (1976a): Altered analgesic responses to morphine in differentially housed mice. *Psychopharmacology,* 49:117–118.

DeFeudis, F. V., Madtes, P., Ojeda, A., and DeFeudis, P. A. (1976b): "Binding" of α-aminobutyric acid and glycine to synaptic particles of the brains of differentially-housed mice; Evidence for morphological changes. *Exp. Neurol.,* 52:285–294.

DeGhett, V. J. (1975): A factor influencing aggression in adult mice: Witnessing aggression when young. *Behav. Biol.,* 13:291–300.

Deiker, T. E. (1973): WAIS characteristics of indicted male murderers. *Psychol. Rep.,* 32:1066.

Deiker, T. E. (1974a): Characteristics of males indicted and convicted for homicide. *J. Soc. Psychol.,* 93:151–152.

Deiker, T. E. (1974b): A cross-validation of MMPI scales of aggression on male criminal criterion groups. *J. Consult. Clin. Psychol.,* 42:196–202.

Delgado, J. M. R. (1955): Cerebral structures involved in transmission and elaboration of noxious stimulation. *J. Neurophysiol.,* 18:261–275.

Delgado, J. M. R. (1961): Chronic implantation of intracerebral electrodes in animals. In: *Electrical Stimulation of the Brain,* edited by D. E. Sheer, pp. 25–36. University of Texas Press, Austin.

Delgado, J. M. R. (1962): Pharmacological modifications of social behavior. In: *Pharmacological Analysis of Central Nervous Action,* edited by W. M. D. Paton and P. Lindgren, pp. 265–292. Pergamon Press, Oxford.

Delgado, J. M. R. (1963a): Cerebral heterostimulation in a monkey colony. *Science,* 141:161–163.

Delgado, J. M. R. (1963b): Telemetry and telestimulation of the brain. In: *Biotelemetry,* edited by L. E. Slater, pp. 231–249. Pergamon Press, New York.

Delgado, J. M. R. (1964): Free behavior and brain stimulation. *Int. Rev. Neurobiol.,* 6:349–449.

Delgado, J. M. R. (1965): Sequential behavior induced repeatedly by stimulation of the red nucleus in free monkeys. *Science,* 148:1361–1363.

Delgado, J. M. R. (1966a): Aggressive behavior evoked by radiostimulation in monkey colonies. *Am. Zool.,* 6:669–681.

Delgado, J. M. R. (1966b): Emotions. In: *Self-Selection Psychology Textbook.* Brown, Dubuque, Iowa.

Delgado, J. M. R. (1967a): Limbic system and free behavior. *Progr. Brain Res.*, 27:48–68.

Delgado, J. M. R. (1967b): Social rank and radio-stimulated aggressiveness in monkeys. *J. Nerv. Ment. Dis.*, 144:383–390.

Delgado, J. M. R. (1969a): Recent advances in neurophysiology. In: *The Present Status of Psychotropic Drugs*, edited by A. Cerletti and F. J. Bové, pp. 36–48. Excerpta Medica, Amsterdam.

Delgado, J. M. R. (1969b): Offensive-defensive behaviour in free monkeys and chimpanzees induced by radio-stimulation of the brain. In: *Aggressive Behaviour*, edited by S. Garattini and E. B. Sigg, pp. 109–119. Excerpta Medica, Amsterdam.

Delgado, J. M. R. (1969c): Radiostimulation of the brain in primates and man. *Anesth. Analg. (Paris)*, 48:529–543.

Delgado, J. M. R. (1969d): *Physical Control of the Mind: Towards a Psychocivilized Society*. Harper and Row, New York.

Delgado, J. M. R. (1974): Communication with the conscious brain by means of electrical and chemical probes. In: *Factors in Depression*, edited by N. S. Kline, pp. 251–268. Raven Press, New York.

Delgado, J. M. R. (1979): Introduction and neuronal and behavioral rhythms. In: *Proceedings of the Second World Congress of Biological Psychiatry*, Barcelona, Aug. 31–Sept. 6, 1978. Elsevier, Amsterdam.

Delgado, J. M. R., and Anand, B. K. (1953): Increase of food intake induced by electrical stimulation of the lateral hypothalamus. *Am. J. Physiol.*, 172:162–168.

Delgado, J. M. R., Fernandez, F., Zaplan, J., and Ruiz, M. L. (1979): Social behaviour, mobility and EEG in Sylvania monkeys. In: *Proceedings of the Second World Congress of Biological Psychiatry*, Barcelona, Aug. 31–Sept. 6, 1978. Elsevier, Amsterdam.

Delgado, J. M. R., and Hamlin, H. (1956): Surface and depth electrography of the frontal lobes in conscious patients. *Electroencephalogr. Clin. Neurophysiol.*, 8:371–384.

Delgado, J. M. R., and Hamlin, H. (1960): Spontaneous and evoked seizures in animals and humans. In: *Electrical Studies on the Unanesthetized Brain*, edited by E. R. Ramey and D. S. O'Doherty, pp. 133–158. Hoeber Publishers, New York.

Delgado, J. M. R., Mark, V., Sweet, W., Ervin, F., Weiss, G., Bach-y-Rita, G., and Hagiwara, R. (1968): Intracerebral radio-stimulation and recording in completely free patients. *J. Nerv. Ment. Dis.*, 147:329–340.

Delgado, J. M. R., Obrador, S., and Martin-Rodriguez, J. C. (1973): Two-way radio communication with the brain in psychosurgical patients. In: *Surgical Approaches in Psychiatry, Pt. VII. Electrophysiological Studies*, edited by L. Laitinen, pp. 215–223. University Park Press, Baltimore.

Delgado, J. M. R., Roberts, W. W., and Miller, N. E. (1954): Learning motivated by electrical stimulation of the brain. *Am. J. Physiol.*, 179:587–593.

Delgado-Garcia, J. M., Grau, C., DeFeudis, P., Del Pozo, F., Jimenez, J. M., and Delgado, J. M. R. (1976): Ultradian rhythms in the mobility and behavior of rhesus monkeys. *Exp. Brain Res.*, 25:79–91.

del Pozo, F., DeFeudis, F. V., and Jimenez, J. M. (1978): Motilities of isolated and aggregated mice. A difference in ultradian rhythmicity. *Experientia*, 34:1302–1304.

DeMaio, D. (1966): Le allucinazioni olfattive dei malati psichici. *Acta Neurol. (Napoli)*, Quaderni N.XXV.

Dembo, R. (1973): Critical factors in understanding adolescent aggression. *Soc. Psychiatry*, 8:212–219.

DeMurua, F. M. (1946): Historia del origen y geneologia real de los reyes Incas del Peru. *Bibl. Missionalia Hispanica.*

De Nelsky, G. Y., and Denenberg, V. H. (1967a): Infantile stimulation and adult exploratory behavior; Effects of handling upon tactual variation seeking. *J. Comp. Physiol. Psychol.*, 63:309–312.

De Nelsky, G. Y., and Denenberg, V. H. (1967b): Infantile stimulation and adult exploratory behaviour in the rat: Effects of handling upon visual variation-seeking. *Anim. Behav.*, 15:568–573.

Denenberg, V. H. (1968): *Psychoneurol. Sci.*, 11:39.

Denenberg, V. H., Gaulin-Kremer, E., Gandelman, R., and Zarrow, M. X. (1973): The development of standard stimulus animals for mouse *(Mus musculus)* aggression testing by means of olfactory bulbectomy. *Anim. Behav.*, 21:590–598.

Denenberg, V. H., and Morton, J. R. C. (1962a): Effects of environmental complexity and social groupings upon the modification of emotional behavior. *J. Comp. Physiol. Psychol.*, 55:242–246.

Denenberg, V. H., and Morton, J. R. C. (1962b): Effects of pre-weaning and post-weaning manipulations upon problem-solving behavior. *J. Comp. Physiol. Psychol.,* 55:1096–1098.

Denenberg, V. H., Morton, J. R., and Haltmeyer, G. C. (1964): Effect of social grouping upon emotional behavior. *Anim. Behav.,* 12:205–208.

Denenberg, V. H., Wehmer, F., Werboff, J., and Zarrow, M. X. (1969): Effects of postweaning enrichment and isolation upon emotionality and brain weight in the mouse. *Physiol. Behav.,* 4:403–406.

Dengering, H. A., and Bertilson, H. S. (1975): Psychopathy and physiological arousal in an aggressive task. *Psychophysiology,* 12:682–684.

De Robertis, E. (1956): Submicroscopic changes of the synapse after nerve section in the acoustic ganglion of the guinea pig: An electron microscope study. *J. Biophys. Biochem. Cytol.,* 2:503–512.

De Robertis, E. (1964): Electron microscope and chemical study of binding sites of brain biogenic amines. *Prog. Brain Res.,* 8:118–136.

De Robertis, E. (1967): Ultrastructure and cytochemistry of the synaptic region. *Science,* 156:907–914.

De Robertis, E. (1971): Molecular biology of synaptic receptors. Synapses of the central nervous system, electric organ, and muscle contain a proteolipid with receptor properties. *Science,* 171:963–971.

De Robertis, E. D. P., and Bennett, H. S. (1954): Submicroscopic vesicular component in the synapse. *Fed. Proc.,* 13:35.

De Robertis, E., and Schmitt, F. O. (1948): An electron microscope analysis of certain nerve axon constituents. *J. Cell. Comp. Physiol.,* 31:1–24.

Descartes, René (Cartesio) (1649): *Les Passions de l'Âme.* Henry Le Gras, Paris.

Desisto, M. J., and Zweig, M. (1974): Differentiation of hypothalamic feeding and killing. *Physiol. Psychol.,* 2:67–70.

Desjardins, C., and Ewing, L. L. (1971): Testicular metabolism in adrenalectomized and corticosterone treated rats. *Proc. Soc. Exp. Biol. Med.,* 137:578–583.

Dessì-Fulgheri, F., Lupo di Prisco, C., and Verdarelli, P. (1975): Influence of long-term isolation on the production and metabolism of gonadal sex steroids in male and female rats. *Physiol. Behav.,* 14:495–499.

Devor, M., and Murphy, M. R. (1973): The effect of peripheral olfactory blockade on the social behavior of the male golden hamster. *Behav. Biol.,* 9:31–42.

De Waal, F. B. M. (1976): Straight-aggression and appeal-aggression in *Macaca fascicularis. Experientia,* 32:1268–1270.

Dewhurst, K., and Beard, A. W. (1970): Sudden religious conversions in temporal lobe epilepsy. *Br. J. Psychiatry,* 117:497–507.

Diamond, M. C., Krech, D., and Rosenzweig, M. R. (1964): The effects of an enriched environment on the histology of the rat cerebral cortex. *J. Comp. Neurol.,* 123:111–120.

Diamond, M. C., Law, F., Rhodes, H., Lindner, B., Rosenzweig, M. R., Krech, D., and Bennett, E. L. (1966): Increases in cortical depth and glia numbers in rats subjected to enriched environment. *J. Comp. Neurol.,* 128:117–126.

Di Chiara, G., Camba, R., and Spano, P. F. (1971): Evidence for inhibition by brain serotonin of mouse killing behavior in cats. *Nature,* 233:272–273.

Dickerson, J. W. T., and Pao, S. K. (1975): The effect of a low protein diet and exogenous insulin on brain tryptophan and its metabolites in the weanling rat. *J. Neurochem.,* 25:559–564.

Didiergeorges, F., and Karli, P. (1966): Privation des afférences olfactives et mise en place d'un mécanisme d'inhibition de l'agressivité interspécifique du rat. *C. R. Soc. Biol. (Paris),* 160:1065–1067.

Didiergeorges, F., and Karli, P. (1967): Hormones stéroides et maturation d'un comportement d'agression interspecifique du rat. *C. R. Soc. Biol. (Paris),* 161:179–181.

Didiergeorges, F., Vergnes, M., and Karli, P. (1966a): Privation des afférences olfactives et agressivité interspécifique du rat. *C. R. Soc. Biol. (Paris),* 160:866–868.

Didiergeorges, F., Vergnes, M., and Karli, P. (1966b): Desafférentation olfactive et agressivité interspecifique du rat. *J. Physiol. (Paris),* 58:510–511.

Dieckmann, G., and Hassler, R. (1975): Unilateral hypothalamotomy in sexual delinquents. Report on six cases. *Confin. Neurol.,* 37:177–186.

Dieterlen, F. (1959): Das Verhalten des synschen Goldhamsters (*Mesocricetus auratus* Waterhouse). *Z. Tierpsychol.,* 16:47–103.

Diez, J. A., Sze, P. Y., and Ginsburg, B. E. (1976): Genetic and developmental variation in mouse brain tryptophan hydroxylase activity. *Brain Res.,* 109:413–417.

Dillon, E. J. (1896): Armenia: An appeal. *Contemp. Rev.,* 69:1–19.

Diserens, C. M., and Vaughn, J. (1931): The experimental psychology of motivation. *Psychol. Bull.,* 28:15–65.

Ditman, K. S., Hayman, M., and Whittlesey, J. R. (1962): Nature and frequency of claims following LSD. *J. Nerv. Ment. Dis.,* 134:346–352.

Ditman, K. S., Tietz, W., Prince, B. S., Forgy, E., and Moss, T. (1968): Harmful aspects of the LSD experience. *J. Nerv. Ment. Dis.,* 145:464–474.

Dixon, A. K., and Mackintosh, J. H. (1971): Effects of female urine upon the social behaviour of adult male mice. *Anim. Behav.,* 19:138–140.

Dohan, F. C. (1966): Cereals and schizophrenia: Data and hypothesis. *Acta Psychiatr. Scand.,* 42:125–152.

Dohan, F. C. (1969a): Schizophrenia: Possible relationship to cereal grains and celiac disease. In: *Schizophrenia. Current Concepts and Research,* edited by D. V. Siva Sankar, pp. 539–551. PJD Publ., Westbury, N.Y.

Dohan, F. C. (1969b): Is celiac disease a clue to the pathogenesis of schizophrenia? *Ment. Hyg.,* 53:525–529.

Dohan, F. C. (1970): Coeliac disease and schizophrenia. *Lancet,* I: 897–898.

Dohan, F. C., and Grasberger, J. C. (1973): Relapsed schizophrenics: Earlier discharge from the hospital after cereal-free, milk-free diet. *Am. J. Psychiatry,* 130:685–688.

Dohan, F. C., Grasberger, J., Lowell, F., Johnston, H., Jr., and Arbegast, A. (1969): Relapsed schizophrenics: More rapid improvement on a milk- and cereal-free diet. *Br. J. Psychiatry,* 115:595–596.

Dolfini, E., Garattini, S., and Valzelli, L. (1969a): Different sensitivity to amphetamine of three strains of mice. *Eur. J. Pharmacol.,* 7:220–223.

Dolfini, E., Garattini, S., and Valzelli, L. (1969b): Activity of (+)-amphetamine at different environmental temperatures in three strains of mice. *J. Pharm. Pharmacol.,* 21:871–872.

Dolfini, E., and Kobayashi, M. (1967): Studies with amphetamine in hyper- and hypothyroid rats. *Eur. J. Pharmacol.,* 2:65–66.

Dolfini, E., Ramirez del Angel, A., Garattini, S., and Valzelli, L. (1970): Brain catecholamine release by dexamphetamine in three strains of mice. *Eur. J. Pharmacol.,* 9:333–336.

Dollard, J., Doob, L. W., Miller, N. E., Mowrer, O. H., and Sears, R. R. (1939): *Frustration and Aggression.* Yale University Press, New Haven.

Dominguez, M., and Longo, V. G. (1969): Taming effects of para-chlorophenylalanine on septal rats. *Physiol. Behav.,* 4:1031–1033.

Dominic, C. J. (1966): Block to pseudopregnancy in mice caused by exposure to male urine. *Experientia,* 22:534–535.

Dominick, J. R., and Greenberg, B. S. (1972): Attitudes toward violence: The interaction of television exposure, family attitudes, and social class. In: *Television and Social Behavior, Vol. 3,* edited by G. A. Comstock and E. A. Rubinstein, pp. 314–335. U.S. Government Printing Office, Washington, D.C.

Donnelly, J. (1978): The incidence of psychosurgery in the United States, 1971–1973. *Am. J. Psychiatry,* 135:1476–1480.

Donnerstein, E., and Donnerstein, M. (1975): The effect of attitudinal similarity on interracial aggression. *J. Pers.,* 43:485–502.

Donnerstein, E., and Donnerstein, M. (1976): Research in the control of interracial aggression. In: *Perspectives on Aggression,* edited by R. G. Geen and E. C. O'Neal. Academic Press, New York.

Donnerstein, E., Donnerstein, M., and Evans, R. (1975): Erotic stimuli and aggression: Facilitation or inhibition. *J. Pers. Soc. Psychol.,* 32:237–244.

Donnerstein, E., Donnerstein, M., Simon, S., and Ditrichs, R. (1972): Variables in interracial aggression: Anonymity, expected retaliation, and a riot. *J. Pers. Soc. Psychol.,* 22:236–245.

Donnerstein, E., and Hallam, J. (1978): Effects of erotica on aggression against women. *J. Pers. Soc. Psychol.,* 36:1270–1277.

Doob, A. N., and Wood, L. E. (1972): Catharsis and aggression: Effects of annoyance and retaliation on aggressive behavior. *J. Pers. Soc. Psychol.,* 22:156–162.

Dow, R. S., Grimm, R. J., and Rushmer, D. S. (1974): Psychosurgery and brain stimulation:

The legislative experience in Oregon in 1973. In: *The Cerebellum, Epilepsy, and Behavior,* edited by I. S. Cooper, M. Biklan, and R. S. Snider, pp. 367–376. Plenum Press, New York.

Drickamer, L. C. (1974): A ten-year summary of reproductive data for free-ranging *Macaca mulatta. Folia Primatol.,* 21:61–80.

Driver, P. M., and Humphries, D. A. (1970): Protean displays as inducers of conflict. *Nature,* 226:968–969.

Drtil, J. (1969a): The disturbances of higher nervous activity preceding aggression of prisoners. *Act. Nerv. Super. (Praha),* 11:297.

Drtil, J. (1969b): Causes of abnormal behavior of prisoners and its prevention. *Czech. Psychol.,* 13:238–248.

Dublin, L. I. (1963): *Suicide: A Sociological and Statistical Study.* Ronald Press, New York.

Duncan, J. W., and Duncan, G. M. (1971): Murder in the family: A study of some homicidal adolescents. *Am. J. Psychiatry,* 127:1498–1502.

Dunham, P. J., and Carr, A. (1976): Pain-elicited aggression in the squirrel monkey: An implicit avoidance contingency. *Anim. Learn. Behav.,* 4:89–95.

Dunn, G. W. (1941): Stilbestrol induced testicular degeneration in hypersexual males. *J. Clin. Endocrinol.,* 1:643–648.

Durell, J., Garland, J. T., and Friedel, R. O. (1969): Acetylcholine action: Biochemical aspects. *Science,* 165:862–866.

Durkheim, E. (1897): *Le Suicide. Etude Sociologique, Vol. I,* Paris.

Eastwood, M. R., and Peacocke, J. (1976): Seasonal patterns of suicide depression and electroconvulsive therapy. *Br. J. Psychiatry,* 129:472–475.

Ebel, A., Mack, G., Stefanovic, V., and Mandel, P. (1973): Activity of choline acetyltransferase and acetyltransferase in the amygdala of spontaneous mouse-killer rats and in rats after olfactory lobe removal. *Brain Res.,* 57:248–251.

Ebert, P. D. (1976): Agonistic behavior in wild and inbred *Mus musculus. Behav. Biol.,* 18:291–294.

Eccles, J. C. (1969): The development of the cerebellum of vertebrates in relation to the control of movement. *Naturwissenschaften,* 56:525–534.

Eccles, J. C. (1970): *Facing Reality. Philosophical Adventures by a Brain Scientist.* Springer-Verlag, Berlin.

Eccles, J. C. (1976): From electrical to chemical transmission in the central nervous system. Proceedings of Sir Henry Dale Centennial Symposium. Cambridge.

Eccles, J. C., Ito, M., and Szentagothai, J. (1967): *The Cerebellum as a Neuronal Machine.* Springer-Verlag, Berlin.

Eccles, J. C., Schmidt, R., and Willis, W. D. (1963): Pharmacological studies of presynaptic inhibition. *J. Physiol.,* 168:500–530.

Eclancher, F., and Karli, P. (1971): Comportement d'agression interspécifique et comportement alimentaire du rat: Effets de lésions des noyaux ventro-médians de l'hypothalamus. *Brain Res.,* 26:71–79.

Eclancher, F., Schmitt, P., and Karli, P. (1975): Effets de lésions précoces de l'amygdale sur le developpement de l'agressivité interspecifique du rat. *Physiol. Behav.,* 14:277–283.

Edgerton, R. B. (1971): *The Individual in Cultural Adaptation: A Study of Four East African Peoples.* Berkeley.

Edwards, D. A. (1970): Effects of cyproterone acetate on aggressive behaviour and the seminal vescicles of male mice. *J. Endocrinol.,* 46:477–481.

Egger, M. D., and Flynn, J. P. (1963): Effect of electrical stimulation of the amygdala on hypothalamically elicited attack behavior in cats. *J. Neurophysiol.,* 26:705–720.

Egger, M. D., and Flynn, J. P. (1967): Further studies on the effects of amygdaloid stimulation and ablation on hypothalamically elicited attack behavior in cats. *Progr. Brain Res.,* 27:165–182.

Ehrenkranz, J., Bliss, E., and Sheard, M. H. (1974): Plasma testosterone: Correlation with aggressive behavior and social dominance in man. *Psychosom. Med.,* 36:469–475.

Ehrlich, A. (1964): Neural control of feeding behavior. *Psychol. Bull.,* 61:100–114.

Eibl-Eibesfeldt, I. (1967): Ontogenetic and maturational studies of aggressive behavior. In: *Aggression and Defense: Neural Mechanisms and Social Patterns,* edited by C. D. Clemente and D. B. Lindsley. University of California Press, Los Angeles.

Eibl-Eibesfeldt, I. (1974a): Phylogenetic adaptation as determinants of aggressive behavior in man.

In: *Determinants and Origins of Aggressive Behavior,* edited by J. de Wit and W. W. Hartup, pp. 29–57. Mouton, The Hague.

Eibl-Eibesfeldt, I. (1974*b*): The myth of the aggression-free hunter and gatherer society. In: *Primate Aggression, Territoriality, and Xenophobia,* edited by R. Halloway, pp. 435–458. Academic Press, New York.

Eichelman, B., Dejong, W., and Williams, R. B. (1973): Aggressive behavior in hypertensive and normotensive rat strains. *Physiol. Behav.,* 10:301–304.

Eichelman, B., Orenberg, E., Hackley, P., and Barchas, J. (1978): Methylxanthine-facilitated shock-induced aggression in the rat. *Psychopharmacology,* 56:305–308.

Eichelman, B. S., Jr., and Thoa, N. B. (1973): The aggressive monoamines. *Biol. Psychiatr.,* 6:143–164.

Eichelman, B. S., Jr., Thoa, N. B., and Ng, K. Y. (1972*a*): Facilitated aggression in the rat following 6-hydroxydopamine administration. *Physiol. Behav.,* 8:1–3.

Eichelman, B., Thoa, N. B., Bugbee, N. M., and Ng, K. Y. (1972*b*): Brain amine and adrenal enzyme levels in aggressive bulbectomized rats. *Physiol. Behav.,* 9:483–485.

Eidelberg, E., and Erspamer, R. (1975): Dopaminergic mechanisms of opiate actions in brain. *J. Pharmacol. Exp. Ther.,* 192:50–57.

Eison, M. S., Stark, A. D., and Ellison, G. (1977): Opposed effects of locus coeruleus and substantia nigra lesions on social behavior in rat colonies. *Pharmacol. Biochem. Behav.,* 7:87–90.

Eleftheriou, B. E., Bailey, D. W., and Denenberg, V. H. (1974): Genetic analysis of fighting behavior in mice. *Physiol. Behav.,* 13:773–777.

Eleftheriou, B. E., and Boehlke, K. W. (1967): Brain monoamine oxidase in mice after exposure to aggression and defeat. *Science,* 155:1693–1694.

Eleftheriou, B. E., and Church, R. L. (1968): Brain levels of serotonin and norepinephrine in mice after exposure to aggression and defeat. *Physiol. Behav.,* 3:977–980.

Elias, J. W., Elias, M. F., and Schlager, G. (1975): Aggressive social interaction in mice genetically selected for blood pressure extremes. *Behav. Biol.,* 13:155–166.

Elkin, A. P. (1964): *The Australian Aborigines.* Angus and Robertson, Sydney.

Ellinwood, E. H., Jr. (1967): Amphetamine psychosis. I. Description of the individuals and process. *J. Nerv. Ment. Dis.,* 144:273–283.

Ellinwood, E. H., Jr. (1971): Assault and homicides associated with amphetamine abuse. *Am. J. Psychiatry,* 127:1170–1175.

Ellinwood, E. (1972): Amphetamine psychosis; individuals, setting and sequences. In: *Current Concepts on Amphetamine Abuse,* edited by E. Ellinwood and S. Cohen, pp. 143–157. National Institute of Mental Health, Rockville, Md.

Elms, A. M., and Milgram, S. (1966): Personality characteristics associated with obedience and defiance toward authoritative command. *J. Exp. Res. Pers.,* 1:282–289.

Elwers, M., and Critchlow, V. (1961): Precocious ovarian stimulation following interruption of stria terminalis. *Am. J. Physiol.,* 201:281–284.

Eng. L. F., Uyeda, C. T., Chao, L. P., and Wolfgram, F. (1974): Antibody to bovine choline acetyltransferase and immunofluorescent localisation of the enzyme in neurones. *Nature,* 250:243–245.

Erickson, T. C. (1945): Erotomania (nymphomania) as an expression of cortical epileptiform discharge. *Arch. Neurol. Psychiatry,* 53:226–231.

Eriksson, K. (1971): Rat strains specially selected for their voluntary alcohol consumption. *Ann. Med. Exp. Biol. Fenn.,* 49:67–72.

Eriksson, K. (1972*a*): Alcohol consumption and blood alcohol in strains selected for their behavior toward alcohol. In: *Biological Aspects of Alcohol Consumption,* edited by O. Forsander and K. Eriksoon, pp. 121–125. Finnish Foundation Alcohol Study no. 20.

Eriksson, K. (1972*b*): Behavioral and physiological differences among rat strains specially selected for their alcohol consumption. *Ann. N.Y. Acad. Sci.,* 197:32–41.

Erlenmeyer-Kimling, L., and Jarvik, L. F. (1963): Genetics and intelligence: A review. *Science,* 142:1477–1479.

Ernst, E. (1976): Biological fundamentals of consciousness. *Acta Biochim. Biophys. Acad. Sci. Hung.,* 11:323–325.

Eron, L. D., Huesmann, L. R., Lefkowitz, M. M., and Walder, L. O. (1974*a*): How learning conditions in early childhood—including mass media—relate to aggression in late adolescence. *Am. J. Orthopsychiatry,* 44:412–423.

Eron, L. D., Walder, L. O., Huesmann, L. P., and Lefkowitz, M. M. (1974b): The convergence of laboratory and field studies of the development of aggression. In: *Determinants and Origins of Aggressive Behavior,* edited by J. de Wit and W. W. Hartup, pp. 347–380. Mouton, The Hague.

Erskine, M. S., and Levine, S. (1973): Suppression of pituitary-adrenal activity and shock-induced fighting in rats. *Physiol. Behav.,* 11:787–790.

Ervin, F. R., and Mark, H. (1969): Behavioral and affective responses to brain stimulation in man. In: *Neurobiological Aspects of Psychopathology,* edited by Y. Zubin and C. Shagass, pp. 54–65. Grune & Stratton, New York.

Essman, W. B. (1968): Differences in locomotor activity and brain-serotonin metabolism in differentially housed mice. *J. Comp. Physiol. Psychol.,* 66:344–346.

Essman, W. B. (1969): "Free" and motivated behaviour and amine metabolism in isolated mice. In: *Aggressive Behaviour,* edited by S. Garattini and E. B. Sigg, pp. 203–208. Excerpta Medica, Amsterdam.

Essman, W. B. (1970): Some neurochemical correlates of altered memory consolidation. *Trans. N.Y. Acad. Sci.,* 32:948–973.

Essman, W. B. (1971a): Isolation-induced behavioral modification: Some neurochemical correlates. In: *Brain Development and Behavior,* pp. 265–276. Academic Press, New York.

Essman, W. B. (1971b): Neurochemical changes associated with isolation and environmental stimulation. *Biol. Psychiatry,* 3:141–147.

Essman, W. B. (1971c): Changes in cholinergic activity and avoidance behavior by nicotine in differentially housed mice. *Int. J. Neurosci.,* 2:199–206.

Essman, W. B. (1971d): The role of biogenic amines in memory consolidation. *Symp. Biol. Hung.,* 10:213–238.

Essman, W. B. (1974): Regional alterations of synaptic O-phosphoryl-lethanolamine in differentially housed mice. *Pharmacol. Res. Commun.,* 6:377–395.

Essman, W. B. (1975a): Diurnal differences in altered brain 5-hydroxytryptamine-related regional protein synthesis. *J. Pharmacol. (Paris),* 6:313–322.

Essman, W. B. (1975b): Lithium. *Lancet,* 2:547.

Essman, W. B. (1978): Benzodiazepines and aggressive behavior. *Mod. Probl. Pharmacopsychiatry,* 13:13–28.

Essman, W. B., and Essman, S. G. (1969): Enhanced memory consolidation with drug-induced regional changes in brain RNA and serotonin metabolism. *Pharmako-Psychiatr. Neuropsychopharmakol.,* 2:28–34.

Essman, W. B., and Frisone, J. F. (1966): Isolation-induced facilitation of gastric ulcerogenesis. *J. Psychosom. Res.,* 10:183–188.

Essman, W. B., Heldman, E., Barker, L. A., and Valzelli, L. (1972): Development of microsomal changes in liver and brain of differentially housed mice. *Fed. Proc.,* 31:121 (abst.).

Evans, C. M., and Brain, P. F. (1978): Effects of age at castration on testosterone induced aggression-promoting cues in groups of male mice. *Physiol. Behav.,* 21:19–23.

Evans, H. L., Ghiselli, W. B., and Patton, R. A. (1973): Diurnal rhythm in behavioral effects of methamphetamine, p-chlormethamphetamine and scopolamine. *J. Pharmacol. Exp. Ther.,* 186:10–17.

Everett, G. (1968): Role of dopamine in aggressive and motor responses in male mice. *Pharmacologist,* 10:181.

Evert, J. (1979): Television: Unbalanced spur to violence and anarchy. *To the Point,* 8:28–29.

Falck, B., Hillarp, N.-Å., Thieme, G., and Torp, A. (1962): Fluorescence of catecholamines and related compounds condensed with formaldehyde. *J. Histochem. Cytochem.,* 10:348–354.

Falconer, M. A., Hill, D., Meyer, A., Mitchell, W., and Pond, D. A. (1955): Treatment of temporal-lobe epilepsy by temporal lobectomy. *Lancet,* I:827–835.

Falconer, M. A., Hill, D., Meyer, A., and Wilson, J. L. (1968): Clinical, radiological and EEG correlations with pathological changes in temporal lobe epilepsy and their significance in surgical treatment. In: *Temporal Lobe Epilepsy,* edited by M. Baldwin and P. Bailey. Charles C Thomas, Springfield, Ill.

Farris, H. E., Gideon, B. E., and Ulrich, R. E. (1970): Classical conditioning of aggression: A developmental study. *Psychol. Rec.,* 20:63–67.

Faull, R. L. M., and Laverty, R. (1969): Changes in dopamine levels in the corpus striatum following lesions in the substantia nigra. *Exp. Neurol.,* 23:332–340.

Faust, V. (1974): The frequency of hospitalization of psychically ill subjects in relation to the time of the year. *J. Interdisciplinary Cycle Res.,* 5:313–319.

Faust, V., and Sarreither, P. (1975): Jahreszeit und psychische krankheit. *Med. Klin.,* 70:467–473.

Fechner, G. T. (1860): *Elemente der Psychophysik, Vol. 2.* Breitkopf & Härtel, Leipzig.

Fechter, J. V. (1971): Modeling and environmental generalization by mentally retarded subjects of televised aggressive or friendly behavior. *Am. J. Ment. Defic.,* 76:266–267.

Felthous, A. R., and Yudowitz, B. (1977): Approaching a comparative typology of assaultive female offenders. *Psychiatry,* 40:270–276.

Fenton, G. W., Tennent, T. G., Fenwick, P. B. C., and Rattray, N. (1974): The EEG in antisocial behaviour: A study of posterior temporal slow activity in special hospital patients. *Psychol. Med.,* 4:181–186.

Fentress, J. C. (1973): Specific and nonspecific factors in the causation of behavior. In: *Perspectives in Ethology,* edited by P. P. G. Bateson and P. H. Klopfer, pp. 155–224. Plenum Press, New York.

Ferchmin, P. A., Eterovic, V. A., and Caputto, R. (1970): Studies on brain weight and RNA content after short periods of exposure to environmental complexity. *Brain Res.,* 20:49–57.

Fernö, A. (1978): The effect of social isolation on the aggressive and sexual behaviour in a cichlid fish, *Haplochromis Bortoni. Behaviour,* 65:43–61.

Fernstrom, J. D., and Wurtman, R. J. (1971): Effect of chronic corn consumption on serotonin content of rat brain. *Nature, [New Biol.],* 234:62–64.

Ferracuti, F., and Wolfgang, M. E. (1963): Design for proposed study of violence. A socio-psychological study of a subculture of violence. *Br. J. Criminol.,* 3:377–380.

Feshbach, S. (1964): The function of aggression and the regulation of aggressive drive. *Psychol. Rev.,* 71:257–272.

Feshbach, S. (1970): Aggression. In: *Carmichael's Manual of Child Psychology, Vol. 3, 3rd Ed.,* edited by P. H. Mussen, pp. 159–259. John Wiley & Sons, New York.

Feshbach, N. D. (1974a): The relationship of child-rearing factors to children's aggression empathy and related positive and negative social behaviors. In: *Determinants and Origins of Aggressive Behavior,* edited by J. de Wit and W. W. Hartup, pp. 427–436. Mouton, The Hague.

Feshbach, S. (1974b): The development and regulation of aggression: Some research gaps and a proposed cognitive approach. In: *Determinants and Origins of Aggressive Behavior,* edited by J. de Wit and W. W. Hartup, pp. 167–181. Mouton, The Hague.

Feshbach, S., and Jaffe, Y. (1970): *Effects of Inhibition of Aggression upon Sexual Responsivity.* Preliminary Report, UCLA, Los Angeles.

Feshbach, S., Stiles, W. B., and Bitter, E. (1967): The reinforcing effect of witnessing aggression. *J. Exp. Res. Pers.,* 2:133–139.

Field, L. H., and Williams, M. (1970): The hormonal treatment of sexual offenders. *Med. Sci. Law,* 10:27–34.

Fisher, S. (1970): Nonspecific factors as determinants of behavioral response to drugs. In: *Clinical Handbook of Psychopharmacology,* edited by A. Di Mascio and R. I. Shader, pp. 17–39. Science House Publ., New York.

Flandera, V. (1977): Stimuli influencing aggressive behaviour of the laboratory rat. *Act. Nerv. Super. (Praha),* 19:247–248.

Flandera, V., and Nováková, V. (1974): Effect of mother on the development of aggressive behavior in rats. *Dev. Psychobiol.,* 8:49–54.

Flannelly, K., and Lore, R. (1977): The influence of females upon aggression in domesticated male rats *(Rattus norvegicus). Anim. Behav.,* 25:654–659.

Flannelly, K. J., and Thor, D. H. (1976): Territorial behavior of laboratory rats under conditions of peripheral anosmia. *Anim. Learn. Behav.,* 4:337–340.

Flannelly, K. J., and Thor, D. H. (1978): Territorial aggression of the rat to males castrated at various ages. *Physiol. Behav.,* 20:785–789.

Flannelly, K. J., Carty, R. W., and Thor, D. H. (1976): Vibrissal anesthesia and the suppression of intruder-elicited aggression in rats. *Psychol. Rec.,* 26:255–261.

Flory, R. (1969): Attack behavior as a function of minimum inter-food interval. *J. Exp. Anal. Behav.,* 12:825–828.

Flory, R. K., Smith, E. L. P., and Ellis, B. B. (1977): The effects of two response-elimination procedures on reinforced and induced aggression. *J. Exp. Anal. Behav.,* 25:5–15.

Flory, R. K., Ulrich, R. E., and Wolff, P. C. (1965): The effects of visual impairment on aggressive behavior. *Psychol. Rec.,* 15:185–190.

Flynn, J. P. (1976): Neural basis of threat and attack. In: *Biological Foundations of Psychiatry*, edited by R. G. Grenell and S. Gabay, pp. 273–295. Raven Press, New York.

Fog, R., Randrup, A., and Pakkenberg, H. (1970): Lesions in corpus striatum and cortex of rat brains and the effect on pharmacologically induced stereotyped, aggressive and cataleptic behaviour. *Psychopharmacologia*, 18:346–356.

Fonberg, E. (1965): Effect of partial destruction of the amygdaloid complex on the emotional-defensive behaviour of dogs. *Bull. Acad. Pol. Sci. (Biol.)*, 13:429–432.

Fonberg, E., and Delgado, J. M. R. (1961a): Avoidance and alimentary reactions during amygdala stimulation. *J. Neurophysiol.*, 24:651–664.

Fonberg, E., and Delgado, J. M. R. (1961b): Inhibitory effects of amygdala on food intake and conditioning. *Fed. Proc.*, 20:335.

Fonnum, F., Walaas, I., and Iversen, E. (1977): Localization of gabaergic, cholinergic and aminergic structures in the mesolimbic system. *J. Neurochem.*, 29:221–230.

Forgays, D. G., and Forgays, J. W. (1952): The nature of the effect of free-environmental experience in the rat. *J. Comp. Physiol. Psychol.*, 45:322–328.

Forssman, H., and Hambert, G. (1963): Incidence of Klinefelter's syndrome among mental patients. *Lancet*, I:1327.

Fortuna, M. B. (1977): Elicitation of aggression by food deprivation in olfactory bulbectomized male mice. *Physiol. Psychol.*, 5:327–330.

Fortuna, M., and Gandelman, R. (1972): Elimination of pain-induced aggression in male mice following olfactory bulb removal. *Physiol. Behav.*, 9:397–400.

Franck, H. (1966a): Ablation des bulbes olfactifs chez la lapine impubère. Répercussions sur le tractus genital et le comportement sexuel. *C.R. Soc. Biol. (Paris)*, 160:389–391.

Franck, H. (1966b): Effets de l'ablation des bulbes olfactifs sur la physiologie génitale chez la lapine adulte. *C.R. Soc. Biol. (Paris)*, 160:863–865.

Frankova, S. (1973): Effect of protein-calorie malnutrition on the development of social behavior in rats. *Dev. Psychobiol.*, 6:33–43.

Franzen, E. A., and Myers, R. E. (1973): Neural control of social behavior: Prefrontal and anterior temporal cortex. *Neuropsychologia*, 11:141–157.

Frederichs, C., and Goodman, H. (1969): *Low Blood Sugar and You*. Constellation International, New York.

Fredericson, E. (1950): The effects of food deprivation upon competitive and spontaneous combat in C57 black mice. *J. Psychol.*, 29:89–100.

Freedman, A. M., Kaplan, H. I., and Sadock, B. J. (1975): *Comprehensive Textbook of Psychiatry, 2nd Ed.* Williams & Wilkins, Baltimore.

Freedman, H. L., and Rockmore, M. J. (1946a): Marihuana; A factor in personality evaluation and Army maladjustment. Part I. *J. Clin. Psychopathol.*, 7:765–782.

Freedman, H. L., and Rockmore, M. J. (1946b): Marihuana; A factor in personality evaluation and Army maladjustment. Part II. *J. Clin. Psychopathol.*, 8:221–236.

Freeman, D. (1964): Human aggression in anthropological perspective. In: *The Natural History of Aggression*, edited by J. D. Carthy and F. J. Ebling, pp. 109–119. Academic Press, London.

Freeman, H. (1978): Mental health and environment. *Br. J. Psychiatry*, 132:113–124.

Freud, A., and Burlingame, D. T. (1944): *Infants without Families*. International University Press, New York.

Freud, S. (1920): *Beyond the Pleasure Principle, Standard Edition*, Hogarth Press, London.

Freud, S. (1923): The ego and the id and other works. In: *Standard Edition of the Complete Psychological Works of Sigmund Freud, Vol. 19*, edited by J. Strachey, 1961. Hogarth Press, London.

Freud, S. (1933): New introductory lectures on psycho-analysis and other works. In: *Standard Edition of the Complete Psychological Works of Sigmund Freud, Vol. 22*, edited by J. Strachey, 1964. Hogarth Press, London.

Freyhan, F. A. (1969): Reevaluation of psychosurgery. *Dis. Nerv. System. [GWAN Suppl.]*, 30:55–56.

Friedman, A. H., and Walker, C. A. (1968): Circadian rhythms in rat mid-brain and caudate nucleus biogenic amine levels. *J. Physiol., (Lond.)*, 197:77–85.

Fröberg, J. E., Karlsson, C. G., Levi, L., and Lidberg, L. (1975): Psychobiological circadian rhythms during a 72 hour vigil. *Försvarsmedicin*, 11:192–201.

Frodi, A. (1977): Sexual arousal, situational restrictiveness and aggressive behavior. *J. Res. Pers.*, 11:48–58.

Frodi, A., Macaulay, J., and Ropert Thome, P. (1977): Are women always less aggressive than men? A review of the experimental literature. *Psychol. Bull.*, 84:634–660.

Frosch, W. A., Robbins, E. S., and Stern, M. (1965): Untoward reactions to lysergic acid diethylamide (LSD) resulting in hospitalization. *N. Engl. J. Med.*, 273:1235–1239.

Fuller, J. L., and Geils, H. D. (1973): Behavioral development in mice selected for differences in brain weight. *Dev. Psychobiol.*, 6:469–474.

Fuller, J. L., Rosvold, H. E., and Pribram, K. M. (1957): The effect on affective and cognitive behavior in the dog of the lesions of pyriform-amygdala-hippocampal complex. *J. Comp. Physiol. Psychol.*, 50:89–96.

Fulton, J. F. (1951): *Frontal Lobotomy and Affective Behavior: A Neurophysiological Approach*, p. 32. Norton Publ., New York.

Funderburk, S. J., and Ferjo, N. (1978): Clinical observations in Klinefelter (47,XXY) syndrome. *J. Ment. Defic. Res.*, 22:207–212.

Futamachi, K. J., and Pedley, T. A. (1976): Glial cells and extracellular potassium: Their relationship in mammalian cortex. *Brain Res.*, 109:311–322.

Fuxe, K., Goldstein, M., Hökfelt, T., and Tong Hyub Joh (1971): Cellular localization of dopamine-β-hydroxylase and phenylethanolamine-N-methyl transferase as revealed by immunohisto-chemistry. *Prog. Brain Res.*, 34:127–138.

Gabel, N. W. (1965): Excitability and the origin of life: A hypothesis. *Life Sci.*, 4:2085–2096.

Gaebelein, J. W. (1973): Third-party instigation of aggression: An experimental approach. *J. Pers. Soc. Psychol.*, 27:289–295.

Gage, F. H., and Olton, D. S. (1976): L-Dopa reduces hyperactivity induced by septal lesions in rats. *Behav. Biol.*, 17:213–218.

Gage, F. H., Olton, D. S., and Bolanowski, D. (1978): Activity, reactivity, and dominance following septal lesions in rats. *Behav. Biol.*, 22:203–210.

Galambos, R. (1961): A glia-neural theory of brain function. *Proc. Natl. Acad. Sci. USA*, 47:129–136.

Galbraith, G. (1978): The economic necessity of marijuana. *Business Soc. Rev.*, 27:58–60.

Galef, B. G., Jr. (1970*a*): Target novelty elicits and directs shock-associated aggression in wild rats. *J. Comp. Physiol. Psychol.*, 71:87–91.

Galef, B. G., Jr. (1970*b*): Aggression and timidity: Responses to novelty in feral Norway rats. *J. Comp. Physiol. Psychol.*, 70:370–381.

Galef, B. G., Jr. (1970*c*): Familiarity of target location as a factor in the shock-associated aggression of wild rats. *Psychonom. Sci.*, 19:299–300.

Galef, B. G., Jr. (1970*d*): Stimulus novelty as a factor in the intraspecific pain-associated aggression of domesticated rats. *Psychosom. Sci.*, 18:21.

Gallager, D. W., and Aghajanian, G. K. (1975): Effects of chlorimipramine and lysergic acid diethylamide on efflux of precursor-formed ^{3}H-serotonin: Correlations with serotonergic impulse flow. *J. Pharmacol. Exp. Ther.*, 193:785–795.

Gallenkamp, C. R., and Rychlak, J. F. (1968): Parental attitudes of sanction in middle-class adolescent male deliquency. *J. Soc. Psychol.*, 75:255–260.

Gandelman, R. (1972*a*): Mice: Postpartum aggression elicited by the presence of an intruder. *Horm. Behav.*, 3:23–28.

Gandelman, R. (1972*b*): Induction of pup killing in female mice by androgenization. *Physiol. Behav.*, 9:101–102.

Gandelman, R., and Svare, B. (1974): Mice: Pregnancy termination, lactation, and aggression. *Horm. Behav.*, 5:397–405.

Gandelman, R., and Vom Saal, F. S. (1975): Pup-killing in mice: The effects of gonadectomy and testosterone administration. *Physiol. Behav.*, 15:647–651.

Gandelman, R., Zarrow, M. X., and Denenberg, V. H. (1971*a*): Stimulus control of cannibalism and maternal behavior in anosmic mice. *Physiol. Behav.*, 7:583–586.

Gandelman, R., Zarrow, M. X., Denenberg, V. H., and Myers, M. (1971*b*): Olfactory bulb removal eliminates maternal behavior in the mouse. *Science*, 171:210–211.

Garattini, S., Giacalone, E., and Valzelli, L. (1967): Isolation, aggressiveness and brain 5-hydroxy-tryptamine turnover. *J. Pharm. Pharmacol.*, 19:338–339.

Garattini, S., Giacalone, E., and Valzelli, L. (1969): Biochemical changes during isolation-induced aggressiveness in mice. In: *Aggressive Behaviour*, edited by S. Garattini and E. B. Sigg, pp. 179–187. Excerpta Medica, Amsterdam.

Garattini, S., and Valzelli, L. (1965): *Serotonin.* Elsevier, Amsterdam.

Garbarg, M., Barbin, G., Feger, J., and Schwartz, J.-C. (1974): Histaminergic pathway in rat brain evidenced by lesions of the medial forebrain bundle. *Science,* 186:833–835.

Gardner, G. E. (1971): Aggression and violence. The enemies of precision learning in children. *Am. J. Psychiatry,* 128:445–450.

Garrod, A. E. (1929): The power of personality. *Br. Med. J.,* 2:509–512.

Gastaut, H., and Collomb, H. (1954): Etude du comportement sexuel chez les épileptiques psychomoteurs. *Ann. Med. Psychol. (Paris),* 112:657–696.

Geen, R. G., and O'Neal, E. C. (Eds.) (1976): *Perspectives on Aggression,* Academic Press, New York.

Geen, R. G., and Stonner, D. (1971): Effects of aggressiveness habit strength on behavior in the presence of aggression-related stimuli. *J. Pers. Soc. Psychol.,* 17:149–153.

Geffen, L. B., and Livett, B. G. (1971): Synaptic vescicles in sympathetic neurons. *Physiol. Rev.,* 51:98–157.

Geissler, K. R., and Melvin, K. B. (1977): Effect of social-housing conditions on aggression in an intruder-resident paradigm in *Cavia procellus. Psychol. Rec.,* 3:537–543.

Gellert, V. F., and Sparber, S. B. (1979): Effects of morphine withdrawal on food competition hierarchies and fighting behavior in rats. *Psychopharmacology,* 60:165–172.

Gelles, R. J. (1973): Child abuse as psychopathology: A sociological critique and reformulation. *Am. J. Orthospychiatry,* 43:611–621.

Genovese, E., Napoli, P. A., and Bolego-Zonta, N. (1969): Self-aggressiveness: A new type of behavioral change induced by pemoline. *Life Sci.,* 8, pt.I:513–515.

Genthner, R. W., and Taylor, S. P. (1973): Physical aggression as a function of racial prejudice and the race of the target. *J. Pers. Soc. Psychol.,* 27:207–210.

Gentry, W. D. (1970): Effects of frustration, attack, and prior aggressive training on overt aggression and vascular processes. *J. Pers. Soc. Psychol.,* 16:718–725.

Gerall, A. A. (1963): An exploratory study of the effect of social isolation variables on the sexual behavior of male guinea pigs. *Anim. Behav.,* 11:274–282.

Gerall, H. D., Ward, I. L., and Gerall, A. A. (1967): Disruption of the male rat's sexual behaviour induced by social isolation. *Anim. Behav.,* 15:54–58.

Gerschenfeld, H. M. (1973): Chemical transmission in invertebrate central nervous systems and neuromuscular junctions. *Physiol. Rev.,* 53:1–119.

Gershoff, S. N., Hegsted, D. M., and Trulson, M. F. (1958): Metabolic studies of mongoloids. *Am. J. Clin. Nutr.,* 6:526–530.

Gerson, L. W. (1978): Alcohol-related acts of violence. Who was drinking and where the acts occurred. *J. Stud. Alcohol,* 39:1294–1296.

Geyer, M. A., and Segal, D. S. (1974): Shock-induced aggression: Opposite effects of intraventricularly infused dopamine and norepinephrine. *Behav. Biol.,* 10:99–104.

Ghiselli, W. B., and Thor, D. H. (1975): Visual, tactual, and olfactory deprivation effects on irritable fighting behavior of male hooded rats. *Physiol. Psychol.,* 3:47–50.

Giacalone, E., Tansella, M., Valzelli, L., and Garattini, S. (1968): Brain serotonin metabolism in isolated aggressive mice. *Biochem. Pharmacol.,* 17:1315–1327.

Giacobini, E. (1961): Localization of carbonic anhydrase in the nervous system. *Science,* 134:1524–1525.

Giacobini, E., and Holmstedt, B. (1958): Cholinesterase content of certain regions of the spinal cord as judged by histochemical and cartesian-diver technique. *Acta Physiol. Scand.,* 42:12–27.

Gianutsos, G., Hynes, M. D., Drawbaugh, R., and Lal, H. (1973): Morphine-withdrawal aggression during protracted abstinence: Role of latent dopaminergic supersensitivity. *Pharmacologist,* 15:348.

Gianutsos, G., Hynes, M. D., Puri, S. K., Drawbaugh, R. B., and Lal, H. (1974): Effect of apomorphine and nigrostriatal lesions on aggression and striatal dopamine turnover during morphine: Evidence for dopaminergic supersensity in protracted abstinence. *Psychopharmacologia,* 34:37–44.

Gianutsos, G., and Lal, H. (1976): Drug-induced aggression. *Curr. Dev. Psychopharmacol.,* 3:199–220.

Gianutsos, G., and Lal, H. (1977): Modification of apomorphine induced aggression by changing central cholinergic activity in rats. *Neuropharmacol.,* 16:7–10.

Gianutsos, G., and Lal, H. (1978): Narcotic analgesics and aggression. *Mod. Probl. Pharmacopsychiatry,* 13:114–138.

Gibbs, F. A. (1951): Ictal and non-ictal psychiatric disorders in temporal lobe epilepsy. *J. Nerv. Ment. Dis.,* 113:522–528.

Gibbs, F. A. (1956): Abnormal electrical activity in the temporal regions and its relationship to abnormalities of behavior. *Res. Publ. Assoc. Res. Nerv. Ment. Dis.,* 36:278–284.

Gil, D. C. (1968): Incidence of child abuse and demographic characteristics of person involved. In: *The Battered Child,* edited by R. E. Helfer and C. H. Kempe. University of Chicago Press, Chicago.

Gil, D. C. (1970): *Violence Against Children.* Harvard University Press, Cambridge, Mass.

Gill, J. R., Reid, L. D., and Porter, P. S. (1966): Effects of restricted rearing on Lashley stand performance. *Psychol. Rep.,* 19:239–242.

Glaser, G. H. (1964): The problem of psychosis in psychomotor temporal lobe epileptics. *Epilepsia,* 5:271–278.

Glees, P., Cole, J., Whitty, C. W. M., and Cairns, H. (1950): The effects of lesions in the cingular gyrus and adjacent areas in monkeys. *J. Neurol. Neurosurg. Psychiatry,* 13:178–190.

Gloor, P. (1960): Amygdala. In: *Handbook of Physiology, Section I.: Neurophysiology, Vol. 2,* edited by J. Field, pp. 1395–1420. American Physiological Society, Washington, D.C.

Gloor, P. (1967): Discussion in the paper: "Brain mechanisms related to aggressive behavior." In: *Brain Function, Vol. V. Aggression and Defence: Neural Mechanisms and Social Patterns,* edited by C. D. Clemente and D. B. Lindsley, pp. 116–133. University of California Press, Los Angeles.

Glueck, S., and Glueck, E. (1950): *Unravelling Juvenile Delinquency.* Commonwealth Fund, New York.

Glusman, M. (1974): The hypothalamic "savage" syndrome. *Aggression,* 52:52–92.

Glusman, M. (1975): Psychoanalytic concepts and brain stimulation: A consideration of relevance. In: *Neurotransmitter Balances Regulating Behavior,* edited by E. F. Domino and J. M. Davis, pp. 1–23. Domino and Davis, Ann Arbor.

Golda, V., Nováková, V., and Šterc, J. (1977): Effects of dorsal, mediobasal and laterobasal lesions of the septal area in the rat. I. The reflexive fighting and the pain threshold. *Acta Nerv. Super. (Praha),* 19:228–229.

Goldberg, M. E., and Horovitz, Z. P. (1978): Antidepressants and aggressive behavior. *Mod. Probl. Pharmacopsychiatry,* 13:29–52.

Goldberg, M. E., and Salama, A. I. (1969): Norepinephrine turnover and brain monoamine levels in aggressive mouse-killing rats. *Biochem. Pharmacol.,* 18:532–534.

Goldberg, M. E., Insalaco, J. R., Hefner, M. A., and Salama, A. I. (1973): Effect of prolonged isolation on learning, biogenic amine turnover and aggressive behaviour in three strains of mice. *Neuropharmacology,* 12:1049–1058.

Goldstein, A., Aronow, L., and Kalman, S. M. (Eds.) (1969): *Principles of Drug Action: The Basis of Pharmacology.* Hoeber Medical Division, New York.

Goldstein, A., and Kaizer, S. (1969): Psychotropic effects of caffeine in man. III. A questionnaire survey of coffee drinking and its effects in a group of housewives. *Clin. Pharmacol. Ther.,* 10:477–488.

Goldstein, A., Kaizer, S., and Whitby, O. (1969): Psychotropic effects of caffeine in man. IV. Quantitative and qualitative differences associated with habituation to coffee. *Clin. Pharmacol. Ther.,* 10:489–497.

Goldstein, A., and Schulz, R. (1973): Morphine-tolerant longitudinal muscle strip from guinea-pig ileum. *Br. J. Pharmacol.,* 48:655–666.

Goldstein, A., Warren, R., and Kaizer, S. (1965a): Psychotropic effects of caffeine in man. I. Individual differences in sensitivity to caffeine-induced wakefulness. *J. Pharmacol. Exp. Ther.,* 149:156–159.

Goldstein, A., Kaizer, S., and Warren, R. (1965b): Psychotropic effects of caffeine in man. II. Alertness, psychomotor coordination, and mood. *J. Pharmacol. Exp. Ther.,* 150:146–151.

Goldstein, M. (1974): Brain research and violent behavior. *Arch. Neurol.,* 30:1–34.

Gonsiorek, J. C., Donovick, P. J., Burright, R. G., and Fuller, J. L. (1974): Aggression in low and high brain weight mice following septal lesions. *Physiol. Behav.,* 12:813–818.

Good, M. I. (1978): Primary affective disorder, aggression, and criminality. *Arch. Gen. Psychiatry,* 35:954–960.

Goodall, J. (1964): Tool-using and aimed throwing in a community of free-living chimpanzees. *Nature (Lond.),* 201:1264–1266.

Goode, E. (1974): The criminogenics of marijuana. *Addict. Dis.*, 1:297–322.

Goode, W. J. (1971): Force and violence in the family. *J. Marriage Fam.*, 33:624–626.

Goodwin, F. K., Murphy, D. L., Brodie, H. K. H., and Bunney, W. E. (1970): L-Dopa, catechol-amines and behavior: A clinical and biochemical study in depressed patients. *Biol. Psychiatry*, 2:341–366.

Goodwin, D. W. (1973): Alcohol in suicide and homicide. *Q. J. Stud. Alcohol*, 34:144–156.

Goodwin, D. W. (1979): Alcoholism and heredity. *Arch. Gen. Psychiatry*, 36:57–61.

Goodwin, D. W., Schulsinger, F., Hermansen, L., Guze, S. B., and Winokur, G. (1973): Alcohol problems in adoptees raised apart from alcoholic biological patients. *Arch. Gen. Psychiatry*, 28:238–243.

Goodwin, D. W., Schulsinger, F., Hermansen, L., Guze, S. B., and Winokur, G. (1975): Alcoholism and the hyperactive child syndrome. *J. Nerv. Ment. Dis.*, 160:349–353.

Goodwin, D. W., Schulsinger, F., Knop, J., Mednick, S., and Guze, S. B. (1977): Alcoholism and depression in adopted-out daughters of alcoholics. *Arch. Gen. Psychiatry*, 34:751–755.

Goodwin, D. W., Schulsinger, F., Møller, N., Hermansen, L., Winokur, G., and Guze, S. B. (1974): Drinking problems in adopted and nonadopted sons of alcoholics. *Arch. Gen. Psychiatry*, 31:164–169.

Goodwin, F. K., Murphy, D. L., Brodie, H. K. H., and Bunney, W. E., Jr. (1971): Levodopa: Alterations in behavior. *Clin. Pharmacol. Ther.*, 12:383–396.

Gorney, R. (1971): Interpersonal intensity, competition, and synergy: Determinants of achievement, aggression, and mental illness. *Am. J. Psychiatry*, 128:436–445.

Gosselin, R. E., Moore, K. E., and Milton, A. S. (1962): Physiological control of molluscan gill cilia by 5-hydroxytryptamine. *J. Gen. Physiol.*, 46:277–296.

Gotsick, J. E., Drew, W. G., and Proctor, D. L. (1975): Apomorphine-induced aggression: An evaluation of possible sensitizing factors in the rat. *Pharmacology*, 13:385–390.

Gottesfeld, Z., Ebsten, B. S., and Samuel, D. (1971): Effect of lithium on concentrations of glutamate and GABA levels in amygdala and hypothalamus of rat. *Nature [New Biol.]*, 234:124–125.

Gottesfeld, Z., and Jacobowitz, D. M. (1978): Further evidence for GABAergic afferents to the lateral habenula. *Brain Res.*, 152:609–613.

Gottesman, I. I., and Shields, J. (1972): *Schizophrenia and Genetics.* Academic Press, New York.

Gottschalk, L. A. (1969): The measurement of hostile aggression through the content analysis of speach, some biologial and interpersonal aspects. In: *Aggressive Behaviour*, edited by S. Garattini and E. B. Sigg, pp. 299–316. Excerpta Medica, Amsterdam.

Goyens, J., and Noirot, E. (1974): Effects of cohabitation with females on aggressive behavior between male mice. *Dev. Psychobiol.*, 8:79–84.

Graff, H., and Stellar, E. (1962): Hyperphagia, obesity and finickiness. *J. Comp. Physiol. Psychol.*, 55:418–424.

Grafstein, B., McEwen, B. S., and Shelanski, M. L. (1970): Axonal transport of neurotubule protein. *Nature*, 227:289–290.

Grant, E. C., and Mackintosh, J. H. (1963): A description of the social postures of some laboratory rodents. *Behaviour*, 21:246–259.

Grant, E. C., Mackintosh, J. H., and Lerwill, C. J. (1970): The effect of a visual stimulus on the agonistic behaviour of the golden hamster. *Z. Thierpsychol.*, 28:73–77.

Grant, L. D., Coscina, D. V., Grossman, S. P., and Freedman, D. X. (1973): Muricide after serotonin depleting lesions of midbrain raphé nuclei. *Pharmacol. Biochem. Behav.*, 1:77–80.

Gray, E. G. (1959): Axo-somatic and axo-dendritic synapses of the cerebral cortex: An electron microscope study. *J. Anat.*, 93:420–433.

Gray, J. A. (1971): *The Psychology of Fear and Stress.* McGraw-Hill, New York.

Graziani, G., and Montanaro, N. (1966): Variazioni della concentrazione di nor-adrenalina cerebrale del ratto albino nel corso della giornata. *Boll. Soc. Ital. Biol. Sper.*, 43:46–48.

Greden, J. F. (1974): Anxiety or caffeinism: A diagnostic dilemma. *Am. J. Psychiatry*, 131:1089–1092.

Green, J. R., Duisberg, R. E. H., and McGrath, W. B. (1951): Focal epilepsy of psychomotor type, a preliminary report of observations on effects of surgical therapy. *J. Neurosurg.*, 8:157–172.

Greenberg, A. S., and Coleman, M. (1976): Depressed 5-hydroxyindole levels associated with hyper-active and aggressive behavior. *Arch. Gen. Psychiatry*, 33:331–336.

Greene, R., and Dalton, K. (1953): The premenstrual syndrome. *Br. Med. J.*, 1:1007–1014.

Greenough, W. T. (1975): Experimental modification of the developing brain. *Am. Sci.,* 63:37–46.

Greenough, W. T., Wood, W. E., and Madden, T. C. (1972): Possible memory storage differences among mice reared in environments varying in complexity. *Behav. Biol.,* 7:717–722.

Griffith, A. W. (1971): Prisoners of XYY constitution: Psychological aspects. *Br. J. Psychiatry,* 119:193–194.

Griffith, W. J., Jr. (1960): Effects of isolation and stress on escape thresholds in albino rats. *Psychol. Rep.,* 6:23–29.

Griffitt, W. (1970): Environmental effects on interpersonal affective behavior: Ambient affective temperature and attraction. *J. Pers. Soc. Psychol.,* 15:240–244.

Griffitt, W., and Veitch, R. (1971): Hot and crowded: Influences of population density and temperature on interpersonal affective behavior. *J. Pers. Soc. Psychol.,* 17:92–98.

Groen, J. J. (1972): The study of human aggression. *Psychother. Psychosom.,* 20:312–315.

Grossman, R. G., and Seregin, A. (1977): Glial-neural interaction demonstrated by the injection of Na^+ and Li^+ into cortical glia. *Science,* 195:196–198.

Gruendel, A. D., and Arnold, W. J. (1969): Effects of early social deprivation on reproductive behavior of male rats. *J. Comp. Physiol. Psychol.,* 67:123–128.

Grundfest, H. (1959): In: *Evolution of Nervous Control from Primitive Organisms to Man,* edited by A. D. Bass, Publication no. 52, p. 43. AAAS, Washington, D.C.

Gumulka, W., Ramirez del Angel, A., Samanin, R., and Valzelli, L. (1970): Lesion of substantia nigra: Biochemical and behavioral effects in rats. *Eur. J. Pharmacol.,* 10:79–82.

Gunn, J. (1977): Criminal behaviour and mental disorder. *Br. J. Psychiatry,* 130:317–329.

Hadaway, C. K. (1978): Life satisfaction and religion. *Soc. Forc.,* 57:636–43.

Haeckel, E. (1868): *Natürliche Schopfungsgeschichte.* Reiner, Berlin.

Hafez, E. S. (1962): *The Behaviour of Domestic Animals.* Williams & Wilkins, Baltimore.

Häfner, H., and Böker, W. (1973): Mentally disordered violent offenders. *Soc. Psychiatry,* 8:220–229.

Hahn, M. E., Haber, S. B., and Fuller, J. L. (1973): Differential agonistic behavior in mice selected for brain weight. *Physiol. Behav.,* 10:759–762.

Haigler, H. J., and Aghajanian, G. K. (1974): Lysergic acid diethylamide and serotonin: A comparison of effects on serotonergic neurons and neurons receiving a serotonergic input. *J. Pharmacol. Exp. Ther.,* 188:688–699.

Halas, E. S., Reynolds, G. M., and Sandstead, H. H. (1977): Intrauterine nutrition and its effects on aggression. *Physiol. Behav.,* 19:653–661.

Haley, T. J. (1957): Intercerebral injection of psychotomimetic and psychotherapeutic drugs into conscious mice. *Acta Pharmacol. Toxicol.,* 13:107–112.

Hall, C. S. (1941): Temperament: A survey of animal studies. *Psychol. Bull.,* 38:909–943.

Hall, E. T. (1959): *The Silent Language.* Doubleday & Co., Garden City, N.Y.

Hall, E. T. (1966): *The Hidden Dimension.* Doubleday & Co., New York.

Halleck, S. L. (1974): Legal and ethical aspects of behavior control. *Am. J. Psychiatry,* 131:381–385.

Hamburg, D. A. (1971): Aggressive behavior of chimpanzees and baboons in natural habitats. *J. Psychiatr. Res.,* 8:385–398.

Hamburg, D. A., and Lunde, D. T. (1966): Sex hormones in the development of sex differences in human behaviour. In: *The Development of Sex Differences in Human Behavior,* edited by E. Maccoby, p. 1. Stanford University Press, Palo Alto.

Hamburg, D. A., Moos, R. H., and Yalom, I. D. (1968): Studies of distress in the menstrual cycle and the postpartum period. In: *Endocrinology and Human Behaviour,* edited by R. P. Michael. Oxford University Press, London.

Hapkiewicz, W. G. (1974): Developmental patterns of aggression. In: *Determinants and Origins of Aggressive Behavior,* edited by J. de Wit and W. W. Hartup, pp. 201–207. Mouton, The Hague.

Harris, G. W. (1964): Sex hormones, brain development and brain function. *Endocrinology,* 75:627–648.

Harris, M. (1971): *Culture, Man, and Nature.* New York.

Harth, E. M., Csermely, T. J., Beek, B., and Lindsay, R. D. (1970): Brain functions and neural dynamics. *J. Theor. Biol.,* 26:93–120.

Hartmann, D. P. (1969): Influence of symbolically modeled instrumental aggression and pain cues on aggressive behavior. *J. Pers. Soc. Psychol.,* 11:280–288.

Hartmann, J. F. (1953): An electron optical study of sections of central nervous system. *J. Comp. Neurol.,* 99:201–225.

Hartup, W. W., and de Wit, J. (1974): The development of aggression: Problems and perspectives. In: *Determinants and Origins of Aggressive Behavior,* edited by J. de Wit and W. W. Hartup, pp. 595–620. Mouton, The Hague.

Hasselager, E., Rolinski, Z., and Randrup, A. (1972): Specific antagonism by dopamine inhibitors of items of amphetamine induced aggressive behaviour. *Psychopharmacologia,* 24:485–495.

Hatch, A., Wiberg, G. S., Balazs, T., and Grice, H. C. (1963): Long-term isolation stress in rats. *Science,* 142:507.

Hatch, A. M., Wiberg, G. S., Zawidzka, Z., Cann, M., Airth, J. M., and Grice, H. C. (1965): Isolation syndrome in the rat. *Toxicol. Appl. Pharmacol.,* 7:737–745.

Haug, M. (1973a): Mise en évidence d'une odeur liée à l'allaitement et stimulant l'agressivité d'un groupe de souris femelles. *C.R. Acad. Sci. [D] (Paris),* 276:3457–3460.

Haug, M. (1973b): L'urine d'une femelle allaitante coutient une phéromone stimulant l'agressivité de petits groupes de souris femelles. *C.R. Acad. Sci. [D] (Paris),* 277:2053–2056.

Haug, M., and Brain, P. F. (1978): Attack directed by groups of castrated male mice towards lactating or non-lactating intruders: A urine-dependent phenomenon? *Physiol. Behav.,* 21:549–552.

Hautojärvi, S., and Lagerspetz, K. (1968): The effects of socially-induced aggressiveness or nonaggressiveness on the sexual behaviour of inexperienced male mice. *Scand. J. Psychol.,* 9:45–49.

Hawke, C. C. (1950): Castration and sex crimes. *Am. J. Ment. Defic.,* 55:220–226.

Hay, D. A. (1975): Y chromosome and aggression in mice. *Nature,* 255:658.

Healy, W., and Bronner, A. T. (1936): *New Light on Delinquency and Its Treatment.* Yale University Press, New Haven, Conn.

Heath, R. G. (1954): Definition of the septal region. In: *Studies in Schizophrenia,* edited by R. G. Heath, pp. 3–5. Harvard University Press, Cambridge, Mass.

Heath, R. G. (1963): Electrical self stimulation of the brain in man. *Am. J. Psychiatry,* 120:571–577.

Heath, R. G. (1964): Plasma response of human subjects to direct stimulation of the brain: Physiologic and psychodynamic considerations. In: *The Role of Plasma in Behavior,* edited by R. G. Heath, pp. 219–243. Hoeber Medical Division, Harper & Row, New York.

Heath, R. G. (1972): Plasma and brain activity in man. Deep and surface electroencephalograms during orgasm. *J. Nerv. Ment. Dis.,* 154:3–18.

Heath, R. G. (1975): Brain function and behavior. I. Emotion and sensory phenomena in psychotic patients and in experimental animals. *J. Nerv. Ment. Dis.,* 160:159–175.

Heath, R. G. (1976): Emotion and sensory perception: Human and animal studies. In: *Biological Foundations of Psychiatry,* edited by R. G. Grenell and S. Gabay, pp. 255–271. Raven Press, New York.

Heath, R. G. (1977): Modulation of emotion with a brain pacemaker. *J. Nerv. Ment. Dis.,* 165:300–317.

Heath, R. G., Dempesy, C. W., Fontan, C. J., and Myers, W. A. (1978): Cerebellar stimulation: Effects on septal region, hippocampus, and amygdala of cats and rats. *Biol. Psychiatry,* 13:501–529.

Heath, R. G., and Harper, J. W. (1974): Ascending projections of the cerebellar fastigial nucleus to the hippocampus, amygdala, and other temporal lobe sites: Evoked potential and histological studies in monkeys and cats. *Exp. Neurol.,* 45:268–287.

Heath, R. G., and Harper, J. W. (1976): Descending projections of the rostral septal region: An electrophysiological-histological study in the cat. *Exp. Neurol.,* 50:536–560.

Heath, R. G., and Mickle, W. A. (1960): Evaluation of seven years' experience with depth electrode studies in human patients. In: *Electrical Studies on the Unanesthetized Brain,* edited by E. R. Ramey and D. S. O'Doherty, p. 214. Hoeber, New York.

Heath, R. G., Monroe, R. R., and Mickle, W. A. (1955): Stimulation of the amygdaloid nucleus in a schizophrenic patient. *Am. J. Psychiatry,* 111:862–863.

Hebb, C. (1970): CNS at the cellular level: Identity of transmitter agents. *Annu. Rev. Physiol.,* 32:165–192.

Hediger, H. (1950): *Wild Animals in Captivity.* Butterworth, London.

Heilingenberger, W., and Kramer, U. (1972): Aggressiveness as a function of external stimulation. *J. Comp. Physiol.*, 77:332–340.

Heimburger, R. F., Whitlock, C. C., and Kalsbeck, J. E. (1966): Stereotaxic amygdalotomy for epilepsy with aggressive behavior. *J.A.M.A.*, 198:741–745.

Heimer, L., and Larsson, K. (1967): Mating behavior of male rats after olfactory bulb lesions. *Physiol. Behav.*, 2:207–209.

Heimstra, N. W. (1965): A further investigation of the development of mouse-killing in rats. *Psychonom. Sci.*, 2:179–180.

Heller, A., Harvey, J. A., and Moore, R. Y.: A demonstration of a fall in brain serotonin following central nervous system lesions in the rat. *Biochem. Pharmacol.*, 11:859–866.

Heller, M. S., and Polsky, S. (1971): Television violence. Guidelines for evaluation. *Arch. Gen. Psychiatry*, 24:279–285.

Henderson, N. D. (1973): Brain weight changes resulting from enriched rearing conditions: A diallel analysis. *Dev. Psychobiol.*, 6:367–376.

Henley, E. D., Moisset, B., and Welch, B. L. (1973): Catecholamine uptake in cerebral cortex: Adaptive change induced by fighting. *Science*, 180:1050–1052.

Henn, F. A., Anderson, D. J., and Sellström, Å. (1977): Possible relationship between glial cells, dopamine and the effects of antipsychotic drugs. *Nature*, 266:637–638.

Henn, F. A., and Hamberger, A. (1971): Glial cell function: Uptake of transmitter substances. *Proc. Natl. Acad. Sci. USA*, 68:2686–2690.

Henry, J. P., Meeham, J. P., and Stephens, P. M. (1967): The use of psychosocial stimuli to induce prolonged systolic hypertension in mice. *Psychosom. Med.*, 29:408–432.

Henry, J. P., Stephens, P. M., Axelrod, J., and Mueller, R. A. (1971): Effect of psychosocial stimulation on the enzymes involved in the biosynthesis and metabolism of noradrenaline and adrenaline. *Psychosom. Med.*, 33:227–237.

Herman, Z. S., Kmieciak-Kołada, K., Słominska-Zurek, J., and Szkilnik, R. (1972): Central effects of acetylcholine. *Psychopharmacologia*, 27:223–232.

Heron, W., Doane, B., and Scott, T. H. (1956): Visual disturbances after prolonged perceptual isolation. *Can. J. Psychol.*, 10:13–16.

Herrmann, W. M., and Beach, R. C. (1978): The psychotropic properties of estrogens. *Pharmakopsychiatr. Neuropsychopharmakol.*, 11:164–176.

Hertwig, O., and Hertwig, R. (1879): *Studien zur Blättertheorie* Heft. I. *Die Actinien.*, G. Fischer, Jena.

Héry, F., Rouer, E., and Glowinski, J. (1972): Daily variations of serotonin metabolism in the rat brain. *Brain Res.*, 43:445–465.

Héry, F., Rouer, E., and Glowinski, J. (1973): Effect of 6-hydroxydopamine on daily variations of 5-HT synthesis in the hypothalamus of the rat. *Brain Res.*, 58:135–146.

Hess, W. R. (1954): *Diencephalon: Autonomic and Extra Pyramidal Function.* Grune Stratton, New York.

Hess, W. R. (1967): Causality, consciousness, and cerebral organization. *Science*, 158:1279–1283.

Hess, W. R., and Brügger, M. (1943): Das subkortikale Zentrum der affektiven Abwehrreaktion. *Helv. Physiol. Pharmacol. Acta*, 1:33–52.

Hewitt, L. E., and Jenkins, R. L. (1946): *Fundamental Patterns of Maladjustment.* State of Illinois, Springfield, Ill.

Hierons, R., and Saunders, M. (1966): Impotence in patients with temporal-lobe lesions. *Lancet*, 2:761–764.

Higgins, J. W., Mahl, G. F., Delgado, J. M. R., and Hamlin, H. (1956): Behavioral changes during intracerebral electrical stimulation. *Arch. Neurol. Psychiatry*, 76:399–419.

Hill, D., Pond, D. A., Mitchell, W., and Falconer, M. A. (1957): Personality changes following temporal lobectomy for epilepsy. *J. Ment. Sci.*, 103:18–27.

Hinde, R. A. (1967): The nature of aggression. *New Soc.*, 9:302–304.

Hodge, G. K., and Butcher, L. L. (1974): 5-Hydroxytryptamine correlates of isolation-induced aggression in mice. *Eur. J. Pharmacol.*, 28:326–337.

Hodge, G. K., and Butcher, L. L. (1975): Catecholamine correlates of isolation-induced aggression in mice. *Eur. J. Pharmacol.*, 31:81–93.

Hoebel, B. G. (1971): Feeding: Neural control of intake. *Annu. Rev. Physiol.*, 33:533–568.

Hoenig, J., and Hamilton, C. M. (1960): Epilepsy and sexual orgasm. *Acta Psychiat. Scand.*, 35:448–456.

Hoffman, H. S., Boskoff, K. J., Eiserer, L. A., and Klein, S. H. (1975): Isolation-induced aggression in newly hatched ducklings. *J. Comp. Physiol. Psychol.,* 89:447–456.

Hoffman, M. L. (1970): Moral development. In: *Carmichael's Manual of Child Psychology, Vol. 2, 3rd ed.,* edited by P. H. Mussen, p. 261. John Wiley & Sons, New York.

Hoffmeister, F., and Wuttke, W. (1969): On the actions of psychotropic drugs on the attack- and the aggressive-defensive behavior of mice and cats. In: *Aggressive Behaviour,* edited by S. Garattini and E. B. Sigg, pp. 273–280. Excerpta Medica, Amsterdam.

Hogan, R. (1973): Moral conduct and moral character: A psychological perspective. *Psychol. Bull.,* 79:217–232.

Hoge, M. A., and Stocking, R. J. (1912): A note on the relative value of punishment and reward as motives. *J. Anim. Behav.,* 2:43–50.

Höhn, R., and Lasagna, L. (1960): Effects of aggregation and temperature on amphetamine toxicity in mice. *Psychopharmacologia,* 1:210–220.

Holloway, R. L. (ed.) (1974): *Primate Agression, Territoriality and Xenophobia. A Comparative Perspective.* Academic Press, New York.

Hong, J. S., Yang, H.-Y. T., Racagni, G., and Costa, E. (1977): Projections of substance P containing neurons from neostriatum to substantia nigra. *Brain Res.,* 122:541–544.

Hook, E. B. (1973): Behavioral implications of the human XYY genotype. *Science,* 179:139–150.

Hook, E. B., and Kim, D.-S. (1971): Height and antisocial behavior in XY and XYY boys. *Science,* 172:284–286.

Horel, J. A., and Misantone, L. J. (1974): The Klüver-Bucy syndrome produced by partial isolation of the temporal lobe. *Exp. Neurol.,* 42:101–112.

Horowitz, M. J. (1969): Flashbacks: Recurrent intrusive images after the use of LSD. *Am. J. Psychiatry,* 126:565–569.

Horowitz, M. J., Duff, D. F., and Stratton, L. O. (1964): Body-buffer zone; Exploration of personal space. *Arch. Gen. Psychiatry,* 11:651–656.

Horvath, F. E. (1963): Effects of basolateral amygdalectomy on three types of avoidance behavior in cats. *J. Comp. Physiol. Psychol.,* 56:380–389.

Hotton, N., III (1976): Origin and radiation of the classes of poikilothermous vertebrates. In: *Evolution of the Brain and Behavior in Vertebrates,* edited by R. B. Masterton, M. E. Bitterman, C. B. G. Campbell, and N. Hotton, pp. 1–24. Lawrence Erlbaum, Hillsdale, N.J.

Howden, J. C. (1872): The religious sentiments in epileptics. *J. Ment. Sci.,* 18:491–497.

Hull, C. D., Buchwald, N. A., and Ling, G. M. (1967): Effects of direct cholinergic stimulation of forebrain structures. *Brain Res.,* 6:22–35.

Hull, E. M., and Homan, H. D. (1975): Olfactory bulbectomy, peripheral anosmia, and mouse killing and eating by rats. *Behav. Biol.,* 14:481–488.

Hull, E. M., Rosselli, L., and Langan, C. J. (1973): Effects of isolation and grouping on guinea pigs. *Behav. Biol.,* 9:493–497.

Humphries, D. A., and Driver, P. M. (1967): Erratic displays as a device against predators. *Science,* 156:1767–1768.

Hunsperger, R. W. (1963): Comportement affectif provoqué par la stimulation électrique du tronc cérébral et du cerveau antérieur. *J. Physiol. (Paris),* 55:45–98.

Hunt, J. McV. (1941): The effect of infant feeding-frustration upon adult hoarding in the albino rat. *J. Abnorm. Soc. Psychol.,* 36:338–360.

Hunt, J. McV., Schlosberg, H., Soloman, R. L., and Stellar, E. (1947): Studies of the effects of infantile experience on adult behavior in rats. I. Effects of infantile feeding frustration on adult hoarding. *J. Comp. Physiol. Psychol.,* 40:291–304.

Hunt, J. McV., and Willoughby, R. R. (1939): The effect of frustration on hoarding in rats. *Psychosom. Med.,* 1:309–310.

Hutchings, B. (1974): Genetic factors in criminality. In: *Determinants and Origins of Aggressive Behavior,* edited by J. de Wit and W. W. Hartup, pp. 255–265. Mouton, The Hague.

Hutchins, D. A., Pearson, J. D. M., and Sharman, D. F. (1974): An altered metabolism of dopamine in the striatal tissue of mice made aggressive by isolation. *Br. J. Pharmacol.,* 51:115P–116P.

Hutchins, D. A., Pearson, J. D. M., and Sharman, D. F. (1975): Striatal metabolism of dopamine in mice made aggressive by isolation. *J. Neurochem.,* 24:1151–1154.

Hutchinson, R. R., and Renfrew, J. W. (1966): Stalking attack and eating behaviors elicited from the same sites in the hypothalamus. *J. Comp. Physiol. Psychol.,* 61:360–367.

Hutchinson, R. R., Renfrew, J. W., and Young, G. A. (1971): Effects of long-term shock and associated stimuli on aggressive and manual responses. *J. Exp. Anal. Behav.*, 15:141–166.

Hutchinson, R. R., Ulrich, R. E., and Azrin, N. H. (1965): Effects of age and related factors on the pain-aggression reaction. *J. Comp. Physiol. Psychol.*, 59:365–369.

Hydén, H. (1960): The neuron. In: *The Cell*, edited by J. Brachet and A. E. Mirsky, *Vol. 4*, pp. 215–323. Academic Press, New York.

Hymovitch, B. (1952): The effects of experimental variations on problem solving in the rat. *J. Comp. Physiol. Psychol.*, 45:313–321.

Igic, R., Stern, P., and Basagic, E. (1970): Changes in emotional behaviour after application of cholinesterase inhibitor in the septal and amygdala region. *Neuropharmacology*, 9:73–75.

Illis, L. (1964): Spinal cord synapses in the cat: The normal appearances by the light microscope. *Brain*, 87:543–554.

Inselman, B. R., and Flynn, J. P. (1972): Modulatory effects of preoptic stimulation on hypothalamically-elicited attack in cats. *Brain Res.*, 42:73–87.

Iorio, L. C., Deacon, M. A., and Ryan, E. A. (1975): Blockade by narcotic drugs of naloxone-precipitated jumping in morphine-dependent mice. *J. Pharmacol. Exp. Ther.*, 192:58–63.

Itil, T. M., and Mukhopadhyay, S. (1978): Pharmacological management of human violence. *Mod. Probl. Pharmacopsychiatry*, 13:139–158.

Iversen, L. L. (1970): Neurotransmitters, neurohormones, and other small molecules in neurons. In: *Neurosciences, Second Study Program*, edited by F. O. Schmitt, pp. 768–782. Rockefeller University Press, New York.

Iwamoto, E. T., Ho, I. K., and Way, E. L. (1973): Elevation of brain dopamine during naloxone precipitated withdrawal in morphine dependent mice and rats. *Proc. West. Pharmacol. Soc.*, 16:14–18.

Jacobs, P. A., Brunton, M., Melville, M. M., Brittain, R. P., and McClemont, W. F. (1965): Aggressive behaviour, mental sub-normality and the XYY male. *Nature*, 208:1351–1352.

Jacobsen, E. (1961): The clinical effects of drugs and their influence on animal behavior. *Rev. Psychol. Appliquée*, 11:421–532.

Jacoby, J. H., Shabshelowitz, H., Fernstrom, J. D., and Wurtman, R. J. (1975): The mechanisms by which methiothepin, a putative serotonin receptor antagonist, increases brain 5-hydroxyindole levels. *J. Pharmacol. Exp. Ther.*, 195:257–264.

Jaffe, Y., Malamuth, N., Feingold, J., and Feshbach, S. (1973): Sexual arousal and behavioral aggression. *Proc. 53rd Western Psychological Association Convention*, Anaheim, Calif.

Jaffe, Y., Malamuth, N., Feingold, J., and Feshbach, S. (1974): Sexual arousal and behavioral aggression. *J. Pers. Soc. Psychol.*, 30:759–764.

James, I. P. (1960): Temporal lobectomy for psychomotor epilepsy. *J. Ment. Sci.*, 106:543–558.

James, W. (1911): *Memories and Studies*, p. 301. Longsman, Green & Co., London.

Jenkins, T. N., Warner, L. H., and Warden, C. J. (1926): Standard apparatus for the study of animal motivation. *J. Comp. Psychol.*, 6:361–382.

Jerison, H. J. (1970): Brain evolution: New light on old principles. *Science*, 170:1224–1225.

Jerôme, H., Lejeune, J., and Turpin, R. (1960): Etude de l'excrétion urinaire de certains métabolites du tryptophane chez les enfants mongoliens. *C. R. Acad. Sci. [D] (Paris)*, 251:474–476.

Johannesson, I. (1974): Aggressive behavior among school children related to maternal practices in early childhood. In: *Determinants and Origins of Aggressive Behavior*, edited by J. de Wit and W. W. Hartup, pp. 413–426. Mouton, The Hague.

Johnson, D. E., and Sellinger, O. Z. (1971): Protein synthesis in neurons and glial cells of the developing rat brain: An *in vivo* study. *J. Neurochem.*, 18:1445–1460.

Johnson, D. N., Funderburk, W. H., Ruckart, R. T., and Ward, J. W. (1972): Contrasting effects of two 5-hydroxytryptamine-depleting drugs on sleep patterns in cats. *Eur. J. Pharmacol.*, 20:80–84.

Johnson, G. A., Kim, E. G., and Boukma, S. J. (1972): 5-Hydroxyindole levels in rat brain after inhibition of dopamine β-hydroxylase. *J. Pharmacol. Exp. Ther.*, 180:539–546.

Johnson, J. (1965): Sexual impotence and the limbic system. *Br. J. Psychiatry*, 111:300–303.

Johnson, R. N., DeSisto, M. J., and Koenig, A. B. (1972): Social and developmental experience and interspecific aggression in rats. *J. Comp. Physiol. Psychol.*, 79:237–242.

Jonas, A. D. (1965): Ictal and subictal neurosis: Diagnosis and treatment. Charles C. Thomas, Springfield, Illinois.

Jones, A. B., Barchas, J. D., and Eichelman, B. (1976): Taming effects of p-chlorophenylalanine on the aggressive behavior of septal rats. *Pharmacol. Biochem. Behav.*, 4:397–400.

Jones, E. G., and Powell, T. P. S. (1970): An anatomical study of converging sensory pathways within the cerebral cortex of the monkey. *Brain,* 93:793–820.

Jones, I. H. (1971): Stereotyped aggression in a group of Australian Western Desert Aborigines. *Br. J. Med. Psychol.,* 44:259–265.

Jones, I. H., and Barraclough, B. M. (1978): Auto-mutilation in animals and its relevance to self-injury in man. *Acta Psychiatr. Scand.,* 58:40–47.

Jones, R. B., and Nowell, N. W. (1973a): Aversive and aggression-promoting properties of urine from dominant and subordinate male mice. *Anim. Learn. Behav.,* 1:207–210.

Jones, R. B., and Nowell, N. W. (1973b): The effect of urine on the investigatory behavior of male albino mice. *Physiol. Behav.,* 11:35–38.

Jones, R. B., and Nowell, N. W. (1974): A comparison of the aversive and female attractant properties of urine from dominant and subordinate male mice. *Anim. Learn. Behav.,* 2:141–144.

Jonsson, G. (1967): Delinquent boys, their parents and grandparents. *Acta Psychiatr. Scand.,* 43:Suppl. 195:1–266.

Jørgensen, F. (1968): Abuse of psychotomimetics. *Acta Psychiatr. Scand.,* Suppl. 203:205–216.

Jouvet, M. (1973): Monoaminergic regulation of the sleep-waking cycle. In: *Pharmacology and the Future of Man, Vol. 4,* edited by G. H. Acheson, pp. 103–107. S. Karger, Basel.

Jung, R., and Hassler, R. (1960): The extrapyramidal motor system. In: *Handbook of Physiology, Sect. 1, Neurophysiology, Vol. 2,* edited by J. Field, H. W. Magoun, and V. E. Hall, pp. 863–927. American Physiological Society, Washington, D.C.

Kaada, B. (1965): Brain mechanisms related to aggressive behavior. *UCLA Forum Med. Sci.,* 7:95–133.

Kaada, B. R., Andersen, P., and Jansen, J., Jr. (1954): Stimulation of the amygdaloid nuclear complex in unanesthetized cats. *Neurology (Minneap.),* 4:48–64.

Kagan, J. (1974): Development and methodological considerations in the study of aggression. In: *Determinants and Origins of Aggressive Behavior,* edited by J. de Wit and W. W. Hartup, pp. 107–114. Mouton, The Hague.

Kahn, M. W. (1961): The effect of socially learned aggression or submission mating behavior of C57 mice. *J. Genet. Psychol.,* 98:211–217.

Kahn, M. W., and Kirk, W. E. (1968): The concepts of aggression: A review and reformulation. *Psychol. Rec.,* 18:559–573.

Kaij, L. (1960): Studies on the etiology and sequels of abuse of alcohol. Thesis, University of Lund, Sweden.

Kalin, R., McClelland, D. C., and Kahn, M. (1965): The effects of male social drinking on fantasy. *J. Pers. Soc. Psychol.,* 1:441–452.

Kalinowsky, L. B. (1969): Psychosurgery today. The present situation. *Dis. Nerv. Syst. GWAN Suppl.,* 30:53.

Kalyanaraman, S. (1975): Some observations during stimulation of the human hypothalamus. *Confin. Neurol.,* 37:189–192.

Kanki, J. P., and Adams, D. B. (1978): Ventrobasal thalamus necessary for visually-released defensive boxing of rat. *Physiol. Behav.,* 21:7–12.

Kant, K. J. (1969): Influences of amygdala and medial forebrain bundle on self-stimulation in the septum. *Physiol. Behav.,* 4:777–784.

Karczmar, A. G., and Scudder, C. L. (1969): Aggression and neurochemical changes in different strains and genera of mice. In: *Aggressive Behaviour,* edited by S. Garattini and E. B. Sigg, pp. 209–227. Excerpta Medica, Amsterdam.

Karli, M. (1961): Rôle des afférences sensorielles dans le déclenchement du comportement d'agression interspécifique rat-souris. *C. R. Soc. Biol. [D] (Paris),* 155:644–646.

Karli, P. (1955): Effects de lésions expérimentales des noyaux amygdaliens et du lobe frontal sur le comportement d'agression du rat vis-à-vis de la souris. *C. R. Soc. Biol. [D] (Paris),* 149:2227–2229.

Karli, P. (1956): The Norway rat's killing-response to the white mouse. An experimental analysis. *Behaviour,* 10:81–103.

Karli, P. (1961): Effets de lésions expérimentales du septum sur l'aggressivité interspécifique rat-souris. *C. R. Soc. Biol. [D] (Paris),* 154:1079–1082.

Karli, P. (1974): Aggressive behavior and its brain mechanisms (as exemplified by an experimental analysis of the rat's mouse-killing behavior). In: *Determinants and Origins of Aggressive Behaviour,* edited by J. de Wit and W. W. Hartup, pp. 277–290. Mouton, The Hague.

Karli, P., and Vergnes, M. (1963): Déclenchement du comportement d'agression interspécifique rat-souris par des lésions experimentales de la bandelette olfactive latérale et du cortex prépyriforme. *C. R. Soc. Biol. [D] (Paris),* 157:373–374.

Karli, P., and Vergnes, M. (1964*a*): Nouvelles donnés sur les bases neurophysiologiques du comportement d'agression intérspecifique rat-souris. *J. Physiol. (Paris),* 56:384.

Karli, P., and Vergnes, M. (1964*b*): Dissociation expérimentale du comportement d'agression interspécifique rat-souris et du comportement alimentaire. *C. R. Soc. Biol. [D] (Paris),* 158:650–653.

Karli, P., and Vergnes, M. (1965*a*): Analyse expérimentale de mécanismes de facilitation et d'inhibition de l'agressivité interspécifique du rat. *J. Physiol. Paris,* 57:637–638.

Karli, P., and Vergnes, M. (1965*b*): Rôle des différentes composantes du complexe nucléaire amygdalien dans la facilitation de l'agressivité interspécifique du rat. *C. R. Soc. Biol. [D] (Paris),* 159:754–756.

Karli, P., Vergnes, M., and Didiergeorges, F. (1969): Rat-mouse interspecific aggressive behaviour and its manipulation by brain ablation and by brain stimulation. In: *Aggressive Behaviour,* edited by S. Garattini and E. B. Sigg, pp. 47–55. Excerpta Medica, Amsterdam.

Karli, P., Vergnes, M., Eclancher, F., and Schmitt, P. (1975): The amygdala and aggressiveness in rodents: Neurochemical correlates. I. Behavioral and neurophysiological data. In: *Neuropsychopharmacology,* edited by J. R. Boissier, H. Hippius, and P. Pichot, pp. 694–697. Excerpta Medica, Amsterdam.

Karli, P., Vergnes, M., Eclancher, F., Schmitt, P., and Chaurand, J. P. (1972): Role of the amygdala in the control of "mouse killing" behavior in the rat. *Adv. Behav. Biol.,* 2:553–580.

Karlsson, J. O. (1977): Is there an axonal transport of amino acids? *J. Neurochem.,* 29:615–617.

Katz, R. J. (1976): Role of the mystacial vibrissae in the control of isolation induced aggression in the mouse. *Behav. Biol.,* 17:399–402.

Katz, R. J., and Thomas, E. (1976): Effects of para-chlorophenylalanine upon brain stimulated affective attack in the cat. *Pharmacol. Biochem. Behav.,* 5:391–394.

Katz, R. J., and Thomas, E. (1977): Effect of food deprivation upon electrically elicited predation in the cat. *Behav. Biol.,* 19:135–140.

Kaufmann, H. (1970): *Aggression and Altruism.* Holt, Rinehart & Winston, New York.

Kavanau, J. L. (1965): *Structure and Function in Biological Membranes, Vol. 1.* Holden-Day, San Francisco.

Kawakami, M., Seto, K., Terasawa, E., and Yoshida, K. (1967): Mechanisms in the limbic system controlling reproductive functions of the ovary with special reference to the positive feedback of progestin to the hippocampus. *Prog. Brain Res.,* 27:69–102.

Keating, E. G., Kormann, L. A., and Horel, J. A. (1970): The behavioral effects of stimulating and ablating the reptilian amygdala *(Caiman sklerops). Physiol. Behav.,* 5:55–59.

Keeler, M. H. (1968): Marihuana induced hallucinations. *Dis. Nerv. Syst.* 29:314–315.

Keeler, M. H., Reifler, C. B., and Liptzin, M. B. (1968): Spontaneous recurrence of marihuana effect. *Am. J. Psychiatry,* 125:384–386.

Kellerman, J., Rigler, D., and Siegel, S. E. (1977): The psychological effects of isolation in protected environments. *Am. J. Psychiatry,* 134:563–565.

Kelly, D. D. (1975): Psychoanalytic concepts and brain stimulation: Further considerations of relevance. In: *Neurotransmitter Balances Regulating Behavior,* edited by E. F. Domino and J. M. Davis, pp. 25–35. Domino & Davis, Ann Arbor.

Kelly, J. F., and Hake, D. F. (1970): An extinction-induced increase in an aggressive response with humans. *J. Exp. Anal. Behav.,* 14:153–164.

Kennard, M. A. (1944): Experimental analysis of the functions of the basal ganglia in monkey and chimpanzees. *J. Neurophysiol.,* 7:127–148.

Kennard, M. A. (1955): Effect of bilateral ablation of cingulate area on behaviour of cats. *J. Neurophysiol.,* 18:159–169.

Kennedy, E. P. (1967): Some recent developments in the biochemistry of membranes. In: *The Neurosciences,* edited by G. C. Quarton, T. Melnechuk, and F. O. Schmitt, pp. 271. Rockefeller University Press, New York.

Kenny, D. A. (1972): Threats to the internal validity of cross-lagged panel inference, as related to "Television violence and child aggression: A followup study." In: *Television and Social Behavior, Vol. 3: Television and Adolescent Aggressiveness,* edited by G. A. Comstock and E. A. Rubinstein, pp. 136–140. U.S. Government Printing Office, Washington, D.C.

Kenyon, J., and Krieckhaus, E. E. (1965): Enhanced avoidance behavior following septal lesions in the rat as a function of lesion size and spontaneous activity. *J. Comp. Physiol. Psychol.*, 59:466–469.

Kephart, N. C. (1940): Influencing the rate of mental growth in retarded children through environmental stimulation. *Yearbook Natl. Soc. Study Educ.*, 39:223–230.

Kesey, K. (1962): *One Flew Over the Cuckoo's Nest.* Viking Press, New York.

Keup, W. (1970): Psychotic symptoms due to cannabis abuse (A survey of newly admitted mental patients). *Dis. Nerv. Syst.* 31:119–126.

Kidd, K. K., and Matthysee, S. (1978): Research designs for the study of gene-environment interactions in psychiatric disorders. *Arch. Gen. Psychiatry*, 35:925–932.

Killackey, H. P., Belford, G., Ryugo, R., and Ryugo, K. (1976): Anomalous organization of thalamocortical projections consequent to vibrissae removal in the newborn rat and mouse. *Brain Res.*, 104:309–315.

Kiloh, L. G., Gye, R. S., Rushworth, R. G., Bell, D. S., and White, R. T. (1974): Stereotactic amygdaloidotomy for aggressive behaviour. *J. Neurol Neurosurg. Psychiatry*, 37:437–444.

Kim, C., Choi, H., Kim, J. K., Kim, M. S., Huh, M. K., and Moon, Y. B. (1971): Sleep pattern of hippocampectomized cat. *Brain Res.*, 29:223–236.

Kimble, C. E., Fitz, D., and Onorad, J. R. (1977): Effectiveness of counteraggression strategies in reducing interactive aggression by males. *J. Pers. Soc. Psychol.*, 35:272–278.

Kimelman, B. R., and Lubow, R. E. (1974): The inhibitory effect of preexposed olfactory cues on intermale aggression in mice. *Physiol. Behav.*, 12:919–922.

King, C. H. (1975): The ego and the integration of violence in homicidal youth. *Am. J. Orthopsychiatry*, 45:134–145.

King, D. L., and Appelbaum, J. R. (1973): Effect of trials on "emotionality" behavior of the rat and mouse. *J. Comp. Physiol. Psychol.*, 85:186–194.

King, F. A. (1958): Effects of septal and amygdaloid lesions on emotional behavior and conditioned avoidance responses in the rat. *J. Nerv. Ment. Dis.*, 126:57–63.

King, H. E. (1961): Psychological effects of excitation in the limbic system. In: *Electrical Stimulation of the Brain,* edited by D. E. Sheer, pp. 477–486. University of Texas Press, Austin, Texas.

King, J. T., Chiung, Puh Lee, Y., and Visscher, M. B. (1955): Single versus multiple cage occupancy and convulsion frequency in C_3H mice. *Proc. Soc. Exp. Biol. Med.*, 88:661–663.

King, M. B., and Hoebel, B. G. (1968): Killing elicited by brain stimulation in rats. *Comm. Behav. Biol., Pat. A*, 2:173–177.

Kinney, D. K., and Matthysese, S. W. (1978): Genetic transmission of schizophrenia. *Annu. Rev. Med.*, 29:459–473.

Kinzel, A. F. (1970): Body-buffer zone in violent prisoners. *Am. J. Psychiatry*, 127:59–64.

Kislak, J. W., and Beach, F. A. (1955): Inhibition of aggressiveness by ovarian hormones. *Endocrinology*, 56:684–692.

Kleinenberg, N. (1872): *Hydra eine Anatomisch-entwicklungsgeschichtliche Untersuchung.* Engelmann, Leipzig.

Klemm, W. R. (1972): *Science, the Brain and Our Future.* Pegasus Publ., New York.

Klerman, G. L. (1963): Assessing the influence of the hospital milieu upon the effectiveness of psychiatric drug therapy: Problems of conceptualization and of research methodology. *J. Ment. Dis.*, 137:143–154.

Kligman, D., and Goldberg, D. A. (1975): Temporal lobe epilepsy and aggression. *J. Nerv. Ment. Dis.*, 160:324–341.

Kling, A. (1965): Behavioral and somatic development following lesions of the amygdala in the cat. *J. Psychiatr. Res.*, 3:263–273.

Kling, A. (1968): Effects of amygdalectomy and testosterone on sexual behavior of male juvenile macaques. *J. Comp. Physiol. Psychol.*, 65:466–471.

Klüver, H., and Bucy, P. C. (1937): "Psychic blindness" and other symptoms following bilateral temporal lobectomy in rhesus monkeys. *Am. J. Physiol.*, 119:352–353.

Klüver, H., and Bucy, P. C. (1938): An analysis of certain effects of bilateral temporal lobectomy in the rhesus monkey, with special reference to "psychic blindness." *J. Psychol.*, 5:33–54.

Klüver, H., and Bucy, P. C. (1939): Preliminary analysis of function of the temporal lobes in monkeys. *Arch. Neurol. Psychiatry*, 42:979–1000.

Knapp, S., and Mandell, A. J. (1973): Short- and long-term lithium administration: Effects on the brain's serotonergic biosynthetic systems. *Science*, 180:645–647.

Knapp, S., and Mandell, A. J. (1975): Effects of lithium chloride on parameters of biosynthetic capacity for 5-hydroxytryptamine in rat brain. *J. Pharmacol. Exp. Ther.*, 193:812–823.

Kniveton, B. H., and Stephenson, G. M. (1973): An examination of individual susceptibility to the influence of aggressive film models. *Br. J. Psychiatry*, 122:53–56.

Knudsen, K. (1964): Homicide after treatment with lysergic acid diethylamide. *Acta Psychiatr. Scand.*, 40:Suppl. 180:389–395.

Knudten, R. D., and Meade, A. C. (1974): Marijuana and social policy. *Addict. Dis.*, 1:323–351.

Knutson, J. F., and Hynan, M. T. (1973): Predatory aggression and irritable aggression: Shock-induced fighting in mouse-killing rats. *Physiol. Behav.*, 11:113–115.

Koe, B., and Weissman, A. (1966): p-Chlorophenylalanine: A specific depletor of brain serotonin. *J. Pharmacol. Exp. Ther.*, 154:499–516.

Koelle, G. B. (1951): The elimination of enzymatic diffusion artifacts in the histochemical localization of cholinesterases and a survey of their cellular distributions. *J. Pharmacol. Exp. Ther.*, 103:153–171.

Koelle, G. B. (1954): The histochemical localization of cholinesterases in the central nervous system of the rat. *J. Comp. Neurol.*, 100:211–235.

Kohlberg, L. (1969): State and sequencies: The cognitive development approach to socialization. In: *Handbook of Socialization Theory and Research*, edited by D. Goslin. Rand McNally, Chicago.

Kohut, H. (1972): Thoughts on narcissism and narcissistic rage. *Psychoanal. Stud. Child*, 27:360–400.

Kolb, B., and Nonneman, A. J. (1974): Frontolimbic lesions and social behavior in the rat. *Physiol. Behav.*, 13:637–643.

Koolhaas, J. M. (1978): Hypothalamically induced intraspecific aggressive behaviour in the rat. *Exp. Brain Res.*, 32:365–375.

Kop, P. P. A. M. (1974): Fluctuations in marriage and divorce frequencies in relation to the month of marriage. *J. Interdisc. Cycle Res.*, 5:327–330.

Kopernik, L. (1964): The family as a breeding ground of violence. *Corr. Psychiatry J. Soc. Ther.*, 10.

Korf, J., Aghajanian, G. K., and Roth, R. H. (1973): Stimulation and destruction of the locus coeruleus: Opposite effects on 3-methoxy-4-hydroxyphenylglycol sulfate levels in the rat cerebral cortex. *Eur. J. Pharmacol.*, 21:305–310.

Kostowski, W. (1966): A note on the effects of some psychotropic drugs on the aggressive behaviour in the ant *Formica rufa*. *J. Pharm. Pharmacol.*, 18:747–749.

Kostowski, W., Członkowski, A., Jerlicz, M., Bidzinski, A., and Hauptmann, M. (1978): Effect of lesions of the locus coeruleus on aggressive behavior in rats. *Physiol. Behav.*, 21:695–699.

Kotowski, W., Członkowski, A., Markowska, L., and Markiewicz, L. (1975): Intraspecific aggressiveness after lesions of midbrain raphe nuclei in rats. *Pharmacology*, 13:81–85.

Kostowski, W., Giacalone, E., Garattini, S., and Valzelli, L. (1968): Studies on behavioural and biochemical changes in rats after lesion of midbrain raphé. *Eur. J. Pharmacol.*, 4:371–376.

Kostowski, W., Samanin, R., Bareggi, S. R., Marc, V., Garattini, S., and Valzelli, L. (1974): Biochemical aspects of the interaction between midbrain raphe and locus coeruleus in the rat. *Brain Res.*, 82:178–182.

Kostowski, W., and Tarchalska, B. (1972): The effects of some drugs affecting brain 5-HT on the aggressive behaviour and spontaneous electrical activity of the central nervous system in the ant, *Formica rufa. Brain Res.*, 38:143–149.

Kostowski, W., and Tarchalska-Krynska, B. (1975): Aggressive behavior and brain serotonin and catecholamines in ants (*Formica rufa*). *Pharmacol. Biochem. Behav.*, 3:717–719.

Kostowski, W., and Valzelli, L. (1974): Biochemical and behavioral effects of lesions of raphe nuclei in aggressive mice. *Pharmacol. Biochem. Behav.*, 2:277–280.

Kostowski, W., Wysokowski, J., and Tarchalska, B. (1962): The effects of some drugs modifying brain 5-hydroxytryptamine on the aggressiveness and spontaneous biocelectrical activity of the central nervous system of the ant *Formica rufa. Diss. Pharm. Pharmacol.*, 24:233–240.

Kracke, W. H. (1967): The maintenance of the ego: Implications of sensory deprivation research for psychoanalytic ego psychology. *Br. J. Med. Psychol.*, 40:17–28.

Kramer J. (1969): Introduction to amphetamine abuse. *J. Psychedelic Drugs*, 2:1–16.

Krames, L., Milgram, N. W., and Christie, D. P. (1973): Predatory aggression: Differential suppression of killing and feeding. *Behav. Biol.*, 9:641–647.

Krech, D. (1967): Psycochemical manipulation and social policy. *Ann. Intern. Med.*, 67:19–24.

Krech, D., Rosenzweig, M. R., and Bennett, E. L. (1962): Relations between brain chemistry and problem-solving among rats raised in enriched and impoverished environments. *J. Comp. Physiol. Psychol.,* 55:801–807.

Krech, D., Rosenzweig, M. R., and Bennett, E. L. (1966): Environmental impoverishment, social isolation and changes in brain chemistry and anatomy. *Physiol. Behav.,* 1:99–104.

Kreiskott, H., and Hofmann, H. P. (1975): Stimulation of a specific drive (predatory behaviour) by p-chlorophenylalanine (pCPA) in the rat. *Pharmakopsychiatrie,* 8:136–140.

Kreschner, M., Bender, M., and Strauss, I. (1936): Mental symptoms in cases of tumor of the temporal lobe. *Arch. Neurol. Psychiatry,* 35:572–596.

Kreuz, L. E., and Rose, R. M. (1972): Assessment of aggressive behavior and plasma testosterone in a young criminal population. *Psychosom. Med.,* 34:321–332.

Krnjević, K. (1974): Chemical nature of synaptic transmission in vertebrates. *Physiol. Rev.,* 54:418–540.

Kršiak, M. (1974): Behavioral changes and aggressivity evoked by drugs in mice. *Res. Commun. Chem. Pathol. Pharmacol.,* 7:237–257.

Kršiak, M. (1975): Timid singly-housed mice: Their value in prediction of psychotropic activity of drugs. *Br. J. Pharmacol.,* 55:141–150.

Kršiak, M., Elis, J., Pöschlová, N., and Mašek, K. (1977): Increased aggressiveness and lower brain serotonin levels in offspring of mice given alcohol during gestation. *J. Stud. Alcohol,* 38:1696–1704.

Kršiak, M., and Janků, I. (1969): The development of aggressive behaviour in mice by isolation. In: *Aggressive Behaviour,* edited by S. Garattini and E. B. Sigg, pp. 101–105. Excerpta Medica, Amsterdam.

Krupp, N. E. (1977): Self-caused skin ulcers. *Psychosomatics,* 18:15–19.

Kruuk, H. (1966): A new view of the hyaena. *New Scientist,* 30:849–851.

Kuffler, S. W. (1967): Neuroglial cells: Physiological properties and a potassium mediated effect of neuronal activity on the glial membrane potential. *Proc. R. Soc. Lond. [Biol.],* 168:1–21.

Kuiper, P. C. (1972): Some theoretical remarks about the theme of aggression. *Psychother. Psychosom.,* 20:260–267.

Kulkarni, A. S. (1968a): Satiation of instinctive mouse killing by rats. *Psychol. Rec.,* 18:385–388.

Kulkarni, A. S. (1968b): Muricidal block produced by 5-hydroxytryptophan and various drugs. *Life Sci.,* 7:125–128.

Kulkarni, A. S., Rahwan, R. G., and Bocknik, S. (1973): Muricidal block induced by 5-hydroxytryptophan in the rat. *Arch. Int. Pharmacodyn. Ther.,* 201:308–313.

Kunkel, B. W. (1919): Instinctive behavior in the white rat. *Science,* 50:305–306.

LaBarba, R. C., Lazar, J. M., and White, J. L. (1972): The effects of maternal separation, isolation and sex on the response to Ehrlich carcinoma in BALB/c mice. *Psychosom. Med.,* 34:557–559.

LaBarba, R. C., and White, J. L. (1971): Maternal deprivation and the response to Ehrlich carcinoma in BALB/c mice. *Psychosom. Med.,* 33:458–460.

Lagerspetz, K. M. J. (1961): Genetic and social causes of aggressive behaviour in mice. *Scand. J. Psychol.,* 2:167–173.

Lagerspetz, K. (1964): Studies on the aggressive behaviour in mice. *Ann. Acad. Sci. Fenn.,* 131:1–131.

Lagerspetz, K. M. J. (1969): Aggression and aggressiveness in laboratory mice. In: *Aggressive Behaviour,* edited by S. Garattini and E. G. Sigg, pp. 77–85. Excerpta Medica, Amsterdam.

Lagerspetz, K., and Hautojärvi, S. (1967): The effect of prior aggressive or sexual arousal on subsequent aggressive or sexual reactions in male mice. *Scand. J. Psychol.,* 8:1–6.

Lagerspetz, K., and Hyvärinen, S. (1971): The effects of conflict and stress on sexual and aggressive behaviour in male mice. *Scand. J. Psychol.,* 12:119–127.

Lagerspetz, K. M. J., and Lagerspetz, K. Y. H. (1971): Changes in the aggressiveness of mice resulting from selective breeding, learning and social isolation. *Scand. J. Psychol.,* 12:241–248.

Lagerspetz, K. Y. H., Tirri, R., and Lagerspetz, K. M. J. (1968): Neurochemical and endocrinological studies of mice selectively bred for aggressiveness. *Scand. J. Psychol.,* 9:157–160.

Lal, H. (1975): Morphine withdrawal aggression. In: *Methods in Narcotic Research,* edited by S. Ehrenpreis and E. Neidle, pp. 149–171. Marcel Dekker, New York.

Lal, H., DeFeo, J. J., Pitterman, A., Patel, G., and Baumel, I. (1972): Effects of prolonged social deprivation or enrichment on neuronal sensitivity for CNS depressants and stimulants. In: *Drug*

Action: Experimental Pharmacology, Vol. 1, pp. 255–266. Futura Publ. Co., Mount Kisco, New York.

Lal, H., O'Brien, J., and Puri, S. K. (1971): Morphine-withdrawal aggression: Sensitization by amphetamines. *Psychopharmacologia,* 22:217–223.

Lamarck, J. B. P. A. (1809): *Philosophie Zoologique.* Paris.

Lamprecht, F., Eichelman, B., Thoa, N. B., Williams, R. B., and Kopin, I. J. (1972): Rat fighting behavior: Serum dopamine-β-hydroxylase and hypothalamic tyrosine hydroxylase. *Science,* 177:1214–1215.

Lange, A. (1972): Possible determinants of aggression. *Psychother. Psychosom.,* 20:241–248.

La Peyronie, F. (1709): Observations par lesquelles on tache de découvrir la partie du cerveau ou l'âme exerce ses fonctions. *Memoires pour l'Histoire des Sciences et des Beaux Arts,* Paris.

La Peyronie, F. (1744): *Histoire de l'Academie Royal des Sciences,* p. 199, Paris.

Lapin, I. P., and Samsonova, M. L. (1964): Vliianie sgruppjrovaniia myshelna ikh vynoslivost'k fenaminu v zabisimosti ot pola, chislennosti gruppy i ploshchadi pomeshcheniia. *Biull. Eksp. Biol. Med.,* 58:66–70.

Laschet, U. (1973): Antiandrogen in the treatment of sex offenders: Mode of action and therapeutic outcome. In: *Contemporary Sexual Behavior,* edited by J. Zulbin and J. Money, p. 311. Johns Hopkins University Press, Baltimore.

Lashley, K. S. (1929): *Brain Mechanisms and Intelligence: A Quantitative Study of Injuries to the Brain.* University of Chicago Press, Chicago.

Latham, E. E., and Thorne, B. M. (1974): Septal damage and muricide: Effects of strain and handling. *Physiol. Behav.,* 12:521–526.

Lau, P., and Miczek, K. A. (1977): Differential effects of septal lesions on attack and defensive-submissive reactions during intraspecies aggression in rats. *Physiol. Behav.,* 18:479–485.

Laver, J. (1964): Costume as a means of social aggression. In: *The Natural History of Aggression,* edited by J. D. Carthy, and F. J. Ebling, pp. 101–108. Academic Press, London.

Leaf, R. C., Lerner, L., and Horowitz, Z. P. (1969): The role of the amygdala in the pharmacological and endocrinological manipulation of aggression. In: *Aggressive Behaviour,* edited by S. Garattini and E. B. Sigg, pp. 120–131. Excerpta Medica, Amsterdam.

Le Beau, J. (1952): The cingular and precingular areas in psychosurgery (agitated behavior, obsessive compulsive states, epilepsy). *Acta Psychiatr. Scand.,* 27:305–316.

Leblond, C. P. (1940): Nervous and hormonal factors in the maternal behavior of the mouse. *J. Gen. Psychol.,* 57:327–344.

Lee, C. T., and Brake, S. C. (1971): Reactions of male fighters to male and female mice untreated or deodorized. *Psychonom. Sci.,* 24:209–211.

Lee, C. T., and Brake, S. C. (1972): Reactions of male mouse fighters to male castrates treated with testosterone propionate or oil. *Psychonom. Sci.,* 27:287–288.

Leff, J. P. (1968): Perceptual phenomena and personality in sensory deprivation. *Br. J. Psychiatry,* 114:1499–1508.

Lefkowitz, M. M., Eron, L. D., Walder, L. O., and Huesmann, L. R. (1972): Television violence and child aggression. A follow-up study. In: *Television and Social Behavior, Vol. 3: Television and Adolescent Aggressiveness,* edited by G. A. Comstock and E. A. Rubinstein, pp. 35–135. U.S. Government Printing Office, Washington, D.C.

Legrand, R., Dahl, G., and Meier, S. (1974): Functional differences between two shock-induced aggressive behaviors of mice. *Anim. Learn. Behav.,* 2:177–180.

Leigh, H., and Hofer, M. A. (1975): Long-term effects of preweaning isolation from littermates in rats. *Behav. Biol.,* 15:173–181.

Leitenberg, H., Rawson, R. A., and Bath, K. (1970): Reinforcement of competing behavior during extinction. *Science,* 169:301–303.

LeMaire, L. (1956): Danish experiences regarding the castration of sexual offenders. *J. Criminal Law Criminol.,* 47:294–310.

Lennenberg, E. H. (Ed.) (1974): Language and brain: Developmental aspects. *Neurosci. Res. Program Bull.,* 12:518–656.

Leonard, C. M. (1972): Effects of neonatal (day 10) olfactory bulb lesions on social behavior of female golden hamsters *(Mesocricetus auratus). J. Comp. Physiol. Psychol.,* 80:208–215.

Leroux, A. G., and Myers, R. D. (1975): Action of serotonin microinjected into hypothalamic sites at which electrical stimulation produced aversive responses in the rat. *Physiol. Behav.,* 14:501–505.

Lesch, M., and Nyhan, W. L. (1964): A familial disorder of uric acid metabolism and central nervous system function. *Am. J. Med.*, 36:561–570.

Lester, D. (1971): Seasonal variation in suicidal deaths. *Br. J. Psychiatry*, 118:627–628.

Lester, D. (1972): Self-mutilating behavior. *Psychol. Bull.*, 73:119–128.

Levi-Montalcini, R. (1975): A new role for the glial cell? *Nature*, 253:687.

Levine, L., Diakow, C. A., and Barsel, G. E. (1965): Interstrain fighting in male mice. *Anim. Behav.*, 13:52–58.

Levine, S., and Otis, L. S. (1958): The effects of handling before and after weaning on the resistance of albino rats to later deprivation. *Can. J. Psychol.*, 12:103–108.

Levitan, I. B., Mushynski, W. E., and Ramirez, G. (1972a): Effects of an enriched environment on amino acid incorporation into rat brain sub-cellular fractions in vivo. *Brain Res.*, 41:498–502.

Levitan, I. B., Mushynski, W. E., and Ramirez, G. (1972b): Effects of environmental complexity on amino acid incorporation into rat cortex and hippocampus in vivo. *J. Neurochem.*, 19:2621–2630.

Levitt, L., and Viney, W. (1973): Inhibition of aggression against the physically disabled. *Percept. Mot. Skills*, 36:255–258.

Libassi, P. T. (1975): Early man, nearly man. *The Sciences*, 15:13–18.

Lichtensteiger, W., Mutzner, U., and Langemann, H. (1967): Uptake of 5-hydroxytryptamine and 5-hydroxytryptophan by neurons of the central nervous system normally containing catecholamines. *J. Neurochem.*, 14:489–497.

Lieber, A. L. (1978): Human aggression and the lunar synodic cycle. *J. Clin. Psychiatry*, 39:385–393.

Lieber, A. L., and Sherin, C. R. (1972): Homicides and the lunar cycle: Toward a theory of lunar influence on human emotional disturbance. *Am. J. Psychiatry*, 129:69–74.

Liebert, R. M. (1974): Television violence and children's aggression: The weight of the violence. In: *Determinants and Origin of Aggressive Behavior*, edited by J. de Wit and W. W. Hartup, pp. 525–531. Mouton, The Hague.

Liebert, R. M., and Poulos, R. W. (1972): TV for kiddies: Truth, goodness, beauty, and a little bit of brainwash. *Psychol. Today*, 6:123–128.

Liebert, R. M., Poulos, R. W., and Strauss, G. D. (1974): *Developmental Psychology*. Prentice-Hall, Englewood Cliffs, N.J.

Lilly, J. C. (1958): Some considerations regarding basic mechanisms of positive and negative types of motivations. *Am. J. Psychiatry*, 115:498–504.

Lilly, J. C., and Miller, A. M. (1962): Operant conditioning of the bottle nose dolphin with electrical stimulation of the brain. *J. Comp. Physiol. Psychol.*, 55:73–79.

Linn, E. (1959): Drug therapy, milieu change and release from a mental hospital. *Arch. Neurol. Psychiatry*, 81:785–794.

Linn, R. L. (1974): Unsquared genetic correlations. *Psychol. Bull.*, 81:203–206.

Lints, C. E., and Harvey, J. A. (1969): Altered sensitivity to foot-shock and decreased brain content of serotonin following brain lesions in the rat. *J. Comp. Physiol. Psychol.*, 67:23–31.

Lion, J. R. (1979): Benzodiazepines in the treatment of aggressive patients. *J. Clin. Psychiatry*, 40:70–71.

Lion, J. R., Madden, D. J., and Christopher, R. L. (1976): A violence clinic: Three years' experience. *Am. J. Psychiatry*, 133:432–435.

Lion, J. R., and Monroe, R. R. (1975): Clinical research of the violent individual. *J. Nerv. Ment. Dis.*, 160:75.

Lion, J. R., and Penna, M. (1974): The study of human aggression. In: *The Neuropsychology of Aggression*, edited by R. E. Whalen, pp. 165–184. Plenum Press, New York.

Lipp, H. P., and Hunsperger, R. W. (1978): Threat, attack and flight elicited by electrical stimulation of the ventromedial hypothalamus of the Marmoset monkey *Callithrix jacchus. Brain Behav. Evol.*, 15:260–293.

Livingston, R. B. (1976): Sensory processing, perception, and behavior. In: *Biological Foundations of Psychiatry, Vol. 1*, edited by R. G. Grenell and S. Gabay, pp. 48–143. Raven Press, New York.

Lloyd, C. W. (1964a): Treatment and prevention of certain sexual behavioral problems. In: *Human Reproduction and Sexual Behavior*, edited by C. W. Lloyd, p. 498. Lea & Febiger, Philadelphia.

Lloyd, C. W. (1964*b*): Problems associated with the menstrual cycle. In: *Human Reproduction and Sexual Behavior,* edited by C. W. Lloyd, p. 490. Lea & Febiger, Philadelphia.

Lloyd, J. A. (1971): Weights of testes, thymi and accessory reproductive glands in relation to rank in paired and grouped house mice *(Mus musculus). Proc. Soc. Exp. Biol. Med.,* 137:19–22.

Loizou, L. A. (1969): Projections of the nucleus locus coeruleus in the albino rat. *Brain Res.,* 15:563–566.

Loney, J., Prinz, R. J., Mishalow, J., and Joad, J. (1978): Hyperkinetic/aggressive boys in treatment: Predictors of clinical response to methylphenidate. *Am. J. Psychiatry,* 135:1487–1491.

Longo, A. M., and Penhoet, E. E. (1974): Nerve growth factor in rat glioma cells. *Proc. Natl. Acad. Sci. U.S.A.,* 71:2347–2349.

Lonowski, D. J., Levitt, R. A., and Larson, S. D. (1973*a*): Mouse killing or carrying by male and female Long-Evans hooded rats. *Bull. Psychonom. Soc.,* 1:349–351.

Lonowski, D. J., Levitt, R. A., and Larson, S. D. (1973*b*): Effects of cholinergic brain injections on mouse killing or carrying by rats. *Physiol. Psychol.,* 1:341–345.

Lorens, S. A., Sorensen, J. P., and Harvey, J. A. (1970): Lesions in the nuclei accumbens septi of the rat: Behavioral and neurochemical effects. *J. Comp. Physiol. Psychol.,* 73:284–290.

Lorens, S. A., Sorensen, J. P., and Yunger, L. M. (1971): Behavioral and neurochemical effects of lesions in the raphe system of the rat. *J. Comp. Physiol. Psychol.,* 77:48–52.

Lorenz, K. (1967): *On Aggression.* Methuen & Co., London.

Lorenz, K. (1973): *Civilized Man's Eight Deadly Sins.* Piper & Co. Verlag, Munich.

Lovaas, O. I. (1961): Interaction between verbal and non verbal behavior. *Child Dev.,* 32:329–336.

Lowry, O. H. (1953): The quantitative histochemistry of the brain. Histological sampling. *J. Histochem. Cytochem.,* 1:420–428.

Luciano, D., and Lore, R. (1975): Aggression and social experience in domesticated rats. *J. Comp. Physiol. Psychol.,* 88:917–923.

Lumia, A. R., Raskin, L. A., and Eckhert, S. (1977): Effects of androgen on marking and aggressive behavior of neonatally and prepubertally bulbectomized and castrated male gerbils. *J. Comp. Physiol. Psychol.,* 91:1377–1389.

Lumia, A. R., Westervelt, M. O., and Rieder, C. A. (1975): Effects of olfactory bulb ablation and androgen on marking and agonistic behavior in male Mongolian gerbils *(Meriones unguiculatus). J. Comp. Physiol. Psychol.,* 89:1091–1099.

Lund, R. (1974): Personality factors and desynchronization of circadian rhythms. *Psychosom. Med.,* 36:224–228.

Lutz, E. G. (1978): Restless legs, anxiety and caffeinism. *J. Clin. Pychiatry,* 39:693–698.

Lycke, E., Modigh, K., and Roos, B.-E. (1969): Aggression in mice associated with changes in the monoamine-metabolism of the brain. *Experientia,* 25:951–953.

Lynn, R. (1971): *Personality and National Character.* Pergamon Press, Oxford.

Lyon, D. O., and Ozolins, D. (1970): Pavlovian conditioning of shock-elicited aggression: A discrimination procedure. *J. Exp. Anal. Behav.,* 13:325–331.

Lystad, M. H. (1975): Violence at home: A review of the literature. *Am. J. Orthopsychiatry,* 45:328–345.

Maas, J. W. (1962): Neurochemical differences between two strains of mice. *Science,* 137:621–622.

Maas, J. W. (1963): Neurochemical differences between two strains of mice. *Nature,* 197:255–257.

Mabille, H. (1899): Hallucinations religieuses et delire religieux transitoire dans l'épilepsie. *Ann. Med. Psychol. (Paris),* Series 8, 9:76–81.

MacDonnell, M. F., and Ehmer, M. (1969): Some effects of ethanol on aggressive behavior in cats. *Q. J. Stud. Alcohol,* 30:312–319.

MacDonnell, M. F., and Flynn, J. P. (1968): Attack elicited by stimulation of the thalamus and adjacent structures of cats. *Behaviour,* 31:185–202.

Mackintosh, J. H. (1970): Territory formation by laboratory mice. *Anim. Behav.,* 18:177–183.

Mackintosh, J. H. (1973): Factors affecting the recognition of territory boundaries by mice *(Mus musculus). Anim. Behav.,* 21:464–470.

MacLean, N., Mitchell, J. M., Harnden, D. G., Williams, J., Jacobs, P. A., Buckton, K. E., Baikie, A. G., Court Brown, W. M., McBride, J. A., Strong, J. A., Close, H. G., and Jones, D. C. (1962): A survey of sex-chromosome abnormalities among 4514 mental defectives. *Lancet,* 1:293–296.

MacLean, P. D. (1949): Psychosomatic disease and the "visceral brain." Recent developments bearing on the Papez theory of emotion. *Psychosom. Med.,* 11:338–353.

MacLean, P. D. (1952): Some psychiatric implications of physiological studies on frontotemporal portion of limbic system (visceral brain). *Electroencephalogr. Clin. Neurophysiol.,* 4:407–418.

MacLean, P. D. (1954): The limbic system and its hippocampal formation. Studies in animals and their possible application to man. *J. Neurosurg.,* 11:29–44.

MacLean, P. D. (1955): The limbic system ('visceral brain') in relation to central gray and reticulum of the brain stem. Evidence of interdependence in emotional processes. *Psychosom. Med.,* 17:355–366.

MacLean, P. D. (1957a): Chemical and electrical stimulation of hippocampus in unrestrained animals. I. Methods and EEG findings. *Arch. Neurol. Psychiatry,* 78:113–127.

MacLean, P. D. (1957b): Chemical and electrical stimulation of hippocampus in unrestrained animals. II. Behavioral findings. *Arch. Neurol. Psychiatry,* 78:128–142.

MacLean, P. D. (1958): The limbic system with respect to self-preservation and the preservation of the species. *J. Nerv. Ment. Dis.,* 127:1–11.

MacLean, P. D. (1962): New findings relevant to the evolution of psychosexual functions of the brain. *J. Nerv. Ment. Dis.,* 135:289–301.

MacLean, P. D. (1964): Mirror display in the squirrel monkey, *Saimiri sciureus. Science,* 146:950–952.

MacLean, P. D. (1965): New findings relevant to the evolution of psychosexual functions of the brain. In: *Sex Research: New Developments,* edited by J. Money. Holt, Rinehart and Winston, New York.

MacLean, P. D. (1970): The triume brain, emotion and scientific bias. In: *The Neurosciences, Second Study Program,* edited by F. O. Schmitt, pp. 336–349. Rockefeller University Press, New York.

MacLean, P. D. (1972a): Cerebral evolution and emotional processes: New findings on the striatal complex. *Ann. N.Y. Acad. Sci.,* 193:137–149.

MacLean, P. D. (1972b): Implications of microelectrode findings on exteroceptive inputs to the limbic cortex. In: *Limbic System Mechanisms and Autonomic Function,* edited by C. H. Hockman, pp. 115–136. Charles C Thomas, Springfield, Ill.

MacLean, P. D. (1973): A triume concept of the brain and behaviour; Lectures I, II, III. In: *The Hincks Memorial Lectures,* edited by T. Boag and D. Campbell. University of Toronto Press, Toronto.

MacLean, P. D. (1976): Sensory and perceptive factors in emotional functions of the triume brain. In: *Biological Foundations of Psychiatry,* edited by R. G. Grenell and S. Gabay, pp. 177–198. Raven Press, New York.

MacLean, P. D., and Delgado, J. M. R. (1953): Electrical and chemical stimulation of frontotemporal portion of limbic system in the waking animal. *Electroencephalogr. Clin. Neurophysiol.,* 5:91–100.

MacLean, P. D., and Ploog, D. W. (1962): Cerebral representation of penile erection. *J. Neurophysiol.,* 25:29–55.

MacLean, P. D., Yokota, T., and Kinnard, M. A. (1968): Photically sustained on-responses of units in posterior hippocampal gyrus of awake monkey. *J. Neurophysiol.,* 31:870–883.

Macovski, E. (1966): The cybernetic functions of living matter components. *Rev. Roum. Biol. Botan. (Bucarest),* 11:333–337.

Madden, D. J. (1977): Voluntary and involuntary treatment of aggressive patients. *Am. J. Psychiatry,* 134:553–555.

Maeder, A. (1909): Sexualität und epilepsie. *Jahrbüch Psychoanal. Psychopathol. Forsch.,* 1:119–155.

Maengwyn-Davies, G. D., Johnson, D. G., Thoa, N. B., Weise, V. K., and Kopin, I. J. (1973): Influence of isolation and of fighting on adrenal tyrosine hydroxylase and phenylethanolamine-N-methyltransferase activities in three strains of mice. *Psychopharmacologia,* 28:339–350.

Mahl, G. F., Rothenberg, A., Delgado, J. M. R., and Hamlin, H. (1964): Psychological responses in the human to intracerebral electrical stimulation. *Psychosom. Med.,* 26:337–368.

Malick, J. B. (1970): A behavioral comparison of three lesion-induced models of aggression in the rat. *Physiol. Behav.,* 5:679–681.

Malick, J. B. (1976): Antagonism of isolation-induced aggression in mice by thyrotropin-releasing hormone (TRH). *Pharmacol. Biochem. Behav.,* 5:665–669.

Malick, J. B., and Barnett, A. (1976): The role of serotonergic pathways in isolation-induced aggression in mice. *Pharmacol. Biochem. Behav.,* 5:55–61.

Malitz, S. (1963): Variables and drug effectiveness. In: *Specific and Nonspecific Factors in Psychopharmacology*, edited by M. Rinkel, pp. 141–148. Philosophical Library Inc., New York.

Mandel, P., Kempf, E., Ebel, A., and Mack, G. (1975): The amygdala and aggressiveness in rodents: Neurochemical correlates. II. Neurochemical and pharmacological data. In: *Neuropsychopharmacology*, edited by J. R. Boissier, H. Hippius, and P. Pichot, pp. 698–703. Excerpta Medica, Amsterdam.

Mandell, A. J. (1978): Toward a psychobiology of transcendence, God in the brain. In: *The Psychobiology of Consciousness*, edited by J. M. Davidson and R. J. Davidson, pp. 1–132. Plenum Press, New York.

Männistö, P. T., and Saarnivaara, L. (1976): Effects of lithium and rubidium on the antinociception and behaviour in mice. II. Studies on three tricyclic antidepressants and pimozide. *Arch. Int. Pharmacodyn.*, 222:293–299.

Manosevitz, M. (1970): Early environmental enrichment and mouse behavior. *J. Comp. Physiol. Psychol.*, 71:459–466.

Manosevitz, M., and Joel, U. (1973): Behavioral effects of environmental enrichment in randomly bred mice. *J. Comp. Physiol. Psychol.*, 85:373–382.

Manosevitz, M., and Montemayor, R. J. (1972): Interaction of environmental enrichment and genotype. *J. Comp. Physiol. Psychol.*, 79:67–76.

Marcucci, F., and Giacalone, E. (1969): N-acetyl aspartic, aspartic and glutamic acid brain levels in aggressive mice. *Biochem. Pharmacol.*, 18:691–692.

Marcucci, F., Mussini, E., Valzelli, L., and Garattini, S. (1968): Decrease in N-acetyl-L-aspartic acid in brain of aggressive mice. *J. Neurochem.*, 15:53–54.

Mark, V. H., and Ervin, F. R. (1970): *Violence and the Brain*. Harper & Row, New York.

Mark, V. H., and Neville, R. (1973): Brain surgery in aggressive epileptics: Social and ethical implications. *J.A.M.A.*, 226:765–772.

Markowitsch, H. J., and Pritzel, M. (1976): Reward related neurons in cat association cortex. *Brain Res.*, 111:185–188.

Marks, P. C., O'Brien, M., and Paxinos, G. (1978): Chlorimipramine inhibition of muricide; The role of the ascending 5-HT projection. *Brain Res.*, 149:270–273.

Marler, P. R., and Hamilton, W. J. (1966): *Mechanisms of Animal Behavior*. John Wiley & Sons, New York.

Marotta, R. F., Logan, N., Potegal, M., Glusman, M., and Gardner, E. L. (1977): Dopamine agonists induce recovery from surgically-induced septal rage. *Nature*, 269:513–515.

Marotta, R. F., Potegal, M., Garnder, E. L., and Glusman, M. (1975): In: Meeting of American Psychological Association, Chicago.

Martinez-Hernandez, A., Bell, K. P., and Noremberg, M. D. (1977): Glutamine synthetase: Glial localization in brain. *Science*, 195:1356–1358.

Maslow, A. H. (1936): The role of dominance in the social and sexual behavior in infrahuman primates. 1. Observations at Villas Park Zoo. *J. Genet. Psychol.*, 48:261–277.

Maslow, A. H. (1940): Dominance-quality and social behavior in infrahuman primates. *J. Soc. Psychol.*, 11:313–324.

Masur, J. (1972): Sex differences in "emotionality" and behavior of rats in the open-field. *Behav. Biol.*, 7:749–754.

Masur, J., and Benedito, M. A. C. (1974): Genetic selection of winner and loser rats in a competitive situation. *Nature*, 249:284.

Masur, J., Czeresnia, S., Skitnevsky, H., and Carlini, E. A. (1974): Brain amine level and competitive behavior between rats in a straight runway. *Pharmacol. Biochem. Behav.*, 2:55–62.

Matte, A. C. (1975): Effect of hashish on isolation induced aggression in wild mice. *Psychopharmacologia*, 45:125–128.

Matte, A. C., and Tornow, H. (1978): Parachlorophenylalanine produces dissociated effects on aggression, "emotionality" and motor activity. *Neuropharmacology*, 17:555–558.

Mawson, A. R., and Jacobs, K. W. (1978): Corn, tryptophan, and homicide. *J. Orthomol. Psychiatry*, 7:227–230.

Maxim, P. E., Bowden, D. M., and Sackett, G. P. (1976): Ultradian rhythms of solitary and social behavior in rhesus monkeys. *Physiol. Behav.*, 17:337–344.

McCabe, T. T., and Blanchard, D. D. (1950): *Three Species of Peromyscus*. Road Associates, Santa Barbara.

McCarthy, D. (1966): Mouse-killing induced in rats treated with pilocarpine. *Fed. Proc.*, 25:385.

McCarthy, D. (1970): Aboriginal antiquities in Australia. Australian Aboriginal Study No. 22. Institute of Aborigine Studies, Canberra.

McClearn, G. E. (1974): Behavioral genetic analyses of aggression. In: *The Neuropsychology of Aggression,* edited by R. E. Whalen, pp. 87–98. Plenum Press, New York.

McCleary, R. A., and Moore, R. Y. (1965): *Subcortical Mechanisms of Behavior.* Basic Books, New York.

McGeer, P. L., Hattori, T., Fibiger, H. C., McGeer, E. G., and Singh, V. H. (1974): Interconnections of dopamine, GABA and acetylcholine containing neurons of the extrapyramidal system. *J. Pharmacol. (Paris),* 5 (Suppl. 1):54.

McGlothlin, W., Cohen, S., and McGlothlin, M. S. (1967): Long lasting effects of LSD on normals. *Arch. Gen. Psychiatry,* 17:521–532.

McHenry, H. M. (1975a): Fossil hominid body weight and brain size. *Nature,* 254:686–688.

McHenry, H. M. (1975b): Fossils and the mosaic nature of human evolution. *Science,* 190:425–431.

McIntosh, J. C., and Cooper, J. R. (1964): Function of N-acetyl aspartic acid in the brain: Effects of certain drugs. *Nature,* 203:658.

McIntosh, J. C., and Cooper, J. R. (1965): Studies on the function of N-acetyl aspartic acid in brain. *J. Neurochem.,* 12:825–835.

McIntyre, D. C. (1978): Amygdala kindling and muricide in rats. *Physiol. Behav.,* 21:49–56.

McKenzie, G. M. (1971): Apomorphine-induced aggression in the rat. *Brain Res.,* 34:323–330.

McKerracher, D. W., Street, D. R. K., and Segal, L. J. (1966): A comparison of the behaviour problems presented by male and female subnormal offenders. *Br. J. Psychiatry,* 112:891–897.

McKinney, W. T. (1974): Primate social isolation. *Arch. Gen. Psychiatry,* 31:422–426.

McKinney, W. T., Jr., Suomi, S. J., and Harlow, H. F. (1972a): Repetitive peer separations of juvenile-age rhesus monkeys. *Arch. Gen. Psychiatry,* 27:200–203.

McKinney, W. T., Jr., Suomi, S. J., and Harlow, H. F. (1972b): Vertical-chamber confinement of juvenile-age rhesus monkeys. *Arch. Gen. Psychiatry,* 26:223–228.

McLain, W. C., III, Cole, B. T., Schrieber, R., and Powell, D. A. (1974): Central catechol- and indolamine systems and aggression. *Pharmacol. Biochem. Behav.,* 2:123–126.

McLennan, H. (1970): *Synaptic Transmission.* W. B. Saunders Co., Philadelphia.

McManamy, M. C., and Schube, P. G. (1936): Caffeine intoxication; Report of a case the symptoms of which amounted to a psychosis. *N. Engl. J. Med.,* 215:616–620.

Mehler, W. R., and Nauta, W. J. H. (1974): Connections of the basal ganglia and of the cerebellum. *Confin. Neurol.,* 36:205–222.

Melander, B. (1960): Psychopharmacodynamic effects of diethylpropion. *Acta Pharmacol. Toxicol. (Copenh.),* 17:182–190.

Meldrum, B. S. (1975): Epilepsy and gamma-aminobutyric acid-mediated inhibition. *Int. Rev. Neurobiol.,* 17:1–36.

Melges, F. T., Tinklenberg, J. R., Deardorff, C. M., Davies, N. H., Anderson, R. E., and Owen, C. A. (1974): Temporal disorganization and delusional-like ideation. Processes induced by hashish and alcohol. *Arch. Gen. Psychiatry,* 30:855–861.

Melges, F. T., Tinklenberg, J. R., Hollister, L. E., and Gillespie, H. K. (1970a): Marihuana and temporal disintegration. *Science,* 168:1118–1120.

Melges, F. T., Tinklenberg, J. R., Hollister, L. E., and Gillespie, H. K. (1970b): Temporal disintegration and depersonalization during marihuana intoxication. *Arch. Gen. Psychiatry,* 23:204–210.

Melvin, K. B., and Ervey, D. H. (1973): Facilitative and suppressive effects of punishment on species-typical aggressive display in *Betta splendens. J. Comp. Physiol. Psychol.,* 83:451–457.

Mempel, E. (1971): Influence of partial amygdalectomy on the emotional disturbances and epileptic seizures. *Neurol. Neurochir. Pol.,* 5:81–86.

Meyer, D. C., and Quay, W. B. (1977): Seasonal changes in the uptake capacity of the suprachiasmatic nucleus for ³H-serotonin. *Experientia,* 33:472–473.

Meyer, T. P. (1972): The effects of sexually arousing and violent films on aggressive behavior. *J. Sex Res.,* 8:324–333.

Meyer-Bahlburg, H. F. L. (1974): Aggression, androgens, and the XYY syndrome. In: *Sex Differences in Behavior,* edited by R. C. Friedman, R. M. Richart, and R. L. Vande Wiele, pp. 433–453. John Wiley & Sons, New York.

Meyer-Bahlburg, H. F. L., Nat, R., Boom, D. A., Sharma, M., and Edwards, J. A. (1974): Aggressiveness and testosterone measures in man. *Psychosom. Med.,* 36:269–274.

Michael, R. P., and Zumpe, D. (1970): Aggression and gonadal hormones in captive rhesus monkeys *(Macaca mulatta)*. *Anim. Behav.,* 18:1–10.

Michael, R. P., and Zumpe, D. (1978): Annual cycles of aggression and plasma testosterone in captive male rhesus monkeys. *Psychoneuroendocrinology,* 3:217–220.

Miczek, K. A. (1976): Does THC induce aggression? Suppression and induction of aggressive reactions by chronic and acute Δ^9-THC treatment in laboratory rats. In: *Pharmacology of Marihuana, Vol. 2,* edited by M. C. Braude and S. Szara, pp. 499–514. Raven Press, New York.

Miczek, K. A. (1977): A behavioral analysis of aggressive behaviors induced and modulated by Δ^9-tetrahydrocannabinol, pilocarpine, d-amphetamine and l-dopa. *Act. Nerv. Super. (Praha),* 19:224–225.

Miczek, K. A. (1979): Chronic Δ^9-tetrahydrocannabinol in rats: Effect on social interactions, mouse killing, motor activity, consummatory behavior, and body temperature. *Psychopharmacology,* 60:137–146.

Miczek, K. A., Altman, J. L., Appel, J. B., and Boggan, W. O. (1975): Para-chlorophenylalanine, serotonin and killing behavior. *Pharmacol. Biochem. Behav.,* 3:355–361.

Miczek, K. A., and Barry, H., III (1976): Pharmacology of sex and aggression. In: *Behavioral Pharmacology,* edited by S. D. Glick and J. Goldfarb, pp. 176–257. C. V. Mosby Co., St. Louis.

Miczek, K. A., Brykczynski, T., and Grossman, S. P. (1974): Differential effects of lesions in the amygdala, periamygdaloid cortex, and stria terminalis on aggressive behaviors in rats. *J. Comp. Physiol. Psychol.,* 87:760–771.

Miczek, K. A., and Grossman, S. P. (1972): Effects of septal lesions on inter- and intraspecies aggression in rats. *J. Comp. Physiol. Psychol.,* 79:37–45.

Midgley, M. (1978): *Beast and Man: The Roots of Human Nature.* Cornell University Press, Ithaca, N.Y.

Miller, B. L., Patcher, J. S., and Valzelli, L. (1979): Brain tryptophan in isolated aggressive mice. *Neuropsychobiology,* 5:11–15.

Miller, M., Leahy, J. P., McConville, F., Morgane, P. J., and Resnick, O. (1977a): Effects of developmental protein malnutrition on tryptophan utilization in brain and peripheral tissues. *Brain Res. Bull.,* 2:347–353.

Miller, M., Leahy, J. P., Stern, W. C., Morgane, P. J., and Resnick, O. (1977b): Tryptophan availability: Relation to elevated brain serotonin in developmentally protein-malnourished rats. *Exp. Neurol.,* 57:142–157.

Miller, M. H. (1976): Dorsolateral frontal lobe lesions and behavior in the macaque: Dissociation of threat and aggression. *Physiol. Behav.,* 17:209–213.

Miller, N. E. (1941): The frustration-aggression hypothesis. *Phsychol. Rev.,* 48:337–342.

Miller, N. E. (1948): Studies of fear as an acquirable drive. I. Fear as motivation and fear-reduction as reinforcement in the learning of new responses. *J. Exp. Psychol.,* 38:89–101.

Miller, N. E. (1957): Experiments on motivation. *Science,* 126:1271–1278.

Miller, N. E. (1961): Implications for theories of reinforcement. In: *Electrical Stimulation of the Brain,* edited by D. E. Sheer, pp. 515–581. University of Texas Press, Austin, Texas.

Miller, N. E. (1965): Chemical coding of behavior in the brain. *Science,* 148:328–338.

Miller, S. C. (1962): Ego-autonomy in sensory deprivation, isolation, and stress. *Int. J. Psychoanaly.,* 43:1–20.

Miller, T.-I., and Miller, P. M. C. (1974): Some physiological correlates of experimentally elicited behavioral aggression and its relationships with fantasy and other paper-and-pencil measure. In: *Determinants and Origins of Aggressive Behavior,* edited by J. de Wit and W. W. Hartup, pp. 579–587. Mouton, The Hague.

Miller, W. B. (1958): Lower class culture as a generating milieu of gang delinquency. *J. Soc. Issues,* 14:5–19.

Miller, W. B., Geertz, H., and Cutter, H. S. G. (1961): Aggression in a boy's street-corner group. *Psychiatry,* 24:283–298.

Milligan, W. L., Powell, D. A., and Borasio, G. (1973): Sexual variables and shock-elicited aggression. *J. Comp. Physiol. Psychol.,* 83:441–450.

Milner, P. M. (1970): *Physiological Psychology.* Holt, Rinehart and Winston, London.

Mishkin, M. (1978): Memory in monkeys severely impaired by combined but not by separate removal of amygdala and hippocampus. *Nature,* 273:297–298.

Missakian, E. A. (1969): Reproductive behavior of socially deprived male rhesus monkeys *(Macaca mulatta)*. *J. Comp. Physiol. Psychol.,* 69:403–407.

Mitchell, G. (1975): What monkeys can tell us about human violence. *Futurist,* April: 75–80.

Mitchell, G. D. (1968): Persistent behavior pathology in rhesus monkeys following early social isolation. *Folia Primatol. (Basel),* 8:132–147.

Mitchell, W., Falconer, M. A., and Hill, D. (1954): Epilepsy with fetishism relieved by temporal lobectomy. *Lancet,* 2:626–630.

Miyamoto, S. (1960): Developmental changes of intelligence in mentally retarded children. *Jpn. J. Child Psychiatry,* 1:114–125.

Mizuno, T., and Yugari, Y. (1975): Prophylactic effect of L-5-hydroxytryptophan on self-mutilation in the Lesch-Nyhan syndrome. *Neuropädiatrie,* 6:13–23.

Moan, C. E., and Heath, R. G. (1972): Septal stimulation for the initiation of heterosexual behavior in a homosexual male. *J. Behav. Ther. Exp. Psychiatry,* 3:23–30.

Modigh, K. (1973): Effects of isolation and fighting in mice on the rate of synthesis of noradrenaline, dopamine and 5-hydroxytryptamine in the brain. *Psychopharmacologia,* 33:1–17.

Modigh, K. (1974): Effects of social stress on the turnover of brain catecholamines and 5-hydroxytryptamine in mice. *Acta Pharmacol. Toxicol. (Copenh.),* 34:97–105.

Money, J. (1956): Mind-body dualism and the unity of body mind. *Behav. Sci.,* 1:212–217.

Montagu, A. (1974): Aggression and the evolution of man. In: *The Neuropsychology of Aggression,* edited by R. E. Whalen, pp. 1–32. Plenum Press, New York.

Montagu, A. (1977): Human aggression. *The Sciences,* 17:6–30.

Montagu, M. F. A. (1968): *Man and Aggression.* Oxford University Press, New York.

Moore, K. E. (1963): Toxicity and catecholamine releasing actions of d- and l-amphetamine in isolated and aggregated mice. *J. Pharmacol. Exp. Ther.,* 142:6–12.

Moore, K. E. (1965): Amphetamine toxicity in hyperthyroid mice: Effects on endogenous catecholamines. *Biochem. Pharmacol.,* 14:1831–1837.

Moore, K. E. (1966): Amphetamine toxicity in hyperthyroid mice: Effects on blood glucose and liver glycogen. *Biochem. Pharmacol.,* 15:353–360.

Mora, G. (1975): Historical and theoretical trends in psychiatry. In: *Comprehensive Textbook of Psychiatry II, Vol. 1,* edited by A. M. Freedman, H. I. Kaplan, and B. J. Sodock, pp. 1–98. Williams & Wilkins, Baltimore.

Morgan, C. T. (1947): The hoarding instinct. *Psychol. Rev.,* 54:335–341.

Morgan, C. T., Stellar, E., and Johnson, O. (1943): Food-deprivation and hoarding in rats. *J. Comp. Psychol.,* 35:275–295.

Morgan, J. J. B. (1923): The measurement of instincts. *Psychol. Bull.,* 20:94.

Morgan, M., and Einon, D. (1975): Incentive motivation and behavioral inhibition in socially-isolated rats. *Physiol. Behav.,* 15:405–409.

Morgan, M. J., Einon, D. F., and Nicholas, D. (1975): The effects of isolation rearing on behavioural inhibition in the rat. *Q. J. Exp. Psychol.,* 27:615–634.

Morgan, W. W., Saldana, J. J., Yndo, C. A., and Morgan, J. F. (1975): Correlations between circadian changes in serum amino acids or brain tryptophan and the contents of serotonin and 5-hydroxyindoleacetic acid in regions of the rat brain. *Brain Res.,* 84:75–86.

Morgan, W. W., and Yndo, C. A. (1973): Daily rhythms in tryptophan and serotonin content in mouse brain: The apparent independence of these parameters from daily changes in food intake and from plasma tryptophan content. *Life Sci.,* 12:395–408.

Morgane, P. J. (1961): Electrophysiological studies of feeding and satiety centers in the rat. *Am. J. Physiol.,* 201:838–844.

Morgane, P. J., and Kosman, A. J. (1960): Relationship of middle hypothalamus to amygdala hyperphagia. *Am. J. Physiol.,* 198:1315–1318.

Morison, R. S., and Dempsey, E. W. (1942): A study of thalamo-cortical relations. *Am. J. Physiol.,* 135:281–292.

Morot-Gaudry, Y., Hamon, M., Bourgoin, S., Ley, J. P., and Glowinski, J. (1974): Estimation of the rate of 5-HT synthesis in the mouse brain by various methods. *Naunyn Schmiedebergs Arch. Pharmacol.,* 282:223–238.

Morris, D. (ed.) (1967): *Primate Ethology.* Weidenfeld and Nicolson, London.

Morris, H. H., Escoll, P. J., and Wexler, R. (1956): Aggressive behavior disorders of childhood: A follow-up study. *Am. J. Psychiatry,* 112:991–997.

Moruzzi, G. (1954): The physiological properties of the brain stem reticular system. In: *Brain Mechanisms and Consciousness,* edited by J. F. Delafresnaye, pp. 21–53. Charles C Thomas, Springfield, Ill.

Moruzzi, G., and Magoun, H. W. (1949): Brain stem reticular formation and activation of the EEG. *Electroencephalogr. Clin. Neurophysiol.,* 1:455–473.

Moss, F. A. (1924): Study of animal drives. *J. Exp. Psychol.,* 7:165–185.

Mountcastle, V. B. (1968): Sleep, wakefulness, and the conscious state: Intrinsic regulatory mechanisms of the brain. In: *Medical Physiology, Vol. 2,* edited by V. B. Mountcastle, pp. 1315–1342. C. V. Mosby Co., St. Louis.

Mowrer, O. H. (1950): *Learning Theory and Personality Dynamics.* Ronald Press, New York.

Mowrer, O. H., and Lamoreaux, R. R. (1946): Fear as an interventing variable in avoidance conditioning. *J. Comp. Psychol.,* 39:29–49.

Moyer, K. E. (1968a): Kinds of aggression and their physiological basis. *Commun. Behav. Biol. [A],* 2:65–87.

Moyer, K. E. (1968b): Brain research must contribute to world peace. *Fiji School Med.,* 3:2–5.

Moyer, K. E. (1969): Internal impulses to aggression. *Trans. N.Y. Acad. Sci.,* 31:104–114.

Moyer, K. E. (1971a): The physiology of aggression and the implications for aggression control. In: *The Control of Aggression and Violence: Cognitive and Physiology Factors,* edited by L. Singer, pp. 61–92. Academic Press, New York.

Moyer, K. E. (1971b): *The Physiology of Hostility.* Markham, Chicago.

Moyer, K. E. (ed.) (1976): *Physiology of Aggression and Implications for Control. An Anthology of Readings.* Raven Press, New York.

Moyer, K. E., and Korn, J. H. (1965): Behavioral effects of isolation in the rat. *Psychonom. Sci.,* 3:503–504.

Mueller, W. J., and Grater, H. A. (1965): Aggression, conflict, anxiety, and ego strength. *J. Consult. Psychol.,* 29:130–134.

Mugford, R. A. (1973): Intermale fighting affected by home-cage odors of male and female mice. *J. Comp. Physiol. Psychol.,* 84:289–295.

Mugford, R. A. (1974): Androgenic stimulation of aggression eliciting cues in adult opponent mice castrated at birth, weaning or maturity. *Horm. Behav.,* 5:93–102.

Mugford, R. A., and Nowell, N. W. (1970a): Pheromones and their effect on aggression in mice. *Nature,* 226:967–968.

Mugford, R. A., and Nowell, N. W. (1970b): The aggression of male mice against androgenized females. *Psychonom. Sci.,* 20:191–192.

Mugford, R. A., and Nowell, N. W. (1971a): Shock-induced release of the preputial gland secretions that elicit fighting in mice. *J. Endocrinol.,* 15:16–17.

Mugford, R. A., and Nowell, N. W. (1971b): The relationship between endocrine status of female opponents and aggressive behaviour of male mice. *Anim. Behav.,* 19:153–155.

Mugford, R. A., and Nowell, N. W. (1971c): Endocrine control over production and activity of the anti-aggression phenomena from female mice. *J. Endocrinol.,* 49:225–232.

Mugford, R. A., and Nowell, N. W. (1971d): The preputial glands as a source of aggression-promoting odors in mice. *Physiol. Behav.,* 6:247–249.

Mugford, R. A., and Nowell, N. W. (1972): The dose-response to testosterone propionate of preputial glands, pheromones and aggression in mice. *Horm. Behav.,* 3:39–46.

Mulder, D., and Daly, D. (1952): Psychiatric symptoms associated with lesions of temporal lobe. *J.A.M.A.,* 150:173–176.

Mullin, W. J., and Phillis, J. W. (1974): Acetylcholine release from the brain of unanaesthetized cats following habituation to morphine and during precipitation of the abstinence syndrome. *Psychopharmacologia,* 36:85–99.

Munn, N. L. (1950): *Handbook of Psychological Research on the Rat.* Houghton Mifflin Co., Boston.

Murphy, M. R. (1970a): Territorial behavior of the caged golden hamster. *Proceedings of the 78th Annual Convention,* pp. 237–238. American Psychological Association.

Murphy, M. R. (1970b): Olfactory bulb removal reduced social and territorial behaviors in the male golden hamster. 41st Annual Meeting of the Eastern Psychological Association, Atlantic City, April 2–4.

Murphy, M. R. (1976a): Blinding increases territorial aggression in male Syrian golden hamsters. *Behav. Biol.,* 17:139–141.

Murphy, M. R. (1976b): Olfactory stimulation and olfactory bulb removal; Effects on territorial aggression in male Syrian golden hamsters. *Brain Res.,* 113:95–110.

Murphy, M. R., and Schneider, G. E. (1970): Olfactory bulb removal eliminates mating behavior in the male golden hamster. *Science,* 167:302–304.

Myer, J. S. (1964): Stimulus control of mouse-killing rats. *J. Comp. Physiol. Psychol.*, 58:112–117.

Myer, J. S. (1966): Punishment of instinctive behaviour: Suppression of mouse-killing by rats. *Psychonom. Sci.*, 4:385–386.

Myer, J. S., and Baenninger, R. (1966): Some effects of punishment and stress on mouse killing by rats. *J. Comp. Physiol. Psychol.*, 62:292–297.

Myer, J. S., and White, R. T. (1965): Aggressive motivation in the rat. *Anim. Behav.*, 13:430–433.

Myers, K. (1964): Influence of density on fecundity, growth rates and mortality in the wild rabbit. *CSIRO Wildlife Res.*, 9:134–137.

Myers, K. (1966): The effects of density on sociality and health in mammals. *Proc. Ecol. Soc. Aust.*, 1:40–64.

Myers, K., Hale, C. S., Mykytowycz, R., and Hughes, R. L. (1971): The effects of varying density and space on sociality and health in animals. In: *Behaviour and Environment*, edited by A. H. Esser, pp. 148–187. Plenum Press, New York.

Myers, R. D. (1964): Emotional and autonomic responses following hypothalamic chemical stimulation. *Can. J. Psychol.*, 18:6–14.

Myers, R. D., and Melchior, C. L. (1975): Alcohol and alcoholism: Role of serotonin. In: *Serotonin in Health and Disease, II*, edited by W. B. Essman, pp. 373–430. Spectrum, New York.

Myers, R. E., Swett, C., and Miller, M. (1973): Loss of social group affinity following prefrontal lesions in free-ranging macaques. *Brain Res.*, 64:257–269.

Mykytowycz, R. (1959): Social behaviour of an experimental colony of wild rabbits *Oryctolagus cuniculus* (L). II. First breeding season. *CSIRO Wildlife Res.*, 4:1–13.

Mykytowycz, R. (1960): Social behaviour of an experimental colony of wild rabbits *Oryctolagus cuniculus* (L). III. Second breeding season. *CSIRO Wildlife Res.*, 5:1–20.

Mykytowycz, R., and Dudzinski, M. L. (1972): Aggressive and protective behaviour of adult rabbits *Oryctolagus cuniculus* (L) towards juveniles. *Behaviour*, 43:97–120.

Nachmansohn, D. (1970): Proteins in excitable membranes. *Science*, 168:1059–1066.

Nageotte, J. (1910): Phénomènes de sécrétion dans le protoplasma des cellules névrogliques de la substance grise. *C. R. Soc. Biol. (Paris)*, 68:1068–1069.

Nansen, F. (1887): *The Structure and Combination of the Histological Elements of the Central Nervous System*. Bergens Mus. Aarsber, Bergen.

Narabayashi, H., and Mizutani, T. (1970): Epileptic seizures and the stereotaxic amygdalotomy. *Confin. Neurol.*, 32:289–297.

Narabayashi, H., Nagao, T., Saito, Y., Yoshida, M., and Nagahata, M. (1963): Stereotaxic amygdalotomy for behavior disorders. *Arch. Neurol.*, 9:1–16.

Narabayashi, H., and Uno, M. (1966): Long range results of stereotaxic amygdalotomy for behavior disorders. *Confin. Neurol.*, 27:168–171.

Nashold, B. S., Jr., and Slaughter, D. G. (1969): Effects of stimulating or destroying the deep cerebellar regions in man. *J. Neurosurg.*, 31:172–186.

Nassi, A. J., and Abramowitz, S. I. (1976): From phrenology to psychosurgery and back again: Biological studies of criminality. *Am. J. Orthopsychiatry*, 46:591–607.

Nath, V. (1957): The Golgi controversy. *Nature*, 180:967–969.

Nauta, W. J. H. (1960): Some neural pathways related to the limbic system. In: *Electrical Studies on the Unanesthetized Brain*, edited by E. R. Rainey and D. S. O'Doherty, pp. 1–16. Hoeber, New York.

Nauta, W. J. H. (1963): Central nervous organization and the endocrine motor system. In: *Advances in Neuroendocrinology*, edited by A. V. Nalbandov, pp. 5–21. University of Illinois Press, Urbana, Ill.

Nauta, W. J. H. (1964): Some efferent connections of the prefrontal cortex in the monkey. In: *The Frontal Granular Cortex and Behavior*, edited by J. M. Warren and K. Akert, pp. 397–409. McGraw-Hill, New York.

Neale, J. M., and Liebert, R. M. (1973): *Science and Behavior: An Introduction to Methods of Research*. Prentice Hall, Englewood Cliffs, N.J.

Neckers, L. M., Biggio, G., Moja, E., and Meek, J. L. (1977): Modulation of brain tryptophan hydroxylase activity by brain tryptophan content. *J. Pharmacol. Exp. Ther.*, 201:110–116.

Neckers, L. M., Zarrow, M. X., Myers, M. M., and Denenberg, V. H. (1975): Influence of olfactory bulbectomy and the serotonergic system upon intermale aggression and maternal behavior in the mouse. *Pharmacol. Biochem. Behav.*, 3:545–550.

Nelson, G. J. (1968): Gill-arch structure in "Acanthodes." In: *Current Problems of Lower Vertebrate Phylogeny,* edited by T. Ørvig, pp. 129–143. John Wiley & Sons, New York.

Newman, J. D., and Symmes, D. (1974): Vocal pathology in socially deprived monkeys. *Dev. Psychobiol.,* 7:351–358.

Nielsen, N. J., and Tsuboi, T. (1970): Correlation between stature, character disorder and criminality. *Br. J. Psychiatry,* 116:145–150.

Nieuwenhuys, R. (1967): Comparative anatomy of the cerebellum. *Prog. Brain Res.,* 25:1–93.

Nishikawa, T., Kajiwara, Y., Kōno, Y., Sano, T., Nagasaki, N., and Tanaka, M. (1976): Different effect of social isolation on the levels of brain monoamines in post-weaning and young-adult rat. *Folia Psychiatr. Neurol. Jpn.,* 30:57–63.

Nishikawa, T., and Tanaka, M. (1978): Altered behavioral responses to intense foot shock in socially-isolated rats. *Pharmacol. Biochem. Behav.,* 8:61–67.

Nissen, H. W. (1930a): A study of exploratory behavior in the white rat by means of the destruction method. *J. Genet. Psychol.,* 37:361–376.

Nissen, H. W. (1930b): A study of maternal behavior in the white rat by means of the obstruction method. *J. Genet. Psychol.,* 37:377–393.

Nobin, A., and Björklund, A. (1973): Topography of the monoamine neuron systems in the human brain as revealed in fetuses. *Acta Physiol. Scand.,* 88 Suppl. 388:1–40.

Noblit, G. W., and Burcart, J. M. (1976): Women and crime: 1960–1970. *Soc. Sci. Q.,* 56:650–657.

Noël, B., and Benezech, M. (1977): YY syndrome in French security settings. *Clin. Genet.,* 12:314–315.

Noirot, E. (1972): Ultrasounds and maternal behavior in small rodents. *Dev. Psychobiol.,* 5:371–387.

Noirot, E., Goyens, J., and Buhot, M.-C. (1975): Aggressive behavior of pregnant mice toward males. *Horm. Behav.,* 6:9–17.

Nuffield, E. J. A. (1961): Neurophysiology and behaviour disorders in epileptic children. *J. Ment. Sci.,* 107:438–458.

Nyby, J., Dizinno, G. A., and Whitney, G. (1976): Social status, and ultrasonic vocalizations of male mice. *Behav. Biol.,* 18:285–289.

Nyström, B., Olson, L., and Ungerstedt, U. (1972): Noradrenaline nerve terminals in human cerebral cortices: First histochemical evidence. *Science,* 176:924–925.

O'Boyle, M. (1974): Rats and mice together: The predatory nature of the rats mouse-killing response. *Psychol. Bull.,* 81:261–269.

O'Brien, D., and Groshek, A. (1962): The abnormality of tryptophane metabolism in children with mongolism. *Arch. Dis. Child.,* 37:17–20.

O'Brien, D., Groshek, A. P., and Streamer, C. W. (1960): Abnormalities of tryptophan metabolism in children with mongolism. *Am. J. Dis. Child.,* 100: 540–541.

O'Connor, N. (1971): Children in restricted environments. *Psychiatr. Neurol. Neurochir.,* 74:71–77.

Oger, J., Arnason, B. G. W., Pantazis, N., Lehrich, J., and Young, M. (1974): Synthesis of nerve growth factor by L and 3T3 cells in culture. *Proc. Natl. Acad. Sci. U.S.A.,* 71:1554–1558.

Ojemann, R. G. (1966): Correlations between specific human brain lesions and memory changes. *Neurosci. Res. Program Bull.,* 4 Suppl.: 1–70.

Oka, H., Yasuda, T., Jinnai, K., and Yoneda, Y. (1976): Reexamination of cerebellar responses to stimulation of sensorimotor areas of the cerebral cortex. *Brain Res.,* 118:312–319.

Oken, D. (1960): An experimental study of suppressed anger and blood pressure. *Arch. Gen. Psychiatry,* 2:441–456.

Okumura, N., Otsuki, S., and Kameyama, A. (1960): Studies on free aminoacids in human brain. *J. Biochem. (Tokyo),* 47:315–320.

Older, J. (1974): Psychosurgery: Ethical issues and a proposal for control. *Am. J. Orthopsychiatry,* 44:661–674.

Olds, J. (1958): Self-stimulation of the brain. Its use to study local effects of hunger, sex, and drugs. *Science,* 127:315–324.

Olds, J. (1962): Hypothalamic substrates of reward. *Physiol. Rev.,* 42:554–604.

Olds, J., and Milner, P. (1954): Positive reinforcement produced by electrical stimulation of septal areas and other regions of the rat brain. *J. Comp. Physiol. Psychol.,* 47:419–427.

Olds, J., Yuwiler, A., Olds, M. E., and Yun, C. (1964): Neurohumors in hypothalamic substrates of reward. *Am. J. Physiol.,* 207:242–254.

Olds, M. E. (1974): Unit responses in the medial forebrain bundle to rewarding stimulation in the hypothalamus. *Brain Res.,* 80:479–495.

Olds, M. E., and Olds, J. (1962): Approach-escape interactions in rat brain. *Am. J. Physiol.,* 203:803–810.

Olds, M. E., and Olds, J. (1969a): Effects of anxiety-relieving drugs on unit discharges in hippocampus, reticular midbrain, and pre-optic area in the freely moving rat. *Int. J. Neuropharmacol.,* 8:87–103.

Olds, M. E., and Olds, J. (1969b): Effects of lesions in medial forebrain bundle on self-stimulation behavior. *Am. J. Physiol.,* 217:1253–1264.

Olson, L. (1974): Post-mortem fluorescence histochemistry of monoamine neuron systems in the human brain: A new approach in the search for a neuropathology of schizophrenia. *J. Psychiatr. Res.,* 11:199–203.

Olson, L., Boréus, L. O., and Seiger, Å. (1973a): Histochemical demonstration and mapping of 5-hydroxytryptamine- and catecholamine-containing neuron systems in the human fetal brain. *Z. Anat. Entwicklungsgeschichte,* 139:259–282.

Olson, L., Nyström, B., and Seiger, Å. (1973b): Monoamine fluorescence histochemistry of human post mortem brain. *Brain Res.,* 63:231–247.

Olson, L., Nyström, B., and Seiger, Å. (1973c): Monoamine neuron systems in the normal and schizophrenic human brain: Fluorescence histochemistry of fetal, neurosurgical and post-mortem material. In: *Frontiers in Catecholamine Research,* edited by E. Usdin and S. H. Snyder, pp. 1097–1100. Pergamon Press, New York.

Olson, R. E., Gursey, D., and Vester, J. W. (1960): Evidence for a defect in tryptophan metabolism in chronic alcoholism. *N. Engl. J. Med.,* 263:1169–1174.

Oomura, Y., Ooyama, H., Naka, F., Yamamoto, T., Ono, T., and Kobayashi, N. (1969): Some stochastical patterns of single unit discharges in the cat hypothalamus under chronic conditions. *Ann. N.Y. Acad. Sci.,* 157:666–689.

Oomura, Y., Ooyama, H., Yamamoto, T., and Naka, F. (1967a): Reciprocal relationships of the lateral and ventromedial hypothalamus in the regulation of food intake. *Physiol. Behav.,* 2:97–115.

Oomura, Y., Ooyama, H., Yamamoto, T., Naka, F., Kobayashi, N., and Ono, T. (1967b): Neuronal mechanism of feeding. *Prog. Brain Res.,* 27:1–33.

Oparin, A. I. (1964): *The Chemical Origin of Life.* Charles C Thomas, Springfield, Ill.

Oparin, A. I. (1976): Evolution of the concepts of the origin of life, 1924–1974. *Orig. Life,* 7:3–8.

Orenberg, E. K., Renson, J., Elliott, G. R., Barchas, J. D., and Kessler, S. (1975): Genetic determination of aggressive behavior and brain cyclic AMP. *Psychopharmacol. Commun.,* 1:99–107.

Orkand, R. K., Nicholls, J. G., and Kuffler, S. W. (1966): Effect of nerve impulses on the membrane potential of glial cells in the central nervous system of amphibia. *J. Neurophysiol.,* 29:788–806.

Oro, J. (1960): Synthesis of adenine from ammonium cyanide. *Biochem. Biophys. Res. Commun.,* 2:407–412.

Oro, J., and Guidry, C. L. (1960): A novel synthesis of polypeptides. *Nature,* 186:156–157.

Oro, J., and Kamat, S. S. (1961): Amino-acid synthesis from hydrogen cyanide under possible primitive earth conditions. *Nature,* 190:442–443.

Oro, J., and Kimball, A. P. (1961): Synthesis of purines under possible primitive earth conditions. I. Adenine from hydrogen cyanide. *Arch. Biochem. Biophys.,* 94:217–227.

Otsuka, M., and Takahashi, T. (1977): Putative peptide neurotransmitters. *Annu. Rev. Pharmacol. Toxicol.,* 17:425–439.

Ounsted, C. (1969): Aggression and epilepsy rage in children with temporal lobe epilepsy. *J. Psychosom. Res.,* 13:237–242.

Oxnard, C. E. (1975): The place of the australopithecines in human evolution: Grounds for doubt? *Nature,* 258:389–395.

Paalzow, G., and Paalzow, L. (1975): Morphine-induced inhibition of different pain responses in relation to the regional turnover of rat brain noradrenaline and dopamine. *Psychopharmacologia,* 45:9–20.

Paasonen, M. K., MacLean, P. D., and Giarman, N. J. (1957): 5-Hydroxytryptamine (serotonin, enteramine) content of structures of the limbic system. *J. Neurochem.,* 1:326–333.

Palade, G. A. (1955): A small particulate component of cytoplasm. *J. Biophys. Biochem. Cytol.,* 1:59–68.

Palade, G. E., and Palay, S. L. (1956): Synapses in the central nervous system. *J. Biophys. Biochem. Cytol.,* 2 Suppl.:193–202.

Palaic, D. J., Desaty, J., Albert, J. M., and Panisset, J. C. (1971): Effect of ethanol on metabolism and subcellular distribution of serotonin in rat brain. *Brain Res.,* 25:381–386.

Palay, S. L., and Palade, G. E. (1955): Fine structure of neurons. *J. Biophys. Biochem. Cytol.,* 1:69–88.

Palermo-Neto, J., and Carlini, E. A. (1972): Aggressive behavior elicited in rats by *Cannabis sativa:* Effects of p-chlorophenylalanine and DOPA. *Eur. J. Pharmacol.,* 17:215–220.

Palkovits, M., Brownstein, M., and Saavedra, J. M. (1974): Serotonin content of the brain stem nuclei in the rat. *Brain Res.,* 80:237–249.

Palkovits, M., Saavedra, J. M., Kobayashi, R. M., and Brownstein, M. (1974): Choline acetyltransferase content of limbic nuclei of the rat. *Brain Res.,* 79:443–450.

Pallaud, B. (1969a): Influence d'un congénère sur l'apprentissage chez la souris. *C. R. Acad. Sci. [D] (Paris),* 268:118–120.

Pallaud, B. (1969b): Influence d'un congénère sur la performance chez la souris. *C. R. Acad. Sci. [D] (Paris),* 269:1101–1104.

Panksepp, J. (1971): Effects of hypothalamic lesions on mouse-killing and shock-induced fighting in rats. *Physiol. Behav.,* 6:311–316.

Pantin, C. F. A. (1956): *Publ. Staz. Zool. Napoli,* 28:171.

Papez, J. W. (1937): A proposed mechanism of emotion. *Arch. Neurol. Psychiatry,* 38:725–743.

Papez, J. W. (1958): Visceral brain, its component parts and their connections. *J. Nerv. Ment. Dis.,* 126:40–56.

Parke, R. D. (1974): A field experimental approach to children's aggression. Some methodological problems and some future trends. In: *Determinants and Origins of Aggressive Behavior,* edited by J. de Wit and W. W. Hartup, pp. 499–508. Mouton, The Hague.

Parker, G. H. (1919): *The Elementary Nervous System.* Lippincott, Philadelphia.

Partanen, J., Bruun, K., and Markkanen, T. (1966): *Inheritance of Drinking Behavior.* Rutgers University Center of Alcohol Studies, New Brunswick, N.J.

Passano, L. M. (1963): Primitive nervous systems. *Proc. Natl. Acad. Sci. U.S.A.,* 50:306–313.

Passano, L. M., and McCullough, C. B. (1962): The light response and the rhythmic potentials of hydra. *Proc. Natl. Acad. Sci. U.S.A.,* 48:1376–1382.

Patni, S. K., and Dandiya, P. C. (1974): Apomorphine induced biting and fighting behaviour in reserpinized rats and an approach to the mechanism of action. *Life Sci.,* 14:737–745.

Paul, L. (1972): Predatory attack by rats: Its relationship to feeding and type of prey. *J. Comp. Physiol. Psychol.,* 78:69–76.

Paul, L., and Kupferschmidt, J. (1975): Killing of conspecific and mouse young by male rats. *J. Comp. Physiol. Psychol.,* 88:755–763.

Paul, L., Miley, W. M., and Baenninger, R. (1971): Mouse killing by rats: Roles of hunger and thirst in its initiation and maintenance. *J. Comp. Physiol. Psychol.,* 76:242–249.

Paul, L., and Posner, I. (1973): Predation and feeding: Comparisons of feeding behavior of killer and non-killer rats. *J. Comp. Physiol. Psychol.,* 81:258–264.

Paxinos, G. (1974): The hypothalamus: Neural systems involved in feeding, irritability, aggression, and copulation in male rats. *J. Comp. Physiol. Psychol.,* 87:110–119.

Paxinos, G., and Atrens, D. M. (1977): 5,7-Dihydroxytryptamine lesions: Effects on body weight, irritability, and muricide. *Aggressive Behav.,* 3:107–118.

Paxinos, G., Burt, J., Atrens, D. M., and Jackson, D. M. (1977): 5-Hyroxytryptamine depletion with para-chlorophenylalanine: Effects on eating, drinking, irritability, muricide, and copulation. *Pharmacol. Biochem. Behav.,* 6:439–447.

Paxinos, G., Emson, P. C., and Cuello, A. C. (1978a): The substance P projections to the frontal cortex and the substantia nigra. *Neurosci. Lett.,* 7:127–131.

Paxinos, G., Emson, P. C., and Cuello, A. C. (1978b): Substance P projections to the entopeduncular nucleus, the medial preoptic area and the lateral septum. *Neurosci. Lett.,* 7:133–136.

Payne, A. P., and Swanson, H. H. (1970): Agonistic behaviour between pairs of hamsters of the same and opposite sex in a neutral observation area. *Behaviour,* 36:260–269.

Payne, A. P., and Swanson, H. H. (1971a): The effect of castration and ovarian implantation on aggressive behaviour of male hamsters. *J. Endocrinol.,* 51:217–218.

Payne, A. P., and Swanson, H. H. (1971b): Hormonal control of aggressive dominance in the female hamster. *Physiol. Behav.,* 6:355–357.

Payne, A. P., and Swanson, H. H. (1972a): The effect of sex hormones on the aggressive behaviour of the female golden hamster (*Mesocricetus auratus* Waterhouse). *Anim. Behav.,* 20:782–787.

Payne, A. P., and Swanson, H. H. (1972b): The effect of sex hormones on the agonistic behavior of the male golden hamster (Mesocricetus auratus Waterhouse). Physiol. Behav., 8:687–691.

Peeke, H. V. S., Ellman, G. E., and Herz, M. J. (1973): Dose dependent alcohol effects on the aggressive behavior of the Convict cichlid (Cichlasoma migrofasciatum). Behav. Biol., 8:115–122.

Pelto, P. (1967): Psychological anthropology. In: Biennial Review of Anthropology, edited by A. Beals and B. Siegel, pp. 151. Stanford.

Penfield, W., and Erickson, T. C. (1941): Epilepsy and Cerebral Localization. Charles C Thomas, Springfield, Ill.

Penfield, W., and Jasper, H. (1954): Epilepsy and the Functional Anatomy of the Human Brain. Little, Brown & Co., Boston.

Pepeu, G., Garau, L., and Mulas, M. L. (1974): Does 5-hydroxytryptamine influence cholinergic mechanisms in the central nervous system? Adv. Biochem. Psychopharmacol., 10:247–252.

Pepeu, G., Mulas, A., Ruffi, A., and Sotgiu, P. (1971): Brain acetylcholine levels in rats with septal lesions. Part I. Life Sci., 10:181–184.

Perry, D. G., and Perry, L. C. (1976): A note on the effects of prior anger arousal and winning or losing a competition on aggressive behaviour in boys. J. Child Psychol. Psychiatry, 17:145–149.

Persky, H., Smith, K. D., and Basu, G. K. (1971): Relation of psychologic measures of aggression and hostility to testosterone production in man. Psychosom. Med., 33:265–277.

Persky, H., Zuckerman, M., Basu, G. K., and Thornton, D. (1966): Psycho-endocrine effects of perceptual and social isolation. Arch. Gen. Psychiatry, 15:499–505.

Peters, D. P. (1978): Effects of prenatal nutritional deficiency on affiliation and aggression in rats. Physiol. Behav., 20:359–362.

Peters, J. M. (1966): Caffeine toxicity in starved rats. Toxicol. Appl. Pharmacol., 9:390–397.

Peters, J. M. (1967): Caffeine-induced hemorrhagic automutilation. Arch. Int. Pharmacodyn. Ther., 169:139–146.

Peters, J. M., and Boyd, E. M. (1966): Diet and caffeine toxicity in rats. Toxicol. Appl. Pharmacol., 8:350–351.

Peters, J. M., and Boyd, E. M. (1967a): The influence of sex and age in albino rats given a daily oral dose of caffeine at a high dose level. Can. J. Physiol. Pharmacol., 45:305–311.

Peters, J. M., and Boyd, E. M. (1967b): The influence of a cachexigenic diet on caffeine toxicity. Toxicol. Appl. Pharmacol., 11:121–127.

Pfaff, D. W., and Pfaffmann, C. (1969): Olfactory and hormonal influences on the basal forebrain in the male rat. Brain Res., 15:137–156.

Pfeiffer, C. J., and Gass, G. H. (1962): Caffeine-induced ulcerogenesis in the rat. Can. J. Biochem. Physiol., 40:1473–6.

Piaget, J. (1932): The Moral Judgement of the Child. Harcourt, Brace and World, New York.

Piggott, L., Purcell, G., Cummings, G., and Caldwell, D. (1976): Vestibular dysfunction in emotionally disturbed children. Biol. Psychiatry, 11:719–729.

Pinderhughes, C. A. (1974): Ego development and cultural differences. Am. J. Psychiatry, 131:171–175.

Pinel, J. P. J., Treit, D., and Rovner, L. I. (1977): Temporal lobe aggression in rats. Science, 197:1088–1089.

Piotrowski, K. W., Losacco, D., and Guze, S. B. (1976): Psychiatric disorders and crime: A study of pretrial psychiatric examinations. Dis. Nerv. Syst., 37:309–311.

Pirch, J. H., and Rech, R. H. (1968): Effect of isolation on α-methyltyrosine-induced behavioral depression. Life Sci., 7:173–182.

Pisano, R., and Taylor, S. P. (1971): Reduction of physical aggression. The effects of four strategies. J. Pers. Soc. Psychol., 19:237–242.

Pitkänen, L. (1973): An aggression machine. II. Interindividual differences in the aggressive defence responses aroused by varying stimulus conditions. Scand. J. Psychol., 14:65–74.

Planansky, K., and Johnston, R. (1977): Homicidal aggression in schizophrenic men. Acta Psychiatr. Scand., 55:65–73.

Plant, T. M., Zumpe, D., Sauls, M., and Michael, R. P. (1974): An annual rhythm in the plasma testosterone of adult male rhesus monkeys maintained in the laboratory. J. Endocrinol., 62:403–404.

Plomin, R., DeFries, J. C., and Loehlin, J. C. (1977): Genotype-environment interaction and correlation in the analysis of human behavior. *Psychol. Bull.,* 84:309–322.

Ploog, D. W., and MacLean, P. D. (1963): Display of penile erection in squirrel monkey *(Saimiri sciureus). Anim. Behav.,* 11:32–39.

Plotnik, R. (1974): Brain stimulation and aggression: Monkeys, apes, and humans. In: *Primate Aggression, Territoriality, and Xenophobia,* edited by R. L. Holloway, pp. 389–415. Academic Press, New York.

Plumer, S. I., and Siegel, J. (1973): Caudate-induced inhibition of hypothalamic attack behavior. *Physiol. Psychol.,* 1:254–256.

Poblete, M., Palestini, M., Figueroa, E., Gallardo, R., Rojas, J., Covarrubias, M. I., and Doyharcabal, Y. (1970): Stereotaxic thalamotomy (lamella medialis) in aggressive psychiatric patients. *Confin. Neurol.,* 32:326–331.

Poeck, K., and Pilleri, G. (1965): Release of hypersexual behaviour due to lesion in the limbic system. *Acta Neurol. Scand.,* 41:233–244.

Pohorecky, L. A., Larin, F., and Wurtman, R. J. (1969): Mechanism of changes in brain norepinephrine levels following olfactory bulb lesions. *Life Sci.,* 8:1309–1317.

Poitou, P., Guerinet, F., and Bohuon, C. (1974): Effect of lithium on central metabolism of 5-hydroxytryptamine. *Psychopharmacologia,* 38:75–80.

Poley, W. (1973): Factor analyses of patterns of alcohol and water consumption in mice. *Q. J. Stud. Alcohol,* 34:202–205.

Poley, W., and Royce, J. R. (1972): Alcohol consumption, water consumption, and emotionality in mice. *J. Abnorm. Psychol.,* 79:195–204.

Pollard, J. C., Uhr, L., and Jackson, C. W., Jr. (1963): Studies in sensory deprivation. *Arch. Gen. Psychiatry,* 8:435–454.

Pollen, D. A., and Trachtenberg, M. C. (1970): Neuroglia: Gliosis and focal epilepsy. *Science,* 167:1252–1253.

Polsky, R. H. (1975): Hunger, prey feeding, and predatory aggression. *Behav. Biol.,* 13:81–93.

Polsky, R. H. (1978): Influence of eating dead prey on subsequent capture of live prey in golden hamsters. *Physiol. Behav.,* 20:677–680.

Pool, J. L. (1954): The visceral brain of man. *J. Neurosurg.,* 11:45–63.

Poole, T. B., and Morgan, H. D. R. (1975): Aggressive behaviour of male mice *(Mus musculus)* towards familiar and unfamiliar opponents. *Anim. Behav.,* 23:470–479.

Pope, A., Hess, A., and Allen, J. (1957): In: *Ultrastructure and Cellular Chemistry of Nervous Tissue,* edited by H. Waelsch, Harper and Row, New York.

Popova, N. K., and Naumenko, E. V. (1972): Dominance relations and the pituitary-adrenal system. *Anim. Behav.,* 20:108–111.

Popova, N. K., Nikulina, E. M., Aruv, V. I., and Kudriavtseva, M. N. (1975): Role of serotonin in mouse-killing behavior in rats. *Fiziol. Zh. SSSR,* 61:183–186.

Popova, N. K., Nikulina, E. M. and Maslova, L. N. (1976): Influence of isolation of the mediobasal hypothalamus of predatory aggression in rats. *Zh. Vyssh. Nerv. Deiat.,* 26:570–572.

Porter, R. W., Conrad, D., and Brady, J. V. (1959): Some neural and behavioral correlates of electrical self-stimulation in the limbic system. *J. Exp. Anal. Behav.,* 2:43–55.

Poschel, B. P. H., and Ninteman, F. W. (1963): Norepinephrine: A possible excitatory neurohormone of the reward system. *Life Sci.,* 2:782–788.

Pöschlová, N., Masĕk, K., and Kršiak, M. (1976): Facilitated intermale aggression in the mouse after 6-hydroxydopamine administration. *Neuropharmacology,* 15:403–407.

Potter, L. T., and Axelrod, J. (1963): Subcellular localization of catecholamines in tissues of the rat. *J. Pharmacol. Exp. Ther.,* 142:291–298.

Powell, D. A., and Creer, T. L. (1969): Interaction of developmental and environmental variables in shock-elicited aggression. *J. Comp. Physiol. Psychol.,* 69:219–225.

Powell, D. A., Milligan, W. L., and Walters, K. (1973): The effects of muscarinic cholinergic blockade upon shock-elicited aggression. *Pharmacol. Biochem. Behav.,* 1:389–394.

Powell, T. P., Cowan, W. M., and Raisman, G. (1965): The central olfactory connexions. *J. Anat.,* 99:791–813.

Power, T. D. (1965): Some aspects of brain-mind relationship. *Br. J. Psychiatry,* 111:1215–1223.

Prescott, J. W. (1971): Early somatosensory deprivation as an ontogenetic process in the abnormal development of the brain and behavior. In: *Proceedings of the 2nd Conference on Experimental Medicine and Surgery in Primates,* New York, 1969, pp. 356–375. S. Karger, Basel.

Prescott, J. W. (1975): Body pleasure and the origins of violence. *The Futurist,* April: 64–74.

Prescott, J. W., and Essman, W. B. (1969): The psychobiology of maternal-social deprivation and the etiology of violent-aggressive behavior: A special case of sensory deprivation. 2nd Annual Winter Conference on Brain Function, Snowmass-at-Aspen, Colo., January 12–17, 1969.

Pribram, K. H., and Bagshaw, M. (1953): Further analysis of the temporal lobe syndrome utilizing frontotemporal ablations. *J. Comp. Neurol.,* 99:347–375.

Price, E. O., Belanger, P. L., and Duncan, R. A. (1976): Competitive dominance of wild and domestic Norway rats *(Rattus norvegicus). Anim. Behav.,* 24:589–599.

Proshansky, E., and Bandler, R. (1975): Midbrain-hypothalamic interrelationships in the control of aggressive behavior. *Aggressive Behav.,* 1:135–155.

Pryse-Phillips, W. (1975): Disturbance in the sense of smell in psychiatric patients. *Proc. R. Soc. Med.,* 68:472–474.

Puri, S. K., and Lal, H. (1974): Reduced threshold to pain induced aggression specifically related to morphine dependence. *Psychopharmacologia,* 35:237–241.

Quay, W. B. (1965): Regional and circadian differences in cerebral cortical serotonin concentrations. *Life Sci.,* 4:379–384.

Quay, W. B. (1967): Twenty-four-hour rhythms in cerebral and brainstem contents of 5-hydroxytryptamine in a turtle, *Pseudemys Scripta elegans. Comp. Biochem. Physiol.,* 20:217–221.

Quay, W. B., Bennett, E. L., Rosenzweig, M. R., and Krech, D. (1969): Effects of isolation and environmental complexity on brain and pineal organ. *Physiol. Behav.,* 4:489–494.

Quay, W. B., and Meyer, D. C. (1978): Rhythmicity and periodic functions of the central nervous system and serotonin. In: *Serotonin in Health and Disease, Vol. 2: Physiological Regulation and Pharmacological Action,* edited by W. B. Essman, pp. 159–204. Spectrum, New York.

Rachman, S. (1965): Pain-elicited aggression and behaviour therapy. *Psychol. Rec.,* 15:465–467.

Rada, R. T., Laws, D. R., and Kellner, R. (1976): Plasma testosterone levels in the rapist. *Psychosom. Med.,* 38:257–268.

Ramón y Cajal, S. (1895): *Les Nouvelles Idées sur la Structure du Système Nerveux chez l'Homme et chez les Vertebrés.* Reinwold ed., Paris.

Ramón y Cajal, S. (1913): Contribucion al conocimiento de la neuroglia del cerebro humano. *Trab. Lab. Biol.,* 11:255–315.

Randolph, T. G. (1962): *Human Ecology and Susceptibility to the Chemical Environment.* Charles C Thomas, Springfield, Ill.

Randrup, A., and Munkvad, I. (1967): Stereotyped activities produced by amphetamine in several animal species and man. *Psychopharmacologia,* 11:300–310.

Randt, C. T., Blizard, D. A., and Friedman, E. (1975): Early life undernutrition and aggression in two mouse strains. *Dev. Psychobiol.,* 8:275–279.

Ransohoff, J. (1969): Current status of psychosurgery. *Dis. Nerv. Syst. GWAN Suppl.,* 30:58–59.

Ranson, S. (1936): Some functions of the hypothalamus. *Harvey Lect.,* 32:93–121.

Ranson, S. W., and Berry, C. (1941): Observations on monkeys with bilateral lesions of the globus pallidus. *Arch. Neurol. Psychiatry,* 46:504–508.

Rawlin, J. (1968): Street level abuse of amphetamine. In: *Amphetamine Abuse,* edited by J. R. Russo, pp. 51–65. Charles C Thomas, Springfield, Ill.

Raynes, A. E., and Ryback, R. S. (1970): Effect of alcohol and congeners on aggressive response in *Betta splendens. Q. J. Stud. Alcohol,* 5, Suppl. 5:130–135.

Raynes, A. E., Ryback, R., and Ingle, D. (1968): The effect of alcohol on aggression in *Betta splendens. Commun. Behav. Biol. [A],* 2:141–146.

Rech, R. H., Borys, H. K., and Moore, K. E. (1966): Alterations in behavior and brain catecholamine levels in rats treated with α-methyltyrosine. *J. Pharmacol. Exp. Ther.,* 153:412–419.

Reimann, H. A. (1967): Caffeinism. A cause of long-continued, low-grade fever. *J.A.M.A.,* 202:1105–1106.

Reis, D. J. (1971): Brain monoamines in aggression and sleep. *Clin. Neurosurg.,* 18:471–502.

Reis, D. J. (1974): Consideration of some problems encountered in relating specific neurotransmitters to specific behaviors or disease. *J. Psychiatr. Res.,* 11:145–148.

Reis, D. J., Corvelli, A., and Conners, J. (1969): Circadian and ultradian rhythms of serotonin regionally in cat brain. *J. Pharmacol. Exp. Ther.,* 167:328–333.

Reis, D. J., Doba, N., and Nathan, M. A. (1973): Predatory attack, grooming and consummatory behaviors evoked by electrical stimulation of cat cerebellar nuclei. *Science,* 182:845–847.

Reis, D. J., and Fuxe, K. (1969): Brain norepinephrine: Evidence that neuronal release is essential for sham rage behavior following brainstem transection in cat. *Proc. Natl. Acad. Sci. U.S.A.,* 64:108–112.

Reis, D. J., and Gunne, L. M. (1965): Brain catecholamines: Relation to the defense reaction evoked by amygdaloid stimulation in cat. *Science,* 149:450–451.

Reis, D. J., Miura, M., Weinbren, M., and Gunne, L.-M. (1967): Brain catecholamines: Relation to defense reaction evoked by acute brainstem transection in cat. *Science,* 156:1768–1770.

Reis, D. J., Weinbren, M., and Corvelli, A. (1968): A circadian rhythm of norepinephrine regionally in cat brain: Its relationships to environmental lighting and to regional diurnal variations in brain serotonin. *J. Pharmacol. Exp. Ther.,* 164:135–145.

Reis, D. J., and Wurtman, R. J. (1968): Diurnal changes in brain noradrenalin. *Life Sci.,* 7:91–98.

Reisz, R. (1972): Pelycosaurian reptiles from the middle pennsylvanian of North America. *Bull. Mus. Comp. Zool.,* 144:27–61.

Resnick, O. (1971): The use of psychotropic drugs with criminals. In: *Psychotropic Drugs in the Year 2000: Use by Normal Humans,* edited by W. O. Evans and N. S. Kline, pp. 109–127. Charles C Thomas, Springfield, Ill.

Revitch, E. (1975): Psychiatric evaluation and classification of antisocial activities. *Dis. Nerv. Syst.,* 36:419–421.

Rewerski, W., Kostowski, W., Piechocki, T., and Rylski, M. (1971): The effect of some hallucinogens on aggressiveness of mice and rats. 1. *Pharmacology,* 5:314–320.

Reynierse, J. H. (1971): Submissive postures during shock-elicited aggression. *Anim. Behav.,* 19:102–107.

Reynolds, H. H. (1963): Effect of rearing and habitation in social isolation on performance of an escape task. *J. Comp. Physiol. Psychol.,* 56:520–525.

Richardson, L. F. (1960): *Statistics of Deadly Quarrels.* Boxwood Press, Pittsburgh.

Richmond, J. S., Young, J. R., and Groves, J. E. (1978): Violent discontrol responsive to d-amphetamine. *Am. J. Psychiatry,* 135:365–366.

Richter, C. P. (1957): On the phenomenon of sudden death in animals and man. *Psychosom. Med.,* 19:191–198.

Rickman, E. E., Williams, E. Y., and Brown, R. K. (1961): Acute toxic psychiatric reactions related to amphetamine medication. *Med. Ann. D.C.,* 30:209–212.

Riege, W. H., and Morimoto, H. (1970): Effects of chronic stress and differential environments upon brain weights and biogenic amine levels in rats. *J. Comp. Physiol. Psychol.,* 71:396–404.

Riegel, K. F. (1972): Influence of economic and political ideologies on the development of developmental psychology. *Psychol. Bull.,* 78:129–141.

Rifkin, R. J., Silverman, J. M., Chavez, F. T., and Frankl, G. (1974): Intensified mouse killing in the spontaneously hypertensive rat. *Life Sci.,* 14:985–992.

Riklan, M., Marisak, K., and Cooper, I. S. (1974): Psychological studies of chronic cerebellar stimulation in man. In: *The Cerebellum Epilepsy, and Behavior,* edited by I. S. Cooper, M. Riklan, and R. S. Snider, pp. 285–342. Plenum Press, New York.

Rioch, D. M. (1967): Discussion of agonistic behavior. In: *Social Communication Among Primates,* edited by S. A. Altmann, p. 115. University of Chicago Press, Chicago.

Roberts, D. R. (1966): Functional organization of the limbic systems. *Int. J. Neuropsychiatry,* 2:279–291.

Roberts, E. (1976): Disinhibition as an organizing principle in the nervous system. The role of the GABA system, application to neurologic and psychiatric disorders. In: *GABA in Nervous System Function,* edited by E. Roberts, T. N. Chase, and D. B. Tower, p. 515. Raven Press, New York.

Roberts, W. W., and Kiess, H. O. (1964): Motivational properties of hypothalamic aggression in cats. *J. Comp. Physiol. Psychol.,* 58:187–193.

Robertson, J. D. (1960): The molecular structure and contact relationships of cell membranes. *Prog. Biophys. Mol. Biol.,* 10:343–418.

Robins, L. N. (1966): *Deviant Children Grown Up.* Williams & Wilkins, Baltimore.

Robinson, E. (1963): Effect of amygdalectomy on fear-motivated behavior in rats. *J. Comp. Physiol. Psychol.,* 56:814–820.

Rodgers, R. J. (1977): Attenuation of morphine analgesia in rats by intra-amygdaloid injection of dopamine. *Brain Res.,* 130:156–162.

Rodgers, R. J., and Brown, K. (1973): *IRCS Med. Sci.,* (73–5) 7–10–5.

Rodgers, R. J., Semple, J. M., Cooper, S. J., and Brown, K. (1976): Shock-induced aggression and pain sensitivity in the rat: Catecholamine involvement in the cortico-medial amygdala. *Aggressive Behaviour,* 2.

Roeder, F. D. (1966): Stereotaxic lesion of the tuber cinereum in sexual deviation. *Confin. Neurol.,* 27:162–163.

Rogers, F. T. (1923): Studies of the brain stem. VI. An experimental study of the corpus striatum of the pigeon as related to various instinctive types of behavior. *J. Comp. Neurol.,* 35:21–60.

Rohles, F. H., Jr., and Wilson, L. M. (1973): Hunger as a catalyst in aggression. *Behaviour,* 48:123–129.

Rolinski, Z. (1974): Analysis of the aggressiveness-stereotypy complex induced in mice by amphetamine or D,L-DOPA. II. *Pol. J. Pharmacol. Pharm.,* 26:369–378.

Rolinski, Z. (1975): Pharmacological studies on isolation-induced aggressiveness in mice in relation to biogenic amines. *Pol. J. Pharmacol. Pharm.,* 27:37–44.

Rollin, H. R. (1973): Deviant behaviour in relation to mental disorder. *Proc. R. Soc. Med.,* 66:99–104.

Romaniuk, A. (1974): Neurochemical bases of defensive behavior in animals. *Acta Neurobiol. Exp.,* 34:205–214.

Romaniuk, A., Brudzyński, S., and Grońska, J. (1973): The effect of chemical blockade of hypothalamic cholinergic system on defensive reactions in cats. *Acta Physiol. Pol.,* 24:809–816.

Romaniuk, A., Brudzyński, S., and Grońska, A. (1974): The effects of intrahypothalamic injections of cholinergic and adrenergic agents on defensive behavior in cats. *Acta Physiol. Pol.,* 25:297–305.

Romaniuk, A., and Gołebiewski, H. (1977): Midbrain interaction with the hypothalamus in expression of aggressive behavior in cats. *Acta Neurobiol. Exp.,* 37:83–97.

Ropartz, P. (1968): The relation between olfactory stimulation and aggressive behaviour in mice. *Anim. Behav.,* 1b:97–100.

Rose, R. M., Bernstein, I. S., and Gordon, T. P. (1975): Consequences of social conflict on plasma testosterone levels in rhesus monkeys. *Psychosom. Med.,* 37:50–61.

Rose, S. (1973): *The Conscious Brain.* Alfred A. Knopf, New York.

Rosecrans, J. A. (1970): Brain serotonin and pituitary-adrenal function in rats of different emotionalities. *Arch. Int. Pharmacodyn. Ther.,* 187:349–366.

Rosecrans, J. A., and Sheard, M. H. (1967): Effects of an acute stressor on the brain amine levels of normal and CNS lesioned rats. *Pharmacologist,* 9:224.

Rosenberg, K. M., Denenberg, V. H., Zarrow, M. X., and Frank, B. L. (1971): Effects of neonatal castration and testosterone on the rat's pup-killing behavior and activity. *Physiol. Behav.,* 7:363–368.

Rosenberg, K. M., and Sherman, G. F. (1975): Influence of testosterone on pup-killing in the rat is modified by prior experience. *Physiol. Behav.,* 15:669–672.

Rosenzweig, M. R. (1964): Effect of heredity and environment on brain chemistry, brain anatomy and learning ability in the rat. *Kans. Stud. Educ.,* 14:3–34.

Rosenzweig, M. R. (1966): Environmental complexity, cerebral change, and behavior. *Am. Psychol.,* 21:321–332.

Rosenzweig, M. R., and Bennett, E. L. (1969): Effects of differential environments on brain weights and enzyme activities in gerbil, rats and mice. *Dev. Psychobiol.,* 2:87–95.

Rosenzweig, N. (1959): Sensory deprivation and schizophrenia: Some clinical and theoretical similarities. *Am. J. Psychiatry,* 116:326–329.

Rosenzweig, N., and Gardner, L. M. (1966): The role of input relevance in sensory isolation. *Am. J. Psychiatry,* 122:920–928.

Ross, S. B., and Ögren, S.-O. (1976): Anti-aggressive action of dopamine-β-hydroxylase inhibitors in mice. *J. Pharm. Pharmacol.,* 28:590–592.

Rossi, A. C. (1975): The "mouse-killing rat": Ethological discussion on an experimental model of aggression. *Pharmacol. Res. Commun.,* 7:199–216.

Rossier, J., Bauman, A., and Benda, P. (1973): Antibodies to rat brain choline acetyltransferase: Species and organ specificity. *FEBS Lett.,* 36:43–48.

Rosvold, H. E. (1968): The prefrontal cortex and caudate nucleus: A system for affecting correction in response mechanisms. In: *Mind as a Tissue,* edited by C. Rupp, pp. 21–38. Hoeber Med. Division, Harper & Row, New York.

Rosvold, H. E., Mirscky, A. F., and Pribram, K. H. (1954): Influence of amygdalectomy on social behavior in monkeys. *J. Comp. Physiol. Psychol.,* 47:173–178.

Roth, M. (1972): Human violence as viewed from the psychiatric clinic. *Am. J. Psychiatry,* 128:1043–1056.

Rowe, F. A., and Edwards, D. A. (1971): Olfactory bulb removal: Influences on the aggressive behaviors of male mice. *Physiol. Behav.,* 7:889–892.

Roy, A. (1978): Self-mutilation. *Br. J. Med. Psychol.,* 51:201–203.

Rule, B. G. (1974): The hostile and instrumental functions of human aggression. In: *Determinants and Origins of Aggressive Behavior,* edited by J. de Wit and W. W. Hartup, pp. 125–145. Mouton, The Hague.

Rule, B. G., and Hewitt, L. S. (1971): Effects of thwarting on cardiac response and physical aggression. *J. Pers. Soc. Psychol.,* 19:181–187.

Rupp, C. (ed.) (1968): *Mind as a Tissue.* Harper & Row, New York.

Rusak, B., and Zucker, I. (1975): Biological rhythms and animal behavior. *Annu. Rev. Psychol.,* 26:137–171.

Russell, P. A., and Williams, D. I. (1973): Effects of repeated testing on rats' locomotor activity in the open-field. *Anim. Behav.,* 21:109–112.

Rutter, M. (1966): *Children of Sick Parents. An Environmental and Psychiatric Study.* Oxford University Press, London.

Rutter, M. (1971): Parent-child separation: Psychological effects on the children. *J. Child Psychol. Psychiatry,* 12:233–260.

Ryall, R. W. (1964): The subcellular distributions of acetylcholine, substance P, 5-hydroxytryptamine, gamma-aminobutyric acid and glutamic acid in brain homogenates. *J. Neurochem.,* 11:131–145.

Ryan, W. (1972): *Blaming the Victim.* Vintage Books, New York.

Ryback, R., Percarpio, B., and Vitale, J. (1969): Equilibration and metabolism of ethanol in the goldfish. *Nature,* 222:1068–1070.

Sadowsky, B. (1972): Intracranial self-stimulation patterns. *Physiol. Behav.,* 8:189–193.

Sainsbury, R. S., and Jason, G. W. (1976): Fimbria-fornix lesions and sexual-social behavior of the guinea pig. *Physiol. Behav.,* 17:963–967.

Saito, Y., Yamashita, I., Yamazaki, K., Okada, F., Satomi, R., and Fujieda, T. (1975): Circadian fluctuation of brain acetylcholine in rats. I. On the variations in the total brain and discrete brain areas. *Life Sci.,* 16:281–288.

Sakata, T., and Fuchimoto, H. (1973a): Stereotyped and aggressive behavior induced by sustained high dose of theophylline in rats. *Jpn. J. Pharmacol.,* 23:781–785.

Sakata, T., and Fuchimoto, H. (1973b): Further aspects of aggressive behavior induced by sustained high dose of theophylline in rats. *Jpn. J. Pharmacol.,* 23:787–792.

Sakata, T., Fuchimoto, H., Kodama, J., and Fukusima, M. (1975): Changes of brain serotonin and muricide behavior following chronic administration of theophylline in rats. *Physiol. Behav.,* 15:449–453.

Salama, A. I., and Goldberg, M. E. (1970): Neurochemical effects of imipramine and amphetamine in aggressive mouse-killing (muricidal) rats. *Biochem. Pharmacol.,* 19:2023–2032.

Salama, A. I., and Goldberg, M. E. (1973): Temporary increase in forebrain norepinephrine turnover in mouse-killing rats. *Eur. J. Pharmacol.,* 21:372–374.

Salmoiraghi, G. C., Costa, E., and Bloom, F. E. (1965): Pharmacology of central synapses. *Annu. Rev. Pharmacol.,* 5:213–234.

Salustiano, J., Hoshino, K., and Carlini, E. A. (1966): Effects of *Cannabis sativa* and chlorpromazine on mice as measured by two methods used for evaluation of tranquilizing agents. *Med. Pharmacol. Exp.,* 15:153–162.

Salzman, C., Van der Kolk, B. A., and Shader, R. I. (1976): Marijuana hostility in a small-group setting. *Am. J. Psychiatry,* 133:1029–1033.

Samanin, R., and Bernasconi, S. (1972): Effect of intraventricularly injected 6-OH dopamine or midbrain raphe lesion on morphine analgesia in rats. *Psychopharmacologia,* 25:175–182.

Samanin, R., Ghezzi, D., Mauron, C., and Valzelli, L. (1973): Effect of midbrain raphe lesion on the antinociceptive action of morphine and other analgesics in rats. *Psychopharmacologia,* 33:365–368.

Samanin, R., Ghezzi, D., Valzelli, L., and Garattini, S. (1972): The effects of selective lesioning of brain serotonin or catecholamine containing neurones on the anorectic activity of fenfluramine and amphetamine. *Eur. J. Pharmacol.,* 19:318–322.

Samanin, R., Gumulka, W., and Valzelli, L. (1970): Reduced effect of morphine in midbrain raphe lesioned rats. *Eur. J. Pharmacol.*, 10:339–343.

Samanin, R., and Valzelli, L. (1971): Increase of morphine-induced analgesia by stimulation of the nucleus raphe dorsalis. *Eur. J. Pharmacol.*, 16:298–302.

Samanin, R., and Valzelli, L. (1972): Serotoninergic neurotransmission and morphine activity. *Arch. Int. Pharmacodyn. Ther.*, 196 (suppl.):138–141.

Sands, D. E. (1954): Further studies on endocrine treatment in adolescence and early adult life. *J. Ment. Sci.*, 100:211–219.

Sands, D. E., and Chamberlain, G. H. A. (1952): Treatment of inadequate personality in juveniles by dehydroisoandrosterone. *Br. Med. J.*, 2:66–68.

Sanides, F. (1969): Comparative architectonics of the neocortex of mammals and their evolutionary interpretation. *Ann. N. Y. Acad. Sci.*, 167:404–423.

Sano, K. (1962): Sedative neurosurgery: With special reference to postero-medial hypothalamotomy. *Neurol. Medico-Chir.*, 4:112–142.

Sano, K., Mayanagi, Y., Sekino, H., Ogashiwa, M., and Ishijima, B. (1970): Results of stimulation and destruction of the posterior hypothalamus in man. *J. Neurosurg.*, 33:689–707.

Sano, K., Yoshioka, M., Ogashiwa, M., Ishijima, B., and Ohye, C. (1966): Postero-medial hypothalamotomy in the treatment of aggressive behaviors. *Confin. Neurol.*, 27:164–167.

Santos, M., Sampaio, M. R. P., Fernandes, N. S., and Carlini, E. A. (1966): Effects of *Cannabis sativa* (marihuana) on the fighting behavior of mice. *Psychopharmacologia*, 8:437–444.

Sargant, W. (1969): The physiology of faith. *Br. J. Psychiatry*, 115:505–518.

Sargent, D. (1971a): The lethal situation: Transmission of urge to kill from parent to child. In: *Dynamics of Violence*, edited by J. Fawcett. American Medical Association, Chicago.

Sargent, D. (1971b): Children who kill—A family conspiracy? In: *Theory and Practice of Family Psychiatry*, edited by J. Howells. Brunner-Mazel, New York.

Sawa, M., Ueki, Y., Arita, M., and Harada, T. (1954): Preliminary report on the amygdaloidectomy on the psychotic patients, with interpretation of oral-emotional manifestation in schizophrenics. *Folia Psychiatr. Neurol. Jpn.*, 7:309–329.

Sbordone, R. J. (1976): A rat model of violent attack behavior. *Dissertation Abst.*, 37:213.

Sbordone, R. J., and Carder, B. (1974): Mescaline and shock induced aggression in rats. *Pharmacol. Biochem. Behav.*, 2:777–782.

Sbordone, R. J., and Garcia, J. (1977): Untreated rats develop "pathological" aggression when paired with a mescaline-treated rat in a shock-elicited aggression situation. *Behav. Biol.*, 21:451–461.

Sbordone, R. J., Wingard, J. A., Elliott, M. L., and Jervey, J. (1978): Mescaline produces pathological aggression in rats, regardless of age or strain. *Pharmacol. Biochem. Behav.*, 8:543–546.

Scaramella, T. J., and Brown, W. A. (1978): Serum testosterone and aggressiveness in hockey players. *Psychosom. Med.*, 40:262–265.

Schachter, J. (1957): Pain, fear, and anger in hypertensives and normotensives. *Psychosom. Med.*, 19:17–29.

Schaefer, K. P. (1970): Unit analysis and electrical stimulation in the optic tectum of rabbits and cats. *Brain Behav. Evol.*, 3:222–240.

Schanberg, S. M., Schildkraut, J. J., and Kopin, I. J. (1967): The effects of psychoactive drugs on norepinephrine-³H metabolism in brain. *Biochem. Pharmacol.*, 16:393–399.

Scheving, L. E., Harrison, W. H., Gordon, P., and Pauly, J. E. (1968): Daily fluctuation (circadian and ultradian) in biogenic amines of the rat brain. *Am. J. Physiol.*, 214:166–173.

Schildkraut, J. J., Logue, M. A., and Dodge, G. A. (1969): The effects of lithium salts on the turnover and metabolism of norepinephrine in rat brain. *Psychopharmacologia*, 14:135–141.

Schildkraut, J. J., Schanberg, S. M., Breese, G. R., and Kopin, I. J. (1967): Norepinephrine metabolism and drugs used in the affective disorders: A possible mechanism of action. *Am. J. Psychiatry*, 124:600–608.

Schipkowensky, N. (1973): Epidemiological aspects of homicide. In: *World Biennial of Psychiatry and Psychotherapy, Vol. 2*, edited by S. Arieti, pp. 192–215. Basic Books, New York.

Schless, A. P., Mendels, J., Kipperman, A., and Cochrane, C. (1974): Depression and hostility. *J. Nerv. Ment. Dis.*, 159:91–100.

Schmidt, R. F. (1971): Presynaptic inhibition in the vertebrate central nervous system. *Ergebm. Physiol.*, 63:20–101.

Schramm, G. (1960): In: *Proceedings of the International Symposium on the Origin of Life on Earth, Moscow*, p. 216. Pergamon Press, London.

Schramm, G., Grötsch, H., and Pollmann, W. (1962): Non-enzymatic synthesis of polysaccharides, nucleosides and nucleic acids and the origin of cells-reproducing systems. *Angew, Chem. (Int. Ed.),* 1:1–7.

Schreiner, L., and Kling, A. (1953): Behavioral changes following rhinencephalic injury in cat. *J. Neurophysiol.,* 16:643–659.

Schreiner, L., and Kling, A. (1956): Rhinencephalon and behavior. *Am. J. Physiol.,* 184:486–490.

Schuck, J. R. (1974): The use of causal nonexperimental models in aggression research. In: *Determinants and Origins of Aggressive Behavior,* edited by J. de Wit and W. W. Hartup, pp. 381–389. Mouton, The Hague.

Schwab, J. J., McGinnis, N. H., and Warheit, G. J. (1973): Social psychiatric impairment: Racial comparisons. *Am. J. Psychiatry,* 130:183–187.

Schwab, R. S., Sweet, W. H., Mark, V. H., Kjellber, R. N., and Ervin, F. R. (1965): Treatment of intractable temporal lobe epilepsy by stereotactic amygdala lesions. *Trans. Am. Neurol. Assoc.,* 90:12–19.

Scott, J. J. (1956): Synthesis of crystallizable porphobilinogen. *Biochem. J.,* 62:6P.

Scott, J. P. (1958a): *Aggression.* University of Chicago Press, Chicago.

Scott, J. P. (1958b): *Animal Behavior.* University of Chicago Press, Chicago.

Scott, J. P. (1962): Hostility and aggression in animals. In: *The Roots of Behavior,* edited by E. L. Bliss, pp. 167–178. Harper & Row, New York.

Scott, J. P. (1965): On the evolution of fighting behavior. *Science,* 148:820–821.

Scott, J. P. (1966a): Fighting. *Science,* 154:636–637.

Scott, J. P. (1966b): Agonistic behavior of mice and rats: A review. *Am. Zool.,* 6:683–701.

Scott, J. P. (1974): Effects of psychotropic drugs on separation distress in dogs. *J. Pharmacol. (Paris),* 5 (suppl. 1):95.

Scott, J. P., and Fredericson, E. (1951): The causes of fighting in mice and rats. *Physiol. Zool.,* 24:273–309.

Scoville, W. B., and Milner, B. (1957): Loss of recent memory after bilateral hippocampal lesions. *J. Neurol. Neurosurg. Psychiatry,* 20:11–21.

Sears, R. R. (1961): Relation of early socialization experiences to aggression in middle childhood. *J. Abnorm. Soc. Psychol.,* 63:466–492.

Sedman, G., and Hopkinson, G. (1966): The psychopathology of mystical and religious conversion experiences in psychiatric patients. A phenomenological study. I and II. *Confin. Psychiatr.,* 9:1–19; 65–77.

Seevers, M. H., and Deneau, G. A. (1963): Physiological aspects of tolerance and physical dependence. In: *Physiological Pharmacology, Vol. 1,* edited by W. S. Root and F. G. Hofmann, pp. 565–640. Academic Press, New York.

Segal, D. S., Knapp, S., Kuczenski, R. T., and Mandell, A. J. (1973): The effects of environmental isolation on behavior and regional rat brain tyrosine hydroxylase and tryptophan hydroxylase activities. *Behav. Biol.,* 8:47–53.

Segawa, T., Bando, S., and Hosokawa, M. (1977): Brain serotonin metabolism and Δ^9-tetrahydrocannabinol induced muricide behavior in rats. *Jpn. J. Pharmacol.,* 27:581–582.

Seixas, F. A., Williams, K., and Eggleston, S. (eds.) (1975): Medical consequences of alcoholism. *Ann. N. Y. Acad. Sci.,* 252 pps.

Selmanoff, M. K., Jumonville, J. E., Maxson, S. C., and Ginsburg, B. E. (1975a): Evidence for a Y chromosomal contribution to an aggressive phenotype in inbred mice. *Nature,* 253:529–530.

Selmanoff, M. K., Jumonville, J. E., Maxson, S. C., and Ginsburg, B. E. (1975b): Reply. *Nature,* 255:658.

Sem-Jacobsen, C. W. (1964): Electrical stimulation of the human brain. *Electroencephalogr. Clin. Neurophysiol.,* 17:211.

Senault, B. (1970): Comportement d'agressivité intraspecifique induit par l'apomorphine chez le rat. *Psychopharmacologia,* 18:271–287.

Senault, B. (1971): Influence de l'isolemont sur le comportement d'agressivité intraspécifique induit par l'apomorphine chez le rat. *Psychopharmacologia,* 20:389–394.

Senault, B. (1972): Influence de la surrénalectomie, de l'hypophysectomie, de la thyroidectomie, de la castration ainsi que de la testostérone sur le comportement d'agressivité intraspécifique induit par l'apomorphine chez le rat. *Psychopharmacologia,* 24:476–484.

Senault, B. (1973): Effets de lésions du septum, de l'amygdale, du striatum de la substantia nigra et de l'ablation des bulbes olfactifs sur le comportement d'agressivité intraspécifique induit par l'apomorphine chez le rat. *Psychopharmacologia,* 28:13–25.

Senault, B. (1974): Amines cérébrales et comportement d'agressivité intraspécifique induit par l'apomorphine chez le rat. *Psychopharmacologia,* 34:143–154.

Senault, B. (1976): Comportement d'agressivité à l'apomorphine: Rélations avec quelques éléments du profil comportemental et avec la sensibilité à l'apomorphine. *Psychopharmacology,* 48:31–35.

Sendi, I. B., and Blomgren, P. G. (1975): A comparative study of predictive criteria in the predisposition of homicidal adolescents. *Am. J. Psychiatry,* 132:423–427.

Serafetinides, E. A. (1965): Aggressiveness in temporal lobe epileptics and its relation to cerebral dysfunction and environmental factors. *Epilepsia,* 6:33–46.

Seward, J. P. (1945): Aggressive behavior in the rat. II. An attempt to establish a dominance hierarchy. *J. Comp. Psychol.,* 38:213–238.

Seymour, R. S. (1976): Dinosaurs, endothermy and blood pressure. *Nature,* 262:207–208.

Shapiro, A. P. (1960): Psychophysiologic mechanisms in hypertensive vascular disease. *Ann. Intern. Med.,* 53:64–83.

Sharma, M., Meyer-Bahlburg, H. F. L., Boon, D. A., Slaunwhite, W. R., Jr., and Edwards, J. A. (1975): Testosterone production by XYY subjects. *Steroids,* 26:175–180.

Shear, C. S., Nyhan, W. L., Kirman, B. H., and Stern, J. (1971): Self-mutilative behavior as a feature of the de Lange syndrome. *J. Pediatr.,* 78:506–509.

Sheard, M. H. (1969): The effect p-chlorophenylalanine on behavior in rats: Relation to brain serotonin and 5-hydroxyindoleacetic acid. *Brain Res.,* 15:524–528.

Sheard, M. H., Astrachan, D. J., and Davis, M. (1977): The effect of D-lysergic acid diethylamide (LSD) upon shock-elicited fighting in rats. *Life Sci.,* 20:427–430.

Sheard, M. H., and Flynn, J. P. (1967): Facilitation of attack behavior by stimulation of the midbrain of cats. *Brain Res.,* 4:324–333.

Sheldon, W. H. (1942): *The Varieties of Temperament; A Psychology of Constitutional Differences.* Arper Publ., New York.

Shen, F. H., Loh, H. H., and Way, E. L. (1970): Brain serotonin turnover in morphine tolerant and dependent mice. *J. Pharmacol. Exp. Ther.,* 175:427–434.

Sherrington, C. S. (1906): *The Integrative Action of the Nervous System.* Yale University Press, New Haven.

Shevitz, S. A. (1976): Psychosurgery: Some current observations. *Am. J. Psychiatry,* 133:266–270.

Shipley, J. E., and Kolb, B. (1977): Neural correlates of species-typical behavior in the Syrian golden hamster. *J. Comp. Physiol. Psychol.,* 91:1056–1073.

Sholl, D. A. (1953): Dendritic organization of the neurons of the visual and motor cortices of the cat. *J. Anat.,* 87:387–406.

Shultz, R. L., Maynard, E. A., and Pease, D. C. (1957): Electron microscopy of neurons and neuroglia of cerebral cortex and corpus callosum. *Am. J. Anat.,* 100:369–408.

Shuntich, R. J., and Taylor, S. P. (1972): The effects of alcohol on human physical aggression. *J. Exp. Res. Pers.,* 6:34–38.

Shupe, L. M. (1954): Alcohol and crime: A study of the urine alcohol concentration found in 882 persons arrested during or immediately after the commission of a felony. *J. Criminal Law, Criminol. Police Sci.,* 44:661–664.

Siegel, A., and Chabora, J. (1971): Effects of electrical stimulation of the cingulate gyrus upon attack behavior elicited from the hypothalamus in the cat. *Brain Res.,* 32:169–177.

Siegel, A., Chabora, J., and Troiano, R. (1972): Effects of electrical stimulation of the pyriform cortex upon hypothalamically-elicited aggression in the cat. *Brain Res.,* 47:497–500.

Siegel, A., Edinger, H., and Dotto, M. (1975): Effects of electrical stimulation of the lateral aspect of the prefrontal cortex upon attack behavior in cats. *Brain Res.,* 93:473–484.

Siegel, A., Edinger, H., and Koo, A. (1977): Suppression of attack behavior in the cat by the prefrontal cortex: Role of the mediodorsal thalamic nucleus. *Brain Res.,* 127:185–190.

Siegel, A., and Flynn, J. P. (1968): Differential effects of electrical stimulation and lesions of the hippocampus and adjacent regions upon attack behavior in cats. *Brain Res.,* 7:252–267.

Siegel, D., and Leaf, R. C. (1969): Effects of septal and amygdaloid brain lesions in rats on mouse killing. Paper presented at the Meeting of the Eastern Psychological Association, Atlantic City, N.J.

Sigal, M. (1976): Psychiatric aspects of temporal lobe epilepsy. *J. Nerv. Ment. Dis.,* 163:348–351.

Sigg, E. B. (1969): Relationship of aggressive behaviour to adrenal and gonadal function in male mice. In: *Aggressive Behaviour,* edited by S. Garattini and E. B. Sigg, pp. 143–149. Excerpta Medica, Amsterdam.

Sigg, E. B., Caprio, G., and Schneider, J. A. (1958): Synergism of amines and antagonism of reserpine to morphine analgesia. *Proc. Soc. Exp. Biol. Med.,* 97:97–100.

Silver, L. B., Dublin, C. C., and Lourie, R. S. (1969): Does violence breed violence? Contributions from a study of the child abuse syndrome. *Am. J. Psychiatry,* 126:404–407.

Silverman, L. H. (1964): Ego disturbance in TAT stories as a function of aggression-arousing stimulus properties. *J. Nerv. Ment. Dis.,* 138:248–254.

Silverman, L. H. (1966): A study of the effects of subliminally presented aggressive stimuli on the production of pathologic thinking in a nonpsychiatric population. *J. Nerv. Ment. Dis.,* 141:443–455.

Silverman, L. H., and Spiro, R. H. (1967): Further investigation of the effects of subliminal aggressive stimulation on the ego functioning of schizophrenics. *J. Consult. Psychol.,* 31:225–232.

Silverman, L. H., and Spiro, R. H. (1968): The effects of subliminal, supraliminal and vocalized aggression on the ego functioning of schizophrenics. *J. Nerv. Ment. Dis.,* 146:50–61.

Silverman, L. H., Spiro, R. H., Weisberg, J. S., and Candell, P. (1969): The effects of aggressive activation and the need to merge on pathological thinking in schizophrenia. *J. Nerv. Ment. Dis.,* 148:39–51.

Simmel, M. L., and Counts, S. (1958): Clinical and psychological results of anterior temporal lobectomy in patients with psychomotor epilepsy. In: *Temporal Lobe Epilepsy,* edited by M. Baldwin and Bailey. Charles C Thomas, Springfield, Ill.

Simpson, G. G. (1953): *The Major Features of Evolution.* Columbia University Press, New York.

Singh, M. M., and Kay, S. R. (1976): Wheat gluten as a pathogenic factor in schizophrenia. *Science,* 191:401–402.

Sjöstrand, F. S. (1956): The ultrastructure of cells as revealed by the electron microscope. *Int. Rev. Cytol.,* 5:455–533.

Skinner, B. F. (1971): *Beyond Freedom and Dignity.* Alfred A. Knopf, New York.

Skultety, F. M., and Gary, T. M. (1962): Experimental hyperphagia in cats following destructive mid-brain lesions. *Neurology (Minneap.),* 12:394–401.

Slotnick, B. M., and McMullen, M. F. (1972): Intraspecific fighting in albino mice with septal forebrain lesions. *Physiol. Behav.,* 8:333–337.

Small, W. S. (1899): Notes on the psychic development of the young white rat. *Am. J. Psychol.,* 2:80–100.

Smith, A. D., and Winkler, H. (1972): Fundamental mechanisms in the release of catecholamines. In: *Catecholamines,* edited by H. Blaschko and E. Muscholl, pp. 538–617. Springer-Verlag, Berlin.

Smith, D. E., King, M. B., and Hoebel, B. G. (1970): Lateral hypothalamic control of killing: Evidence for a cholinoceptive mechanism. *Science,* 167:900–901.

Smith, M. O., and Holland, R. C. (1975): Effects of lesions of the nucleus accumbens on lactation and postpartum behavior. *Physiol. Psychol.,* 3:331–336.

Smith, O. A. (1956): Stimulation of lateral and medial hypothalamus and food intake in the rat. *Anat. Rec.,* 124:363–364.

Smith, R. (1969): The world of the Haight Ashbury speed freak. *J. Psychedel. Drugs,* 2:172–188.

Smith, R. L., Mosko, S., and Lynch, G. (1974): Role of various brain areas in recovery from partial cerebellar lesions in the adult rat. *Behav. Biol.,* 12:165–176.

Smythies, J. R. (1970): *Brain Mechanisms and Behavior: An Outline of the Mechanisms of Emotion, Memory, Learning and the Organization of Behavior, with Particular Regard to the Limbic System.* Academic Press, New York.

Snider, R. S., and Maiti, A. (1976): Cerebellar contributions to the papez circuit. *J. Neurosci. Res.,* 2:133–146.

Snider, St. R., and Snider, R. S. (1977): Alterations in forebrain catecholamine metabolism produced by cerebellar lesions in the rat. *J. Neural Transm.,* 40:115–128.

Snyder, D. R. (1970): Fall from social dominance following orbital frontal ablation in monkeys. *Proc. Annual Convention of the American Psychiatric Association,* pp. 235–236.

Snyder, S. H. (1973): Amphetamine psychosis: A "model" schizophrenia mediated by catecholamines. *Am. J. Psychiatry,* 130:61–67.

Snyder, S. H., and Meyerhoff, J. L. (1973): How amphetamine acts in minimal brain disfunction. *Ann. N.Y. Acad. Sci.,* 205:310–320.

Sodetz, F. J., Matalka, E. S., and Bunnell, B. N. (1967): Septal ablation and affective behavior in the golden hamster. *Psychonom. Sci.,* 7:189–190.

Sodetz, K., and Bunnell, B. N. (1970): Septal ablation and the social behavior of the golden hamster. *Physiol. Behav.,* 5:79–88.

Sofia, D. R., and Salama, A. I. (1970): Circadian rhythm for experimentally-induced aggressive behavior in mice. *Life Sci.,* 9:331–338.

Somers, A. R. (1976): Violence, television and the health of American youth. *N. Engl. J. Med.,* 294:811–817.

Sorensen, J. P., Jr., and Harvey, J. A. (1971): Decreased brain acetylcholine after septal lesions in rats: Correlation with thirst. *Physiol. Behav.,* 6:723–725.

Sorensen, C. A., and Gordon, M. (1975): Effects of 6-hydroxydopamine on shock-elicited aggression, emotionality, and maternal behavior in female rats. *Pharmacol. Biochem. Behav.,* 3:331–335.

Southern, H. H. (1948): Sexual and aggressive behaviour in the wild rabbit. *Behaviour,* 1:173–194.

Southwick, C. H. (1955): The population dynamics of confined house mice supplied with unlimited food. *Ecology,* 36:212–225.

Southwick, C. H., and Clark, L. H. (1968): Interstrain differences in aggressive behavior and exploratory activity in inbred mice. *Commun. Behav. Biol. [A],* 1:49–59.

Southwick, C., Siddiqi, M. F., Farooqui, M. Y., and Pal, B. C. (1974): Xenophobia among free-ranging rhesus groups in India. In: *Primate Aggression, Territoriality, and Xenophobia,* edited by R. L. Holloway, pp. 185–210. Academic Press, New York.

Spector, S. A., and Hull, E. M. (1972): Anosmia and mouse-killing by rats: A nonolfactory role for the olfactory bulbs. *J. Comp. Physiol. Psychol.,* 80:354–356.

Spencer, J., Gray, J., and Dalhouse, A. (1973): Social isolation in the gerbil: Its effect on exploratory or agonistic behavior and adrenocortical activity. *Physiol. Behav.,* 10:231–237.

Sperry, R. W. (1969): A modified concept of consciousness. *Psychol. Rev.,* 76:532–536.

Sperry, R. W., Gazzaniga, M. S., and Bogen, J. E. (1969): Interhemispheric relationships: The neocortical syndromes and hemisphere disconnection. In: *Handbook of Clinical Neurology, Vol. 4,* edited by P. J. Vinken, and G. W. Bruyn. North-Holland, Amsterdam.

Spevak, A. M., Quadagno, D. M., Knoeppel, D., and Poggio, J. P. (1973): The effects of isolation on sexual and social behavior in the rat. *Behav. Biol.,* 8:63–73.

Spiegel, E. A., Wycis, H. T., Freed, H., and Orchinik, C. (1951): The central mechanism of the emotions. *Am. J. Psychiatry,* 108:426–532.

Sprague, J. M., Chambers, W. W., and Stellar, E. (1961): Attentive, affective, and adaptive behavior in the cat. *Science,* 133:165–173.

Šramka, M., and Nádvornik, P. (1975): Surgical complication of posterior hypothalamotomy. *Confin. Neurol.,* 37:193–194.

Srebro, B., and Lorens, S. A. (1975): Behavioral effects of selective midbrain raphe lesions in the rat. *Brain Res.,* 89:303–325.

Stachnik, T. J., Ulrich, R. E., and Mabry, J. H. (1966a): Reinforcement of aggression through intracranial stimulation. *Psychonom. Sci.,* 5:101–102.

Stachnik, T. J., Ulrich, R., and Mabry, J. H. (1966b): Reinforcement of intra- and inter-species aggression with intracranial stimulation. *Am. Zool.,* 6:663–668.

Stark, P., Fazio, G., and Boyd, E. S. (1962): Monopolar and bipolar stimulation of the brain. *Am. J. Physiol.,* 203:371–373.

Stark, P., and Henderson, J. K. (1966): Increased reactivity in rats caused by septal lesions. *Int. J. Neuropharmacol.,* 5:379–383.

Staub, E. (1952): Instigation to goodness: The role of social norms and interpersonal influence. *J. Soc. Issues,* 47:463–474.

Steadman, H. J., Vanderwyst, D., and Ribner, S. (1978): Comparing arrest rates of mental patients and criminal offenders. *Am. J. Psychiatry,* 135:1218–1220.

Stein, L. (1968): Chemistry of reward and punishment. In: *Psychopharmacology, A Review of Progress 1957–1967,* edited by D. M. Efron, pp. 105–123. U.S.G.P.O., Washington, D.C.

Stein, L., and Wise, C. D. (1971): Possible etiology of schizophrenia: Progressive damage to the noradrenergic reward system by 6-hydroxydopamine. *Science,* 171:1032–1036.

Steinmetz, S. K., and Straus, M. A. (eds.) (1973): *Violence in the Family.* Dodd, Mead, New York.

Stellar, E., and Morgan, C. T. (1943): The roles of experience and deprivation on the onset of hoarding behavior in the rat. *J. Comp. Psychol.*, 36:47–55.

Stensiö, E. A. (1963): Anatomical studies on the arthrodiran head. 1. *Kungliga Svenska Vetenskapsakad. Handlingar Series 4*, 9:1–419.

Stensiö, E. A. (1968): The cyclostomes with special reference to the petronyzontida and myxinoidea. In: *Current Problems of Lower Vertebrate Phylogeny*, edited by T. Ørvig, pp. 13–71. John Wiley & Sons, New York.

Stephan, H., and Andy, O. J. (1969): Quantitative comparative neuroanatomy of primates: An attempt at a phylogenetic interpretation. *Ann. N.Y. Acad. Sci.*, 167:370–387.

Stern, D. N., Fieve, R. R., Neff, N. H., and Costa, E. (1969): The effect of lithium chloride administration on brain and heart norepinephrine turnover rates. *Psychopharmacologia*, 14:315–322.

Sternlicht, M., and Silver, E. F. (1965): The relationship between fantasy aggression and overt hostility in mental retardates. *Am. J. Ment. Defic.*, 70:486–488.

Steuer, F. B., Applefield, J. M., and Smith, R. (1971): Televised aggression and the interpersonal aggression of preschool children. *J. Exp. Child Psychol.*, 11:442–447.

Stevens, C. F. (1966): *Neurophysiology. A Primer*. John Wiley & Sons, New York.

Stevens, J. R. (1973): An anatomy of schizophrenia? *Arch. Gen. Psychiatry*, 29:177–189.

Stevens, J. R., Kim, C., and MacLean, P. D. (1961): Stimulation of caudate nucleus. *Arch. Neurol.*, 4:47–54.

Stevenson, J. A. F. (1969): Neural control of food and water intake. In: *The Hypothalamus*, edited by W. Haymaker, E. Anderson, and W. J. H. Nauta, Charles C. Thomas, Springfield, Ill.

Stewart, M. A., Cummings Adams, C., and Meardon, J. K. (1978): Unsocialized aggressive boys: A follow-up study. *J. Clin. Psychiatry*, 39:797–799.

Stewart, M. A., and Leone, L. (1978): A family study of unsocialized aggressive boys. *Biol. Psychiatry*, 13:107–117.

Stiglick, A., and White, N. (1977): Effects of lesions of various medial forebrain bundle components on lateral hypothalamic self-stimulation. *Brain Res.*, 133:45–63.

Still, G. F. (1902): Some abnormal psychical conditions in children. *Lancet*, 1:1163–1168.

Stille, G., Ackermann, H., Eichenberger, E., and Lauener, H. (1963): Vergleichende pharmakologische Untersuchung eines neuen zentralen Stimulans, 1-p-Tolyl-1-oxo-2-pyrrolidino-n-pentan-HC1. *Arzneim. Forsch.*, 13:871–877.

Stokes, A. W., and Cox, L. M. (1970): Aggressive man and aggressive beast. *Bioscience*, 20:1092–1095.

Stolk, J. M., Conner, R. L., Levine, S., and Barchas, J. D. (1974): Brain norepinephrine metabolism and shock-induced fighting behavior in rats: Differential effects of shock and fighting on the neurochemical response to a common footshock stimulus. *J. Pharmacol. Exp. Ther.*, 190:193–209.

Stolurow, L. M. (1948): Rodent behavior in the presence of barriers. I. Apparatus and methods. *J. Comp. Physiol. Psychol.*, 41:219–231.

Stone, C. P. (1937): A paper-window obstruction apparatus. *J. Genet. Psychol.*, 50:206–209.

Stone, C. P. (1942): Maturation and instinctive functions and motivations. In: *Comparative Psychology, Rev. Ed.*, edited by F. A. Moss, pp. 32–97. Prentice-Hall, New York.

Stone, W. E. (1969): Action of convulsants: Neurochemical aspects. In: *Basic Mechanisms of the Epilepsies*, edited by H. H. Jasper, A. A. Ward, Jr., and A. Pope, pp. 184–193. Churchill, London.

Story, I. (1976): Caricature and impersonating the other: Observations from the psychotherapy of anorexia nervosa. *Psychiatry*, 39:176–188.

Straus, M. A. (1973): A general systems theory approach to a theory of violence between family members. *Soc. Sci. Inform.*, 12:105–125.

Strauss, E. B., Sands, D. E., Robinson, A. M., Tindall, W. J., and Stevenson, W. A. H. (1952): Use of dehydroisoandrosterone in psychiatric treatment. *Br. Med. J.*, 2:64–66.

Strauss, I., and Kreschner, M. (1935): Mental symptoms in cases of tumor of the frontal lobe. *Arch. Neurol. Psychiatry*, 33:986–1005.

Suchowsky, G. K., Pegrassi, L., and Bonsignori, A. (1969): The effect of steroids on aggressive behaviour in isolated male mice. In: *Aggressive Behaviour*, edited by S. Garattini and E. G. Sigg, pp. 164–171. Excerpta Medica, Amsterdam.

Suchowsky, G. K., Pegrassi, L., and Bonsignori, A. (1971): Steroids and aggressive behaviour in isolated male and female mice. *Psychopharmacologia*, 21:32–38.

Sudak, H. S., and Maas, J. W. (1964a): Central nervous system serotonin and norepinephrine localization in emotional and nonemotional strains in mice. *Nature*, 203:1254–1256.

Sudak, H. S., and Maas, J. W. (1964b): Behavioral-neurochemical correlation in reactive and non-reactive strains of rats. *Science*, 146:418–420.

Suedfeld, P. (1975): The clinical relevance of reduced sensory stimulation. *Can. Psychol. Rev.*, 16:88–103.

Summers, T. B., and Kaelber, W. W. (1962): Amygdalectomy: Effects in cats and a survey of its present status. *Am. J. Physiol.*, 203:1117–1119.

Suomi, S. J., Harlow, H. F., and Kimball, S. D. (1971): Behavioral effects of prolonged partial social isolation in the rhesus monkey. *Psychol. Rep.*, 29:1171–1177.

Suomi, S. J., Harlow, H. F., and Lewis, J. K. (1970): Effect of bilateral frontal lobectomy on social preferences of rhesus monkeys. *J. Comp. Physiol. Psychol.*, 70:448–453.

Šváb, L., Gross, J., and Langová, J. (1972): Stuttering and social isolation. *J. Nerv. Ment. Dis.*, 155:1–5.

Svare, B. B. (1977): Maternal aggression in mice: Influence of the young. *Biobehav. Rev.*, 1:151–164.

Svare, B., and Gandelman, R. (1973): Postpartum aggression in mice: Experiential and environmental factors. *Horm. Behav.*, 4:323–334.

Svare, B., and Gandelman, R. (1974): Stimulus control of aggressive behavior in androgenized female mice. *Behav. Biol.*, 10:447–459.

Svare, B., and Gandelman, R. (1975): Postpartum aggression in mice: Inhibitory effect of estrogen. *Physiol. Behav.*, 14:31–35.

Svare, B., and Gandelman, R. (1976a): A longitudinal analysis of maternal aggression in Rockland-Swiss albino mice. *Dev. Psychobiol.*, 9:437–446.

Svare, B., and Gandelman, R. (1976b): Suckling stimulation induces aggression in virgin female mice. *Nature*, 260:606–608.

Svare, B. B., and Leshner, A. I. (1973): Behavioral correlates of intermale aggression and grouping in mice. *J. Comp. Physiol. Psychol.*, 85:203–210.

Sweet, W. H., Ervin, F., and Mark, V. H. (1969): The relationship of violent behaviour to focal cerebral disease. In: *Aggressive Behaviour*, edited by S. Garattini and E. B. Sigg, pp. 336–352. Excerpta Medica, Amsterdam.

Sylvester-Bradley, P. C. (1976): Evolutionary oscillation in prebiology: Igneous activity and the origins of life. *Orig. Life*, 7:9–18.

Szasz, T. S. (1968): *Law, Liberty and Psychiatry*. Collier Books, New York.

Tagliamonte, A., Biggio, G., Vargiu, L., and Gessa, G. L. (1973): Free tryptophan in serum controls brain tryptophan level and serotonin synthesis. *Life Sci.*, 12:277–287.

Tagliamonte, A., Gessa, R., Biggio, G., Vargiu, L., and Gessa, G. L. (1974): Daily changes of free serum tryptophan in humans. *Life Sci.*, 14:349–354.

Tagliamonte, A., Tagliamonte, P., Perez-Cruet, J., and Gessa, G. L. (1971a): Increase of brain tryptophan caused by drugs which stimulate serotonin synthesis. *Nature [New Biol.]*, 229:125–126.

Tagliamonte, A., Tagliamonte, P., Perez-Cruet, J., Stern, S., and Gessa, G. L. (1971b): Effect of psychotropic drugs on tryptophan concentration in the rat brain. *J. Pharmacol. Exp. Ther.*, 177:475–480.

Takemoto, T.-I., Suzuki, T., and Miyama, T. (1975): An unexpected incidence of convulsive attack in male mice after long-term isolated condition. *Tohoku J. Exp. Med.*, 115:97–98.

Talbott, J. A., and Teague, J. W. (1969): Marihuana psychosis: Acute toxic psychosis associated with the use of cannabis derivatives. *J.A.M.A.*, 210:299–302.

Talkington, L. W., and Altman, R. (1973): Effects of film-mediated aggressive and affectual models on behavior. *Am. J. Ment. Defic.*, 77:420–425.

Talkington, L. W., Hall, S., and Altman, R. (1971): Communication deficits and aggression in the mentally retarded. *Am. J. Ment. Defic.*, 76:235–237.

Tallan, H. H. (1957): Studies on the distribution of N-acetyl-l-aspartic acid in brain. *J. Biol. Chem.*, 224:41–45.

Tannenbaum, G. A., Paxinos, G., and Bindra, D. (1974): Metabolic and endocrine aspects of the ventromedial hypothalamic syndrome in the rat. *J. Comp. Physiol. Psychol.*, 86:401–413.

Tarchalska, B., Kostowski, W., Markowska, L., and Markiewicz, L. (1975): On the role of serotonin in aggressive behaviour of ants genus *formica*. *Pol. J. Pharmacol. Pharm.*, 27 suppl.:237–239.

Tarr, R. S. (1977): Role of the amygdala in the intraspecies aggressive behavior of the Ignanid lizard, *Sceloporus occidentalis. Physiol. Behav.,* 18:1153–1158.

Tata, J. R. (1966): Hormones and the synthesis and utilization of ribonucleic acids. *Prog. Nucleic Acid Res.,* 5:191–250.

Taylor, D. C. (1969): Sexual behavior and temporal lobe epilepsy. *Arch. Neurol.,* 21:510–516.

Taylor, L. R., and Costanzo, D. J. (1975): Social dominance, adrenal weight, and the reticuloendothelial system in rats. *Behav. Biol.,* 13:167–174.

Taylor, S. P., and Gammon, C. B. (1975): Effects of type and dose of alcohol on human physical aggression. *J. Pers. Soc. Psychol.,* 32:169–175.

Taylor, S. P., and Pisano, R. (1971): Physical aggression as a function of frustration and physical attack. *J. Soc. Psychol.,* 84:261–267.

Taylor, S. P., Vardaris, R. M., Rawtich, A. B., Gammon, C. B., Cranston, J. W., and Lubetkin, A. I. (1976): The effects of alcohol and Δ^9-tetrahydrocannabinol on human physical aggression. *Aggressive Behav.,* 2:153–161.

Tedeschi, D. H., Fowler, P. J., Miller, R. B., and Macko, E. (1969): Pharmacological analysis of footshock-induced fighting behaviour. In: *Aggressive Behaviour,* edited by S. Garattini and E. D. Sigg. pp. 245–252. Excerpta Medica, Amsterdam.

Tedeschi, J. T., Smith, R. B., III, and Brown, R. C., Jr. (1974): A reinterpretation of research on aggression. *Psychol. Bull.,* 81:540–562.

Templer, D. I. (1974): The efficacy of psychosurgery. *Biol. Psychiatry,* 9:205–209.

Tenen, S. S. (1967): The effects of p-chlorophenylalanine, a serotonin depletor, on avoidance acquisition, pain sensitivity and related behavior in the rat. *Psychopharmacologia,* 10:204–219.

Tennant, F. S., Jr., and Groesbeck, C. J. (1972): Psychiatric effects of hashish. *Arch. Gen. Psychiatry,* 27:133–136.

Terada, C. W., and Masur, J. (1973): Amphetamine- and apomorphine-induced alteration of the behavior of rats submitted to a competitive situation in a straight runway. *Eur. J. Pharmacol.,* 24:375–380.

Terzian, H. (1958): Observations on the clinical symptomatology of bilateral partial or total removal of the temporal lobes in man. In: *Temporal Lobe Epilepsy,* edited by M. Baldwin, pp. 510–529. Charles C Thomas, Springfield, Ill.

Terzian, H., and Dalle Ore, G. (1955): Syndrome of Klüver and Bucy reproduced in man by bilateral removal of temporal lobes. *Neurology (Minneap.),* 5:373–380.

The National Commission for the Protection of Human Subjects of Biomedical and Behavioral Research (1977): *Report and Recommendations.* Psychosurgery DHEW Publ. 77-0001, U.S.G.P.O., Washington, D.C.

Thierry, A.-M., Javoy, F., Glowinski, J., and Kety, S. S. (1968): Effects of stress on the matabolism of norepinephrine, dopamine and serotonin in the central nervous system of the rat. I. Modifications of norepinephrine turnover. *J. Pharmacol. Exp. Ther.,* 163:163–171.

Thiessen, D. D., Friend, H. C., and Lindzey, G. (1968): Androgen control of territorial marking in the mongolian gerbil. *Science,* 160:432–434.

Thiessen, D. D., Lindzey, G., and Nyby, J. (1970): The effects of olfactory deprivation and hormones on territorial marking in the male mongolian gerbil. *Horm. Behav.,* 1:315–325.

Thiessen, D. D., Yahr, P. I., and Owen, K. (1973): Regulatory mechanisms of territorial marking in the mongolian gerbil. *J. Comp. Physiol. Psychol.,* 82:382–393.

Thoa, N. B., Eichelman, B., and Ng, L. K. Y. (1972a): Shock-induced aggression: Effects of 6-hydroxydopamine and other pharmacological agents. *Brain Res.,* 43:467–475.

Thoa, N. B., Eichelman, B., and Ng, L. K. Y. (1972b): Aggression in rats treated with dopa and 6-hydroxydopamine. *J. Pharm. Pharmacol.,* 24:337–338.

Thoa, N. B., Eichelman, B., Richardson, J. S., and Jacobowitz, D. (1972c): 6-Hydroxydopa depletion of brain norepinephrine and the facilitation of aggressive behavior. *Science,* 178:75–77.

Thomas, E. M. (1959): *The Harmless People.* Alfred A. Knopf, New York.

Thomas, G. S., Caccamise, D. J., and Clark, D. L. (1978): Aggression increase and water competition decrease in squirrel monkeys given physostigmine injections. *Pharmacol. Biochem. Behav.,* 8:633–639.

Thompson, M. E., and Thorne, B. M. (1975): The effects of colony differences and olfactory bulb lesions on muricide in rats. *Physiol. Psychol.,* 3:285–289.

Thompson, R. F. (1975): *Introduction to Physiological Psychology.* Harper & Row, New York.

Thompson, T., and Bloom, W. (1966): Aggressive behavior and extinction-induced response-rate increase. *Psychonom. Sci.,* 5:335–336.

Thor, D. H., and Flannelly, K. J. (1976a): Age of intruder and territorial-elicited aggression in male Long-Evans rats. *Behav. Biol.,* 17:237–241.

Thor, D. H., and Flannelly, K. J. (1976b): Intruder gonadectomy and elicitation of territorial aggression in the rat. *Physiol. Behav.,* 17:725–727.

Thor, D. H., and Ghiselli, W. B. (1973a): Prolonged suppression of irritable aggression in rats by facial anesthesia. *Psychol. Rep.,* 33:815–820.

Thor, D. H., and Ghiselli, W. B. (1973b): Suppression of shock-elicited aggression in rats by facial anesthesia. *Proceedings of the 81st Annual Convention of the American Psychiatric Association,* pp. 1025–1026.

Thor, D. H., and Ghiselli, W. B. (1974): Visual and social determinants of shock-elicited aggressive responding in rats. *Anim. Learn, Behav.,* 2:74–76.

Thor, D. H., and Ghiselli, W. B. (1975a): Vibrissal anesthesia and suppression of irritable fighting in rats: A temporary duration of effect in experienced fighters. *Physiol. Psychol.,* 3:1–3.

Thor, D. H., and Ghiselli, W. B. (1975b): Suppression of mouse killing and apomorphine-induced social aggression in rats by local anesthesia of the mystacial vibrissae. *J. Comp. Physiol. Psychol.,* 88:40–46.

Thor, D. H., Ghiselli, W. B., and Lambelet, D. C. (1974): Sensory control of shock-elicited fighting in rats. *Physiol. Behav.,* 13:683–686.

Thor, D. H., Hoats, D. L., and Thor, C. J. (1970): Morphine-induced fighting and prior social experience. *Psychonom. Sci.,* 18:137–139.

Thorne, B. M., Aaron, M., and Latham, E. E. (1973): Effects of olfactory bulb ablation upon emotionality and muricidal behavior in four rat strains. *J. Comp. Physiol. Psychol.,* 84:339–344.

Thorne, B. M., Aaron, M., and Latham, E. E. (1974): Olfactory system damage in rats and emotional, muricidal, and rat pup killing behavior. *Physiol. Psychol.,* 2:157–163.

Thorne, B. M., and Thompson, M. E. (1976): The effect of different types of mice upon muricidal behavior in the Long-Evans rat. *Physiol. Psychol.,* 4:238–246.

Thorne, B. M., Wallace, T., and Danzig, I. (1978): A comparison of killer and nonkiller rats. *Physiol. Psychol.,* 6:43–47.

Thornton, W. E., and Pray, B. J. (1975): The portrait of a murderer. *Dis. Nerv. Syst.,* 36:176–178.

Thurmond, J. B. (1975): Technique for producing and measuring territorial aggression using laboratory mice. *Physiol. Behav.,* 14:879–881.

Thurmond, J. B., Lasley, S. M., Conkin, A. L., and Brown, J. W. (1977): Effects of dietary tyrosine, phenylalanine, and tryptophan on aggression in mice. *Pharmacol. Biochem. Behav.,* 6:475–478.

Tilly, C. (1969): Collective violence in European perspective. In: *Violence in America: Historical and Comparative Perspectives,* edited by H. D. Graham and T. R. Gurr, pp. 5–34. U.S.G.P.O., Washington, D.C.

Tilson, H. A., Rech, R. H., and Stolman, S. (1973): Hyperalgesia during withdrawal as a means of measuring the degree of dependence in morphine dependent rats. *Psychopharmacologia,* 28:287–300.

Tinbergen, N. (1953): *Social Behavior in Animals,* John Wiley & Sons, New York.

✓Tinbergen, N. (1973): On war and peace in animals and man. In: *Readings in Animal Behavior, 2nd Ed.,* edited by T. E. McGill, 2nd. ed., pp. 453–468. Holt, Rinehart and Winston, London.

Tinklenberg, J. R. (1973): Alcohol and violence. In: *Alcoholism: Progress in Research and Treatment,* edited by P. G. Bourne and R. Fox, pp. 195–210. Academic Press, New York.

Tinklenberg, J. R., and Murphy, D. (1972): Marihuana and crime: A survey report. *J. Psychedelic Drugs,* 5:183–191.

Tinklenberg, J. R., Murphy, P. L., Murphy, P., Darley, C. F., Roth, W. T., and Kopell, B. S. (1974): Drug involvement in criminal assaults by adolescents. *Arch. Gen. Psychiatry,* 30:685–689.

Tinklenberg, J. R., and Woodrow, K. M. (1974): Drug use among youthful assaultive and sexual offenders. *Aggression, Res. Publ. Assoc. Res. Nerv. Ment. Dis.,* 52:209–224.

Tondat, L. M. (1974): Is the effect of preshock treatment on shock-elicited aggression independent of situational stimuli? *Psychol. Rec.,* 24:409–417.

Tondat, L. M., and Daly, H. B. (1972): The combined effects of frustrative nonreward and shock on aggression between rats. *Psychonom. Sci.,* 28:25–28.

Tonini, G., Riccioni, M. L., Babbini, M., and Missere, G. (1963): Evaluation of central pharmacological actions in rats with septal lesions. In: *Psychopharmacological Methods,* edited by Z. Votava, M. Horvath, and O. Vinar, pp. 106–114. Pergamon Press, Oxford.

Tow, P. M., and Whitty, C. W. M. (1953): Personality changes after operations on the cingulate gyrus in man. *J. Neurol. Neurosurg. Psychiatry,* 16:186–193.

Trachtenberg, M. C., and Pollen, D. A. (1970): Neuroglia: Biophysical properties and physiologic function. *Science,* 167:1248–1252.

Tryon, R. C., Tryon, C. M., and Kuznets, G. (1941*a*): Studies in individual differences in maze ability. IX. Ratings of hiding, avoidance, escape and vocalization responses. *J. Comp. Psychol.,* 32:407–435.

Tryon, R. C., Tryon, C. M., and Kuznets, G. (1941*b*): Studies in individual differences in maze ability. X. Ratings and other measures of initial emotional responses of rats to novel inanimate objects. *J. Comp. Psychol.,* 32:447–473.

Tsuang, M. T. (1975): Genetics of affective disorder. In: *The Psychobiology of Depression,* edited by J. Mendels, pp. 85–100. Spectrum, New York.

Tsuang, M. T. (1977): Genetic factors in suicide. *Dis. Nerv. Syst.,* 38:498–501.

Tucker, I. F. (1970): *Adjustment, Models and Mechanisms.* Academic Press, New York.

Tulunay, F. C., Sparber, S. B., and Takemori, A. E. (1975): The effect of dopaminergic stimulation and blockade on the nociceptive and antinociceptive responses of mice. *Eur. J. Pharmacol.,* 33:65–70.

Tulunay, F. C., and Takemori, A. E. (1974): Dopaminergic system and analgesia. *Pharmacologist,* 16:248.

Tupin, J. P., Mahar, D., and Smith, D. (1973): Two types of violent offenders with psychosocial descriptors. *Dis. Nerv. Syst.* 34:356–363.

Turnbull, F. (1969): Neurosurgery in the control of unmanageable affective reactions: A critical review. *Clin. Neurosurg.,* 16:218–233.

Turner, B. H. (1970): Neural structures involved in the rage syndrome of the rat. *J. Comp. Physiol. Psychol.,* 71:103–113.

Turner, B. N., and Iverson, S. L. (1973): The annual cycle of aggression in male *Microtus pennsylvanicus* and its relation to population parameters. *Ecology,* 54:967–981.

Turner, C. H., Davenport, R. K., Jr., and Rogers, C. M. (1969): The effect of early deprivation on the social behavior of adolescent chimpanzees. *Am. J. Psychiatry,* 125:1531–1536.

Turner, J. L., Boice, R., and Powers, P. C. (1973): Behavioral components of shock-induced aggression in ground squirrels *(Citellus tridecemlineatus). Anim. Learn. Behav.,* 1:254–262.

Turner, J. W., Jr. (1975): Influence of neonatal androgen on the display of territorial marking behavior in the gerbil. *Physiol. Behav.,* 15:265–270.

Tytell, M., and Myers, R. D. (1973): Metabolism of [^{14}C]-serotonin in the caudate nucleus, hypothalamus and reticular formation of the rat after ethanol administration. *Biochem. Pharmacol.,* 22:361–372.

Ueki, S., Fujiwara, M., and Ogawa, N. (1972): Mouse-killing behavior (muricide) induced by Δ^9-tetrahydrocannabinol in the rat. *Physiol. Behav.,* 9:585–587.

Ulehla, Z. J., and Adams, D. K. (1973): Detection theory and expectations for social reinforcers: An application to aggression. *Psychol. Rev.,* 80:439–445.

Ulrich, R. (1966): Pain as a cause of aggression. *Am. Zool.,* 6:643–662.

Ulrich, R. E., and Azrin, N. H. (1962): Reflexive fighting in response to aversive stimulation. *J. Exp. Anal. Behav.,* 5:511–520.

Ulrich, R., Wolfe, M., and Dulaney, S. (1969): Punishment of shock-induced aggression. *J. Exp. Anal. Behav.,* 12:1009–1015.

Ulrich, R. E., Wolff, P. C., and Azrin, N. H. (1964): Shock as an elicitor of intra- and inter-species fighting behaviour. *Anim. Behav.,* 12:14–15.

Ungerlieder, J. T., Fisher, D. D., and Fuller, M. (1966): The danger of LSD. *J.A.M.A.,* 197:389–392.

Ungerleider, J. T., Fisher, D. D., Fuller, M., and Caldwell, A. (1968*a*): The "bad trip." The etiology of the adverse LSD reaction. *Am. J. Psychiatry,* 124:1483–1490.

Ungerleider, J. T., Fisher, D. D., Goldsmith, S. R., Fuller, M., and Forgy, E. (1968*b*): A statistical survey of adverse reactions to LSD in Los Angeles county. *Am. J. Psychiatry,* 125:352–357.

Ungerstedt, U. (1971): Stereotaxic mapping of the monoamine pathways in the rat brain. *Acta Phsyiol. Scand.,* 82, suppl. 367:1–48.

Uretsky, N. J., and Iversen, L. L. (1970): Effects of 6-hydroxydopamine on catecholamine containing neurones in the rat brain. *J. Neurochem.,* 17:269–278.

Ursin, H. (1960): The temporal lobe substrate of fear and anger. A review of recent stimulation and ablation studies in animals and humans. *Acta Psychiatr. Neurol. Scand.,* 35:378–396.

Ursin, H. (1965): The effect of amygdaloid lesions on flight and defense behavior in cats. *Exp. Neurol.*, 11:61–79.

Ursin, H., and Kaada, B. R. (1960): Functional localization within the amygdaloid complex in the cat. *Electroencephalogr. Clin. Neurophysiol.*, 12:1–20.

U.S. Riot Commission (1968): *Report of the National Advisory Commission on Civil Disorders.* Bantam Books, New York.

Uyeno, E. T. (1966a): Effects of d-lysergic acid diethylamide and 2-brom-lysergic acid diethylamide on dominance behavior of the rat. *Int. J. Neuropharmacol.*, 5:317–322.

Uyeno, E. T. (1966b): Inhibition of isolation-induced attack behavior of mice by drugs. *J. Pharm. Sci.*, 55:215–216.

Uyeno, E. T. (1978): Effect of psychodysleptics on aggressive behavior of animals. *Mod. Probl. Pharmacopsychiatry*, 13:103–113.

Uyeno, E. T., and Benson, W. M. (1965): Effects of lysergic acid diethylamide and attack behavior of male albino mice. *Psychopharmacologia*, 7:20–26.

Vaernet, K., and Madsen, A. (1970): Stereotaxic amygdalotomy and basofrontal tractotomy in psychotics with aggressive behaviour. *J. Neurol. Neurosurg. Psychiatry*, 33:858–863.

Vale, J. R., Ray, D., and Vale, C. A. (1972): The interaction of genotype and exogenous neonatal androgen: Agonistic behavior in female mice. *Behav. Biol.*, 7:321–334.

Vale, J. R., Ray, D., and Vale, C. A. (1973): Interaction of genotype and exogenous neonatal estrogen: Aggression in female mice. *Physiol. Behav.*, 10:181–183.

Vale, J. R., Vale, C. A., and Harley, J. P. (1971): Interaction of genotype and population number with regard to aggressive behavior, social grooming and adrenal and gonadal weight in male mice. *Commun. Behav. Biol.*, 6:209–221.

Valenstein, E. (1974): *Brain Control.* Wiley-Interscience, New York.

Valle, F. P. (1970): Effects of strain, sex, and illumination on open-field behavior of rats. *Am. J. Psychol.*, 83:103–111.

Valverde, F. (1965): *Studies of Piriform Lobe.* Harvard University Press, Cambridge.

Valzelli, L. (1967a): Biological and pharmacological aspects of aggressiveness in mice. In: *Neuropsychopharmacology*, edited by H. Brill, J. O. Cole, P. Deniker, H. Hippius, and P. B. Bradley, pp. 781–788. Excerpta Medica, Amsterdam.

Valzelli, L. (1967b): Drugs and aggressiveness. *Adv. Pharmacol.*, 5:79–108.

Valzelli, L. (1969): The exploratory behaviour in normal and aggressive mice. *Psychopharmacologia*, 15:232–235.

Valzelli, L. (1970): Variazioni biochimiche cerebrali nel ratto muricida. In: *Atti della II Riunione Nazionale della Società Italiana di Neuropsicofarmacologia*, Tirrenia, Pisa, June, 1969, pp. 20–25. Pacini-Mariotti, Pisa.

Valzelli, L. (1971a): Agressivité chez le rat et la souris: aspects comportementaux et biochimiques. *Actual. Pharmacol. (Paris)*, 24:133–152.

Valzelli, L. (1971b): Further aspects of the exploratory behaviour in aggressive mice. *Psychopharmacologia*, 19:91–94.

Valzelli, L. (1973a): The "isolation syndrome" in mice. *Psychopharmacologia*, 31:305–320.

Valzelli, L. (1973b): Environmental influences upon neurometabolic processes in learning and memory. In: *Current Biochemical Approaches to Learning and Memory*, edited by W. B. Essman and S. Nakajima, pp. 29–47. Spectrum, New York.

Valzelli, L. (1975): Pharmacology and biochemistry of anxiety and aggressiveness: Introductory remarks. In: *Neuropsychopharmacology*, edited by J. R. Boissier, H. Hippius, and P. Pichot, pp. 691–693. Excerpta Medica, Amsterdam.

Valzelli, L. (1977a): Social experience as a determinant of normal behavior and drug effect. In: *Handbook of Psychopharmacology, Vol. 7*, edited by L. L. Iversen, S. D. Iversen, and S. H. Snyder, pp. 369–392. Plenum Press, New York.

Valzelli, L. (1977b): About a "specific" neurochemistry of aggressive behavior. In: *Behavioral Neurochemistry*, edited by J. M. R. Delgado and F. V. De Feudis, pp. 113–132. Spectrum, New York.

Valzelli, L. (1977c): The behavioral and neurochemical correlates of central stimulants. In: *Therapy and Psychosomatic Medicine, Pharmacotherapeutic Tribune*, edited by F. Antonelli, pp. 639–651. L. Pozzi, Rome.

Valzelli, L. (1978a): Affective behavior and serotonin. In: *Serotonin in Health and Disease, Vol. 3, The Central Nervous System*, edited by W. B. Essman, pp. 145–201. Spectrum, New York.

Valzelli, L. (1978b): Clinical pharmacology of serotonin. In: *Serotonin in Health and Disease, Vol.*

4, Physiological Regulation and Pharmacological Action, edited by W. B. Essman, pp. 295–339. Spectrum, New York.

Valzelli, L. (ed.) (1978c): Psychopharmacology of Aggression. *Mod. Probl. Pharmacopsychiatry,* Vol. 13, Karger, Basel.

Valzelli, L. (1978d): Human and animal studies on the neurophysiology of aggression. *Prog. Neuropsychopharmacol.,* 2:591–611.

Valzelli, L. (1979a): *An Approach to Neuroanatomical and Neurochemical Psychophysiology.* Granata Ed., Geneva.

Valzelli, L. (1979b): Effect of sedatives and anxiolytics on aggressivity. *Mod. Probl. Pharmacopsychiatry,* 14:143–156.

Valzelli, L., and Bernasconi, S. (1971): Differential activity of some psychotropic drugs as a function of emotional level in animals. *Psychopharmacologia,* 20:91–96.

Valzelli, L., and Bernasconi, S. (1973): Behavioral and neurochemical aspects of caffeine in normal and aggressive mice. *Pharmacol. Biochem. Behav.,* 1:251–254.

Valzelli, L., and Bernasconi, S. (1976): Psychoactive drug effect on behavioural changes induced by prolonged socio-environmental deprivation in rats. *Psychol. Med.,* 6:271–276.

Valzelli, L., and Bernasconi, S. (1978): Alcohol, prolonged isolation and barbiturate sedation in two strains of mice. *Neuropsychobiology,* 4:86–92.

Valzelli, L., and Bernasconi, S. (1979): Aggressiveness by isolation and brain serotonin turnover changes in different strains of mice. *Neuropsychobiology,* 5:129–135.

Valzelli, L., Bernasconi, S., Coen, E., and Petkov, V. V. (1979): Effect of different psychoactive drugs on serum and brain tryptophan levels. *Neuropsychobiology,* 6:224–229.

Valzelli, L., Bernasconi, S., and Cusumano, G. (1977): Annual and daily changes in brain serotonin content in differentially housed mice. *Neuropsychobiology,* 3:35–41.

Valzelli, L., Bernasconi, S., and Gomba, P. (1974): Effect of isolation on some behavioral aspects of three strains of mice. *Biol. Psychiatry,* 9:329–334.

Valzelli, L., and Garattini, S. (1968a): Biogenic amines in discrete brain areas after treatment with monoamineoxidase inhibitors. *J. Neurochem.,* 15:259–261.

Valzelli, L., and Garattini, S. (1968b): Behavioral changes and 5-hydroxytryptamine turnover in animals. *Adv. Pharmacol.,* 6B:249–260.

Valzelli, L., and Garattini, S. (1972): Biochemical and behavioural changes induced by isolation in rats. *Neuropharmacology,* 11:17–22.

Valzelli, L., Giacalone, E., and Garattini, S. (1967): Pharmacological control of aggressive behavior in mice. *Eur. J. Pharmacol.,* 2:144–146.

Valzelli, L., and Pawłowski, L. (1979): Effect of p-chlorophenylalanine (PCPA) on avoidance learning of two differentially housed mouse strains. *Neuropsychobiology,* 5:121–128.

Valzelli, L., and Sarteschi, P. (1977): *Considerazioni in Tema di Schizofrenia.* Edizioni Medico Scientifiche, Turin.

Van der Kooy, D., and Phillips, A. G. (1977): Trigeminal substrates of intracranial self-stimulation in the brainstem. *Science,* 196:447–449.

VanderWende, C., and Spoerlein, M. T. (1972): Antagonism by dopa of morphine analgesia. A hypothesis for morphine tolerance. *Res. Commun. Chem. Pathol. Pharmacol.,* 3:37–45.

VanderWende, C., and Spoerlein, M. T. (1973): Role of dopaminergic receptors in morphine analgesia and tolerance. *Res. Commun. Chem. Pathol. Pharmacol.,* 5:35–43.

Vanderwolf, C. H., Kolb, B., and Cooley, R. K. (1978): Behavior of the rat after removal of the neocortex and hippocampal formation. *J. Comp. Physiol. Psychol.,* 92:156–175.

Van Hemel, P. E. (1972): Aggression as a reinforcer: Operant behavior in the mouse-killing rat. *J. Exp. Anal. Behav.,* 17:237–245.

Van Hemel, P. E., and Colucci, V. M. (1973): Effects of target movement on mouse-killing attack by rats. *J. Comp. Physiol. Psychol.,* 85:105–110.

Van Hemel, P., and Myer, J. S. (1970): Satiation of mouse killing by rats in an operant situation. *Psychonom. Sci.,* 21:129–130.

Van Hoesen, G. W., Pandya, D. N., and Butters, N. (1972): Cortical afferents to the entorhinal cortex of the rhesus monkey. *Science,* 175:1471–1473.

Van Kreveld, D. (1970): A selective review of dominance-subordination relations in animals. *Genet. Psychol. Monogr.,* 81:143–173.

Van Reeth, P. C., Dierkens, J., and Luminet, D. (1958): L'hypersexualité dan l'épilepsie et les tumeurs du lobe temporal. *Acta Neurol. Belg.,* 58:194–218.

Van Wazer, J. R. (1961): *Phosphorus and Its Compounds*, p. 961. Interscience Publ., New York.

Vergnes, M. (1975): Déclenchement de réactions d'agression interspécifique après lésion amygdalienne chez le rat. *Physiol. Behav.*, 14:271–276.

Vergnes, M., Boehrer, A., and Karli, P. (1974a): Interspecific aggressiveness and reactivity in mouse-killing and nonkilling rats: Compared effects of olfactory bulb removal and raphe lesions. *Aggressive Behav.*, 1:1–16.

Vergnes, M., and Karli, P. (1963a): Effets de lésions expérimentales du néocortex frontal et du noyau caudé sur l'agressivité interspécifique rat-souris. *C. R. Soc. Biol. [D] (Paris)*, 157:176–178.

Vergnes, M., and Karli, P. (1963b): Déclenchement du comportement d'agression interspécifique rat-souris par ablation bilaterale des bulbes olfactifs. Action de l'hydroxyzine sur cette agressivité provoquée. *C. R. Soc. Biol. [D] (Paris)*, 157:1061–1063.

Vergnes, M., and Karli, P. (1964): Etude des voies nerveuses de l'influence facilitatrice exercée par les noyaux amygdaliens sur le comportement d'agression interspécifique rat-souris. *C. R. Soc. Biol. [D] (Paris)*, 158:856–858.

Vergnes, M., and Karli, P. (1965): Etude des voies nerveuses d'une influence inhibitrice s'exerçant sur l'agressivité interspécifique du rat. *C. R. Soc. Biol. [D] (Paris)*, 159:972–975.

Vergnes, M., and Karli, P. (1968): Activité électrique de l'hippocampe et comportement d'agression interspécifique rat-souris. *C. R. Soc. Biol. [D] (Paris)*, 162:555–558.

Vergnes, M., and Karli, P. (1969a): Effets de l'ablation des bulbes olfactifs et de l'isolement sur le développement de l'agressivité interspécifique du rat. *C. R. Soc. Biol. [D] (Paris)*, 163:2704–2706.

Vergnes, M., and Karli, P. (1969b): Effets de la stimulation de l'hypothalamus latéral, de l'amygdale et de l'hippocampe sur le comportement d'agression interspécifique rat-souris. *Physiol. Behav.*, 4:889–894.

Vergnes, M., and Karli, P. (1970): Déclenchement d'un comportement d'agression par stimulation électrique de l'hypothalamus médian chez le rat. *Physiol. Behav.*, 5:1427–1430.

Vergnes, M., and Karli, P. (1972): Stimulation électrique du thalamus dorsomédian et comportement d'agression interspécifique du rat. *Physiol. Behav.*, 9:889–892.

Vergnes, M., Mack, G., and Kempf, E. (1973): Lésions du raphé et réaction d'agression interspécifique rat-souris. Effets comport mentaux et biochimiques. *Brain Res.*, 57:67–74.

Vergnes, M., Mack, G., and Kempf, E. (1974b): Contrôle inhibiteur du comportement d'agression interspécifique du rat: Système sérotoninergique du raphé et afférences olfactives. *Brain Res.*, 70:481–491.

Vergnes, M., and Penot, C. (1976): Agression intraspécifique induite par chocs électriques et réactivité après lésion du raphé chez le rat. Effets de la physostigmine. *Brain Res.*, 104:107–119.

Vergnes, M., Penot, C., Kempf, E., and Mack, G. (1977): Lésion sélective des neurones sérotoninergiques du raphé par la 5,7-dihydroxytryptamine: Effets sur le comportement d'agression interspécifique du rat. *Brain Res.*, 133:167–171.

Vernon, W. M. (1969): Animal aggression: Review of research. *Genet. Psychol. Monogr.*, 80:3–28.

Victor, M., Adams, R. D., and Collins, G. H. (1971): *The Wernicke-Korsakoff Syndrome*. Davis Books, Philadelphia.

Virchow, R. (1859): *Cellulapathologie*. Hirchwald, Berlin.

Visser, P. (1972): Some remarks about psychophysiological aspects of aggressive behaviour. *Psychother. Psychosom.*, 20:249–256.

Vizi, E. S., Illés, P. Rónai, A., and Knoll, J. (1972): The effect of lithium on acetylcholine release and synthesis. *Neuropharmacology*, 11:521–530.

Vogel, J. R., and Leaf, R. C. (1972): Initiation of mouse killing in non-killer rats by repeated pilocarpine treatment. *Physiol. Behav.*, 8:421–424.

Vogel, W. H., Ahlberg, C. D., Di Carlo, V., and Horwitt, M. K. (1967): Pink spot, p-tyramine and schizophrenia. *Nature*, 216:1038–1039.

Vonderahe, A. R. (1944): The anatomic substratum of emotion. *The New Scholasticism*, 18:76–95.

Vonnegut, K., Jr. (1959): *The Sirens of Titan*. Dell, New York.

Vorhees, C. V., Barrett, R. J., and Schenker, S. (1975): Increased muricide and decreased avoidance and discrimination learning in thiamine deficient rats. *Life Sci.*, 16:1187–1200.

Waldbillig, R. J. (1979): The role of the dorsal and median raphe in the inhibition of muricide. *Brain Res.*, 160:341–346.

Waldeck, B. (1973): Sensitization by caffeine of central catecholamine receptors. *J. Neural Transm.,* 34:61–72.

Walker, A. E. (1973): Man and his temporal lobes. Johns Hughlings Jackson Lecture. *Surg. Neurol.,* 1:69–79.

Walker, T. A. (1899): *A History of the Law of Nations. Vol. I. From Earliest Times to the Peace of Westphalia, 1648.* Cambridge University Press, Cambridge.

Wallace, T., and Thorne, B. M. (1978): The effect of lesions in the septal region on muricide, irritability, and activity in the Long-Evans rat. *Physiol. Psychol.,* 6:36–42.

Walters, R. H., and Brown, M. (1963): Studies of reinforcement of aggression. III. Transfer of responses to an interpersonal situation. *Child Dev.,* 34:563–571.

Ward, A. A., Jr. (1948): The cingular gyrus: Area 24. *J. Neurophysiol.,* 11:13–23.

Warden, C. J., and Aylesworth, M. (1927): The relative value of reward and punishment in the formation of a visual discrimination habit in the white rat. *J. Comp. Psychol.,* 7:117–127.

Warner, W. L. (1937): *A Black Civilization.* Harper, New York.

Wasman, M., and Flynn, J. P. (1962): Directed attack elicited from hypothalamus. *Arch. Neurol.,* 6:220–227.

Wasman, M., and Flynn, J. P. (1966): Directed attack behavior during hippocampal seizures. *Arch. Neurol.,* 14:408–414.

Watson, J. B. (1913): Psychology as the behaviorist views it. *Psychol. Rev.,* 20:158–177.

Weichert, P., and Herbst, A. (1966): Provocation of cerebral seizures by derangement of the natural balance between glutamic acid and γ-aminobutyric acid. *J. Neurochem.,* 13:49–64.

Weigert, C. (1895): *Beiträge zur Kenntniss der normalen menschilechen Neuroglia.* Frankfurt am Main, Frankfurt.

Weil, A. T. (1970): Adverse reactions to marihuana. Classification and suggested treatment. *N. Engl. J. Med.,* 282:997–1000.

Weingarten, H., and White, N. (1978): Exploratory behavior evoked by electrical stimulation of the amygdala of rats. *Physiol. Psychol.,* 6:229–235.

Weininger, O. (1953): Mortality of albino rats under stress as a function of early handling. *Can. J. Psychol.,* 7:111–114.

Weininger, O. (1956): The effects of early experience on behavior and growth characteristics. *J. Comp. Physiol. Psychol.,* 49:1–9.

Weinstock, R. (1976): Capgras syndrome: A case involving violence. *Am. J. Psychiatry,* 133:855.

Weiskrantz, L. (1956): Behavioral changes associated with ablation of the amygdaloid complex in monkeys. *J. Comp. Physiol. Psychol.,* 49:381–391.

Weiss, B., Laties, V. G., and Blanton, F. L. (1961): Amphetamine toxicity in rats and mice subjected to stress. *J. Pharmacol. Exp. Ther.,* 132:366–371.

Weissman, M., Fox, K., and Klerman, G. L. (1973): Hostility and depression associated with suicide attempts. *Am. J. Psychiatry,* 130:450–455.

Welch, A. S., and Welch, B. L. (1971): Isolation, reactivity and aggression: Evidence for an involvement of brain catecholamines and serotonin. In: *The Physiology of Aggression and Defeat,* edited by B. E. Eleftheriou and J. P. Scott, pp. 91–142. Plenum Press, New York.

Welch, B. L. (1965): Psychophysiological response to the mean level of environmental stimulation: A theory of environmental integration. In: *Symposium of Medical Aspects of Stress in the Military Climate,* edited by D. Mck.Rioch, pp.39–96. U.S.G.P.O., Washington, D.C.

Welch, B. L., and Welch, A. S. (1966): Differential effect of chronic grouping and isolation on the metabolism of brain biogenic amines. *Fed. Proc.,* 25:623.

Welch, B. L., and Welch, A. S. (1968): Greater lowering of brain and adrenal catecholamines in group-housed than in individually-housed mice administered DL-α-methyltyrosine. *J. Pharm. Pharmacol.,* 20:244–246.

Welch, B. L., and Welch, A. S. (1969): Aggression and the biogenic amine neurohumors. In: *Aggressive Behaviour,* edited by S. Garattini and E. B. Sigg, pp. 188–202. Excerpta Medica, Amsterdam.

Welch, B. L., Brown, D. G., Welch, A. S., and Lin, D. C. (1974): Isolation, restrictive confinement or crowding of rats for one year. I. Weight, nucleic acids and protein of brain regions. *Brain Res.,* 75:71–84.

Welch, J. P., Borgaonkar, D. S., and Herr, H. M. (1967): Psychopathy, mental deficiency, aggressiveness and the XYY syndrome. *Nature,* 214:500–501.

Welsh, J. H., and Williams, L. D. (1970): Monoamine-containing neurons in planaria. *J. Comp. Neurol.,* 138:103–115.

Wescott, R. W. (1976): Protolinguistics: The study of protolanguages as an aid to glossogonic research. *Ann. N.Y. Acad. Sci.,* 280:104–116.

West, D. J., and Farrington, D. P. (1973): *Who Becomes Delinquent?* Heinemann, London.

Westermeyer, J., and Kroll, J. (1978): Violence and mental illness in a peasant society: Characteristics of violent behaviours and "Folk" use of restraints. *Br. J. Psychiatry,* 133:529–541.

Whalen, R. E., and Fehr, M. (1964): The development of the mouse-killing response in rats. *Psychonom. Sci.,* 1:77–78.

Whelton, J. P., Jr., and O'Boyle, M. (1977): Early experience and the development of predatory and intraspecific aggression in mice. *Anim. Learning Behav.,* 5:291–296.

White, E. I. (1935): The ostracoderm *Pteraspis* knerr and the relationships of the agnathous vertebrates. *Philos. Trans. R. Soc. Lond. [Biol.],* 225:381–458.

White, M., Mayo, S., and Edwards, D. A. (1969): Fighting in female mice as a function of the size of the opponent. *Psychonom. Sci.,* 16:14–15.

Whitsett, J. M. (1975): The development of aggressive and marking behavior in intact and castrated male hamsters. *Horm. Behav.,* 6:47–57.

Whittaker, V. P. (1964): Investigations on the storage sites of biogenic amines in the central nervous system. *Prog. Brain Res.,* 8:90–117.

Whittaker, V. P., Essman, W. B., and Dowe, G. H. C. (1972): The isolation of pure cholinergic synaptic vesicles from the electric organs of Elasmobranch fish of the family of Torpedimidae. *Biochem. J.,* 128:833–846.

Widdowson, E. M. (1951): Mental contentment and physical growth. *Lancet,* 1:1316–1318.

Wilcock, J. (1968): Strain differences in response to shock in rats selectively bred for emotional elimination. *Anim. Behav.,* 16:294–297.

Wilder, J. (1947): Sugar metabolism in its relation to criminology. In: *Handbook of Correctional Psychology,* edited by R. Lindner, and R. Seliger, pp. 98–129. Philosophical Library, New York.

Wilder, J. (1967): *Stimulus and Response: The Law of Initial Volume,* p. viii. Idmwrigh and Sons, Bristol.

Wilkinson, G. S. (1975): Isolation and psychological disorder. *Psychol. Rep.,* 36:631–634.

Williams, D. I., and Russell, P. A. (1972): Open-field behaviour in rats: Effects of handling, sex and repeated testing. *Br. J. Psychol.,* 63:593–596.

Wilson, A. P., and Boelkins, R. C. (1970): Evidence for seasonal variation in aggressive behaviour by *Macaca mulatta. Anim. Behav.,* 18:719–724.

Wilson, A. T. (1960): Synthesis of macromolecules under possible primeval earth conditions. *Nature,* 188:1007–1009.

Wilson, L., and Rogers, R. W. (1975): The fire this time: Effects of race of target, insult, and potential retaliation on block aggression. *J. Pers. Soc. Psychol.,* 32:857–864.

Wimer, C., and Prater, L. (1966): Some behavioral differences in mice genetically selected for high and low brain weight. *Psychol. Rep.,* 19:675–681.

Wimer, C., Roderick, T. H., and Wimer, R. E. (1969): Supplementary report: Behavioral differences in mice genetically selected for brain weight. *Psychol. Rep.,* 25:363–368.

Wimer, R. E., Norman, R., and Eleftheriou, E. (1973): Serotonin levels in hippocampus: Striking variations associated with mouse strain and treatment. *Brain Res.,* 63:397–401.

Winokur, G., and Tsuang, M. (1975): The Iowa 500: Suicide in mania, depression, and schizophrenia. *Am. J. Psychiatry,* 132:650–651.

Wise, C. D., Baden, M. M., and Stein, L. (1974): Post-mortem measurement of enzymes in human brain: Evidence of a central noradrenergic deficit in schizophrenia. *J. Psychiatr. Res.,* 11:185–198.

Wise, D. A., and Pryor, T. L. (1977): Effects of ergocornine and prolactin on aggression in the postpartum golden hamster. *Horm. Behav.,* 8:30–39.

Witkin, H. A., Goodenough, D. R., and Hirschhorn, K. (1977): XYY Men: Are they criminally aggressive? *The Sciences,* 17:n.6:10–13.

Wnek, D. J., and Leaf, C. (1973): Effects of cholinergic drugs on prey-killing by rodents. *Physiol. Behav.,* 10:1107–1113.

Wolfe, J. B. (1939): An exploratory study of food-storing in rats. *J. Comp. Physiol.,* 28:97–108.

Wolff, H. H. (1973): Aggression in relation to health and illness. *Br. J. Med. Psychol.,* 46:23–27.

Wolfgang, M. E. (1966): *Patterns in Criminal Homicide,* Science editions, pp. 134–166. John Wiley & Sons, New York.

Wolfgang, M. E., and Ferracuti, F. (1967): *The Sub-culture of Violence: Towards an Integrated Theory in Criminology.* Tavistock, London.

Wolfgang, M. E., and Strohm, R. B. (1956): The relationship between alcohol and criminal homicide. *Q. J. Stud. Alcohol,* 17:411–425.

Wood, C. D. (1958): Behavioral changes following discrete lesions of temporal lobe structures. *Neurology (Minneap.),* 8 suppl. 1:215–220.

Woods, J. W. (1956): "Taming" of the wild Norway rat by rhinencephalic lesions. *Nature,* 178:869.

Woodworth, C. H. (1971): Attack elicited in rats by electrical stimulation of the lateral hypothalamus. *Physiol. Behav.,* 6:345–353.

Wooten, G. F., and Coyle, J. T. (1973): Axonal transport of catecholamine synthesizing and metabolizing enzymes. *J. Neurochem.,* 20:1361–1371.

Worrall, E. P., Moody, J. P., and Naylor, G. J. (1975): Lithium in non-manic-depressives: Antiaggressive effect and red blood cell lithium values. *Br. J. Psychiatry,* 126:464–468.

Wright, J. J., and Craggs, M. D. (1977): Arousal and intracranial self-stimulation in split-brain monkeys. *Exp. Neurol.,* 55:295–303.

Wurtman, R. J., and Fernstrom, D. (1974): Effects of the diet on brain neurotransmitters. *Nutr. Rev.,* 32:193–200.

Wynne-Edwards, V. C. (1963): Intergroup selection in the evolution of social systems. *Nature,* 200:623–626.

Wynne-Edwards, V. C. (1965): Self-regulating systems in populations of animals. *Science,* 147:1543–1548.

Yakovlev, P. I. (1970): The structural and functional "trinity" of the body, brain and behavior. *Top. Probl. Psychiatry Neurol.,* 10:197–208.

Yamamoto, T., and Ueki, S. (1977): Characteristics in aggressive behavior induced by midbrain raphe lesions in rats. *Physiol. Behav.,* 19:105–110.

Yarrow, L. J. (1961): Maternal deprivation: Toward an empirical and conceptual re-evaluation. *Psychol. Bull.,* 58:459–490.

Yen, H. C. Y., Day, C. A., and Sigg, E. B. (1962): Influence of endocrine factors on development of fighting behavior in rodents. *Pharmacologist,* 4:173.

Yen, H. C. Y., Katz, M. H., and Krop, S. (1970): Effects of various drugs on 3,4-dihydroxyphenylalanine (DL-DOPA)-induced excitation (aggressive behavior) in mice. *Toxicol. Appl. Pharmacol.,* 17:597–604.

Yen, C. Y., Stanger, R. L., and Millman, N. (1959): Ataractic suppression of isolation-induced aggressive behavior. *Arch. Int. Pharmacodyn. Ther.,* 123:179–185.

Yoshimura, H., Gomita, Y., and Ueki, S. (1974): Changes in acetylcholine content in rat brain after bilateral olfactory bulbectomy in relation to mouse-killing behavior. *Pharmacol. Biochem. Behav.,* 2:703–705.

Yoshimura, H., and Ueki, S. (1977): Biochemical correlates in mouse-killing behavior of the rat: Prolonged isolation and brain cholinergic function. *Pharmacol. Biochem. Behav.,* 6:193–196.

Young, J. K. (1975): A possible neuroendocrine basis of two clinical syndromes: Anorexia nervosa and the Kline-Levin syndrome. *Physiol. Psychol.,* 3:322–330.

Young, L. D., Suomi, S. S., Harlow, H. F., and McKinney, W. T., Jr. (1973): Early stress and later response to separation in rhesus monkeys. *Am. J. Psychiatry,* 130:400–405.

Young, W. C., Goy, R. W., and Phoenix, C. H. (1964): Hormones and sexual behavior. *Science,* 143:212–218.

Zalba, S. R. (1966): The abused child. I. A survey of the problem. *Social Work,* 11:3–16.

Zangwill, O. L. (1976): Thought and the brain. *Br. J. Psychol.,* 67:301–314.

Zeman, W., and Innes, J. R. M. (1963): *Craigie's Neuroanatomy of the Rat.* Academic Press, New York.

Zigler, E., and Williams, J. (1963): Institutionalization and the effectiveness of social reinforcement: A three-year follow-up study. *J. Abnorm. Soc. Psychol.,* 66:197–205.

Zillmann, D. (1971): Excitation transfer in communication-mediated aggressive behavior. *J. Exp. Soc. Psychol.,* 7:419–434.

Zillmann, D. (1978): *Hostility and Aggression.* Lawrence Earlbaum Associates, Hillsdale, N.J.

Ziskind, E., and Augsburg, T. (1967): Hallucinations in sensory deprivation. (Method of madness?). *Dis. Nerv. Syst.,* 28:721–726.

Zook, J. M., and Adams, D. B. (1975): Competitive fighting in the rat. *J. Comp. Physiol. Psychol.,* 88:418–423.

Zucker, I., and Stephan, F. K. (1973): Light-dark rhythms in hamster eating, drinking and locomotor behaviors. *Physiol. Behav.,* 11:239–250.

Zuckerman, M., Persky, H., Hopkins, T. R., Murtaugh, T., Pasu, G. K., and Schilling, M. (1966): Comparison of stress effects of perceptual and social isolation. *Arch. Gen. Psychiatry,* 14:356–365.

Zumpe, D., and Michael, R. P. (1970): Redirected aggression and gonadal hormones in captive rhesus monkeys *(Macaca mulatta). Anim. Behav.,* 18:11–19.

SUBJECT INDEX